Behavioral Measures
OF
Neurotoxicity

Report of a Symposium

Roger W. Russell, Pamela Ebert Flattau,
and Andrew M. Pope, editors

U.S. National Committee for the International
Union of Psychological Science

Commission on Behavioral and
Social Sciences and Education

National Research Council

NATIONAL ACADEMY PRESS
Washington, D.C.   1990

National Academy Press • 2101 Constitution Avenue, NW • Washington, DC 20418

NOTICE: The project that is the subject of this report was approved by the Governing Board of the National Research Council, whose members are drawn from the councils of the National Academy of Sciences, the National Academy of Engineering, and the Institute of Medicine. The members of the committee responsible for the report were chosen for their special competences and with regard for appropriate balance.

This report has been reviewed by a group other than the authors according to procedures approved by a Report Review Committee consisting of members of the National Academy of Sciences, the National Academy of Engineering, and the Institute of Medicine.

Library of Congress Cataloguing-in-Publication Data

Behavioral measures of neurotoxicity : report of a symposium / Roger
   W. Russell, Pamela Ebert Flattau, and Andrew M. Pope, editors.
      p. cm.
   Based on the Workshop on Behavioral Aspects of Neurotoxicity,
Critical Issues, held in Aug. 1988 at the Australian National
University in Canberra; sponsored by the U.S. National Committee for
the International Union of Psychological Science (USNC/IUPsyS).
   Includes bibliographical references.
   Includes index.
   ISBN 0-309-04047-7
   1. Neurotoxicology—Congresses. 2. Pollutants—Toxicology—
Congresses. 3. Nervous system—Diseases—Environmental aspects—
Congresses. 4. Neuropsychological tests—Congresses. I. Russell,
Roger W. II. Flattau, Pamela Ebert. III. Pope, Andrew MacPherson,
1950-   . IV. U.S. National Committee for the International Union
of Psychological Science. V. Workshop on Behavioral Aspects of
Neurotoxicity: Critical Issues (1988 : Australian National
University)
   [DNLM: 1. Behavior—drug effects—congresses. 2. Nervous System
Diseases—chemically induced—congresses. 3. Neurotoxins—adverse
effects—congresses.   WL 103 B419 1988]
RC347.5.B44   1990
616.8—dc20
DNLM/DLC
for Library of Congress   90-6565
                                                                              CIP

Copyright © 1990 by the National Academy of Sciences

Printed in the United States of America

U.S. NATIONAL COMMITTEE FOR THE INTERNATIONAL UNION OF PSYCHOLOGICAL SCIENCE

JAMES L. MCGAUGH *(Chair)*, Center for the Neurobiology of Learning and Memory, University of California, Irvine

DOROTHEA JAMESON *(Vice Chair)*, Department of Psychology, University of Pennsylvania

ALBERT BANDURA, Department of Psychology, Stanford University

ROCHEL GELMAN, Department of Psychology, University of California, Los Angeles

ROBERT GLASER, Learning Research and Development Center, University of Pittsburgh

WILLARD W. HARTUP, Institute of Child Development, University of Minnesota

GEORGE MANDLER, Department of Psychology, University of California, San Diego

MARK R. ROSENZWEIG, Department of Psychology, University of California, Berkeley

ROBERT B. ZAJONC, Research Center for Group Dynamics, University of Michigan

ROBERT MCC. ADAMS *(ex officio)*, Chair, Commission on Behavioral and Social Sciences and Education

WILLIAM E. GORDON *(ex officio)*, Foreign Secretary, National Academy of Sciences

WAYNE H. HOLTZMAN *(ex officio)*, President, International Union of Psychological Science

PAMELA EBERT FLATTAU, Staff Officer

CAROL METCALF, Administrative Secretary

The National Academy of Sciences is a private, nonprofit, self-perpetuating society of distinguished scholars engaged in scientific and engineering research, dedicated to the furtherance of science and technology and to their use for the general welfare. Upon the authority of the charter granted to it by the Congress in 1863, the Academy has a mandate that requires it to advise the federal government on scientific and technical matters. Dr. Frank Press is president of the National Academy of Sciences.

The National Academy of Engineering was established in 1964, under the charter of the National Academy of Sciences, as a parallel organization of outstanding engineers. It is autonomous in its administration and in the selection of its members, sharing with the National Academy of Sciences the responsibility for advising the federal government. The National Academy of Engineering also sponsors engineering programs aimed at meeting national needs, encourages education and research, and recognizes the superior achievements of engineers. Dr. Robert M. White is president of the National Academy of Engineering.

The Institute of Medicine was established in 1970 by the National Academy of Sciences to secure the services of eminent members of appropriate professions in the examination of policy matters pertaining to the health of the public. The Institute acts under the responsibility given to the National Academy of Sciences by its congressional charter to be an adviser to the federal government and, upon its own initiative, to identify issues of medical care, research, and education. Dr. Samuel O. Thier is president of the Institute of Medicine.

The National Research Council was organized by the National Academy of Sciences in 1916 to associate the broad community of science and technology with the Academy's purposes of furthering knowledge and advising the federal government. Functioning in accordance with general policies determined by the Academy, the Council has become the principal operating agency of both the National Academy of Sciences and the National Academy of Engineering in providing services to the government, the public, and scientific and engineering communities. The Council is administered jointly by both Academies and the Institute of Medicine. Dr. Frank Press and Dr. Robert M. White are chairman and vice chairman, respectively, of the National Research Council.

SYMPOSIUM CONTRIBUTORS*

W. KENT ANGER, National Institute of Occupational Safety and Health, Cincinnati, Ohio, U.S.A.

ELAINE L. BAILEY, School of Biological Science, The Flinders University, Australia

JOHN GRAHAM BEAUMONT, Department of Psychology, University College of Swansea, United Kingdom

M.G. CASSITTO, Institute of Occupational Health, Milan, Italy

DEBORAH A. CORY-SLECHTA, School of Medicine and Dentistry, University of Rochester, New York, U.S.A.

PAMELA EBERT FLATTAU, National Research Council, Washington, D.C., U.S.A.

FRANCESCO GAMBERALE, Division of Psychophysiology, National Institute of Occupational Safety and Health, Solna, Sweden

RENATO GILIOLI, Institute of Occupational Health, Milan, Italy

HELENA HANNINEN, Institute of Occupational Health, Helsinki, Finland

WAYNE H. HOLTZMAN, Hogg Foundation for Mental Health, University of Texas, Austin, U.S.A.

ANDERS IREGREN, Division of Psychophysiology, National Institute of Occupational Safety and Health, Solna, Sweden

ANDERS KJELLBERG, Division of Psychophysiology, National Institute of Occupational Safety and Health, Solna, Sweden

NORMAN KRASNEGOR, National Institute of Child Health and Human Development, Bethesda, Maryland, U.S.A.

BEVERLY M. KULIG, Medical Biological Laboratories, TNO, Rijswijk, The Netherlands

ROBERT MACPHAIL, Environmental Protection Agency, Research Triangle Park, North Carolina, U.S.A.

HANNA MICHALEK, Laboratory of Pharmacology, Instituto Superiore di Sanita, Rome, Italy

DAVID H. OVERSTREET, School of Biological Science, The Flinders University, Australia

---

*Affiliations of contributors are as of 1988.

ANNITA PINTOR, Laboratory of Pharmacology, Instituto Superiore di Sanita, Rome, Italy

ANDREW M. POPE, National Research Council, Washington, D.C., U.S.A.

DEBORAH C. RICE, Toxicology Research Division, Bureau of Chemical Safety, Ottawa, Canada

ROGER W. RUSSELL, Department of Pharmacology, University of California, Los Angeles, U.S.A.

VINOD BEHARI SAXENA, Department of Psychology, P.P.N. College, Kanpur University, India

GEORGE SINGER, Brain-Behavior Research Institute, La Trobe University, Australia

PETER S. SPENCER, Center for Research on Occupational and Environmental Toxicology, Oregon Health Sciences University, Portland, U.S.A.

BERNARD WEISS, School of Medicine and Dentistry, University of Rochester, New York, U.S.A.

ANN M. WILLIAMSON, National Institute of Occupational Health and Safety, Sydney, Australia

GERHARD WINNEKE, Medical Institute for Environmental Hygiene, Dusseldorf, West Germany

LIANG YOU-XIN, Department of Occupational Health, Shanghai Medical University, The People's Republic of China

Preface

The U.S. National Committee for the International Union of Psychological Science (USNC/IUPsyS) is a standing committee of the National Research Council's (NRC) Commission on Behavioral and Social Sciences and Education. Established in 1985, the committee promotes the advancement of the science of psychology in the United States and throughout the world. It does so by advising the President of the National Academy of Sciences on matters pertaining to U.S. participation in the Union and by planning and sponsoring scientific meetings in the United States and abroad in accordance with the objectives of the International Union of Psychological Science.

A continuing concern of the committee has been U.S. participation in the International Congress of Psychology. Although the USNC/IUPsyS is a relatively young committee by NRC standards (the oldest were established in the early 1920s), U.S. representatives have been active in the affairs of the Union since its inception in 1952. Thus, in its establishment by the NRC in 1985, the role of the committee was to continue U.S. participation in Union affairs. One of the first items of committee business was the exchange of correspondence with the Australian scientific program committee regarding possible topics for inclusion in the XXIV International Congress of Psychology held in Sydney in August 1988. The committee recommended that U.S. psychologists both organize and participate in symposia in scientific psychology.

One area that emerged during committee deliberations as a topic of significant scientific interest was behavioral analysis of the action of environmental toxicants on neural tissue. The committee noted that numerous researchers in the United States and abroad were engaged in the development of neurobehavioral tests of toxic exposure. With funds provided by the U.S. Environmental Protection Agency (EPA), the committee organized a meeting in 1987 to develop the agenda for an invitational workshop to be held in conjunction with the XXIV International Congress, drawing speakers from the diverse array of behavioral scientists planning to attend the congress.

In August 1988, the committee convened the "Workshop on Behavioral Aspects of Neurotoxicity: Critical Issues." Workshop participants, drawn from 12 countries, met for three days at the Australian National University in Canberra; Vice Chancellor L.W. Nichol kindly welcomed participants on behalf of the University. The 20 specialists participating in the workshop met to review progress toward the development of behavioral tests of neurotoxicity. The meeting began with a report of the findings from field work studies by Professor Peter Spencer. Spencer's research has led to a new understanding of the direct relationship between environmental toxicants and the occurrence of neurodegenerative disease arising from the health and social practices of Pacific Islanders. Papers by the remaining workshop participants provided a state-of-the-art review of behavioral measures of neurotoxicity and suggested new directions, given emerging knowledge of the interactive effects of environmental toxicants and neurodegenerative disorders.

Five panels were assembled to address separate but interrelated topics: (1) assessment of neurobehavioral tests now in use; (2) animal models: what has worked and what is needed; (3) "chemical time bombs": environmental causes of neurodegenerative diseases; (4) regional issues in neurobehavioral toxicity testing; and (5) recommendations for further research and testing. The Appendix contains a list of conference participants and the detailed program. Background papers were prepared by each participant and circulated well in advance of the meeting. Discussants similarly prepared a response to panelists' papers, with the preliminary responses also circulated prior to the meeting. This report presents those papers in a collected volume. Funding for the workshop was provided by several U.S. sources, including EPA, the National Institute for Occupational Safety and Health, the National Institute of Child Health and Human Development, and a consortium of private contributors (Rohm & Haas Co., Abbott Laboratories, Lilly Research Laboratories, Burroughs Wellcome Co., Hoffmann-La Roche).

In addition to the workshop participants, a number of people con-

PREFACE   ix

tributed in important ways to the success of the workshop and to this report. Roger Russell, Emeritus Professor, University of California, Los Angeles, has served as a wise and enthusiastic consultant to the committee, helping to shape the workshop program, meeting logistics, and the final report. The committee is deeply indebted to Professor Russell for these contributions. Andrew M. Pope, staff officer with the NRC Board on Environmental Studies and Toxicology, provided valuable technical assistance in organizing the workshop. Pamela Ebert Flattau, the committee's study director, planned the workshop and played a significant role in overseeing the development of this volume. Carol Metcalf, the committee's administrative secretary, worked tirelessly to coordinate the production of the workshop papers. To all these people, we express our gratitude for their efforts.

> James L. McGaugh, *Chair*
> U.S. National Committee for the International
> Union of Psychological Science

Contents

Introduction 1

Are Neurotoxins Driving Us Crazy? Planetary
Observations on the Causes of Neurodegenerative
Diseases of Old Age 11
   Peter S. Spencer

PART I. ASSESSMENT OF NEUROBEHAVIORAL TESTS NOW IN USE

Methods in Behavioral Toxicology: Current Test
Batteries and Need for Development 39
   Helena Hanninen

The Current Status of Test Development in
Neurobehavioral Toxicology 56
   Ann M. Williamson

Human Neurobehavioral Toxicology Testing 69
   W. Kent Anger

Neurobehavioral Tests: Problems, Potential, and
Prospects 86
   J. Graham Beaumont

PART II. ASSESSMENT OF ANIMAL MODELS: WHAT HAS WORKED AND WHAT IS NEEDED

Exposure to Neurotoxins Throughout the Life Span: Animal Models for Linking Neurochemical Effects to Behavioral Consequences 101
Hanna Michalek and Annita Pintor

Animal Models of Dementia: Their Relevance to Neurobehavioral Toxicology Testing 124
David H. Overstreet and Elaine L. Bailey

Bridging Experimental Animal and Human Behavioral Toxicology Studies 137
Deborah A. Cory-Slechta

Methods and Issues in Evaluating the Neurotoxic Effects of Organic Solvents............................... 159
Beverly M. Kulig

Animal Models: What Has Worked and What Is Needed... 184
Robert C. MacPhail

PART III. CHEMICAL TIME BOMBS: ENVIRONMENTAL CAUSES OF NEURODEGENERATIVE DISEASES

On the Identification and Measurement of Chemical Time Bombs: A Behavior Development Perspective 191
Norman A. Krasnegor

Neurobehavioral Time Bombs: Their Nature and Their Mechanisms.. 206
Roger W. Russell

Neurobehavioral Toxicity of Selected Environmental Chemicals: Clinical and Subclinical Aspects 226
Gerhard Winneke

The Health Effects of Environmental Lead Exposure: Closing Pandora's Box.. 243
Deborah C. Rice

Chemical Time Bombs: Environmental Causes of Neurodegenerative Diseases 268
Peter S. Spencer

PART IV. BEHAVIORAL ASPECTS OF NEUROTOXICITY: REGIONAL ISSUES

Neurobehavioral Toxicity Testing in China 287
  Liang You-Xin

Regional Issues in the Development of the
Neurobehavioral Core Test Battery 312
  Renato Gilioli and Maria G. Cassitto

Issues in the Development of Neurobehavioral Toxicity
Tests in India 322
  Vinod Behari Saxena

Regional Issues in Neurobehavioral Testing:
An Overview 337
  Ann M. Williamson

PART V. RECOMMENDATIONS FOR FURTHER RESEARCH AND TESTING

Environmental Modulation of Neurobehavioral
Toxicity 347
  Robert C. MacPhail

Computerized Performance Testing in Neurotoxicology:
Why, What, How, and Whereto? 359
  Francesco Gamberale, Anders Iregren, and Anders Kjellberg

The Scope and Promise of Behavioral Toxicology 395
  Bernard Weiss

Appendix: Symposium Agenda 415

Index 419

Behavioral Measures
OF
Neurotoxicity

Introduction

Behavior is a basic property of living organisms as they adjust to their physical and psychological environments. It is dependent upon a multitude of biochemical and electrophysiological functions at various morphological sites within the body, particularly within the nervous system. Behavioral analysis is a process of "sampling" the integrated outputs of these various functions. The major objectives of neurobehavioral toxicity testing are to determine whether exposure to a potentially toxic chemical produces effects on these functions that exceed the organism's ability to adjust, to assist in identifying the nature of such malfunctions, and to provide information about the effects of extended exposure (both dose and time effects) that may be used in establishing threshold limit values ("no-observed-adverse-effect levels") for purposes of human safety.

Behavior is a dynamic process, however, and exposure to a toxicant may or may not result in behavioral change. This situation arises from the fact that toxic agents do not affect behavior directly (Russell, 1979). In general terms, toxicants enter the body through one or more of several routes, and once within the body they combine with receptor molecules to produce an agent-receptor complex. These interactions, which may or may not be reversible, initiate a series of biochemical and electrophysiological changes that characterize the overall response to the agent. Biochemical events occurring after exposure to a toxicant have significant influence on subsequent behavioral effects and constitute the "mechanisms of action" underlying those effects. If the linkage is direct, changes in biochemical events are related to specific changes in behavior; if the linkage is diffuse, changes in biochemical events may affect a variety of behavior patterns.

In 1975, the National Research Council (NRC) report *Evaluating Chemicals in the Environment,* was one of the first reports to formalize the concept of "behavioral toxicology." The report identified two levels at which the effects of toxicants on the nervous system could be tested by using a variety of assays including behavioral measures. The first level of investigation involves studies designed to screen for possible behavioral abnormalities produced by chemical substances. A battery of tests is often used to assess possible effects on a wide range of behavioral processes (for example, gross and fine motor functions, vision, hearing, cognition). Studies at this level involve large numbers of subjects and use simple, rapid, and inexpensive behavior technologies. Screening strategies are suited for investigating large numbers of putative toxicants in a short time. These procedures are generally not appropriate to reveal basic mechanisms, nor are they suited to assess more subtle or complex behavioral functions. The second level of analysis focuses on characterizing in more detail the basic mechanisms that are altered by substances to which the organism has been exposed. Typically, tests at this level use a smaller number of subjects who are often studied in greater detail over some lengthy period of time. The behavioral technologies that are involved are more elaborate and time-consuming, and frequently require periods of retraining.

Behavioral analysis thus offers useful tools for measuring the direct action of a chemical on neural tissue. Such measures of behavior are often particularly sensitive to toxic agents, showing significant effects when no overt symptomatology is observable. They are "noninvasive," requiring no sampling of body fluids or tissues. They may be applied repeatedly to study progressive effects of contacts with toxicants, for monitoring the effectiveness of regulations established to control adverse exposures, and to assist in the management of patients under treatment for toxic illnesses.

IDENTIFICATION OF BEHAVIORAL TOXIC AGENTS

It is obviously desirable that potential dangers resulting from chemical agents in the environment be identified as early as possible. One category of critical issues certain to continue to demand attention into the next decade—and, indeed, well beyond—is that encompassing the identification of toxic agents that affect behavior. Given the existence of a behavioral abnormality, what chemical agent, if any, is its antecedent? Given a chemical agent, what may be its behavioral consequences?

Six centuries ago, Paracelsus said, "All things are poisons, for there is nothing without poisonous qualities. It is only the dose which makes a thing a poison" (Sigerist, 1958). If Paracelsus was correct, the number of chemicals having potentially toxic effects on behavior corresponds to the number of chemicals in the environment.

A first step in the risk assessment process and the establishment of threshold limit values for purposes of regulation is to define a dose-effect relationship for the particular substance under consideration. A "critical issue" needing attention arises from the fact that the "effective" dose often is not directly proportional to the "administered" dose. Methods have been suggested that enable dose-effect relations to be constructed on the basis of the former rather than the latter, thus increasing ". . . confidence in the reliability of the observed relationship as a starting point for extrapolation outside the experimental dose range" (O'Flaherty, 1986).

Another series of questions related to dose-effect relationships centers on increasing appreciation of the fact that different behaviors have different toxic thresholds. For example, untoward effects on cognitive processes may occur without immediately obvious behavioral symptomatologies. This puts pressure on decisions about the breadth of behavioral testing and when during the life span to test.

Publication of the *Proceedings of the Third International Conference on Indoor Air and Climate* (Berglund et al., 1984) made it convincingly clear that potentially toxic chemicals are to be found in all human environments. Discovering these "ecological traps," defining their effects on behavior, and inventing methods for preventing them from being sprung is yet another area of critical issues for the foreseeable future.

MECHANISMS AND SITES OF ACTION

Behaviorally, toxic chemicals may be identified solely on the basis of their measured effects on behavior, without knowledge of the physiological mechanisms or sites of action within the body. Indeed, it is a safe generalization to state that these properties are not fully understood for most of the agents of current concern. Seeking the mechanism of action of a chemical affecting the central nervous system (CNS) is not like looking for a needle in a haystack, but rather like looking for a needle in a heap of needles. What have we to lose if we do not search for mechanisms and sites of action? Clearly there are losses for the development of our knowledge in the basic biobehavioral sciences. However, let us consider three specific questions of a more applied nature, all of which document the kinds of "critical issues" within this general area of mechanisms by which chemicals act.

Can an understanding of mechanisms of action help in synthesizing new compounds for specific purposes? Yes. For example, knowledge about the nature of receptors in the CNS is being used to create molecules that are less toxic and others that are more effective as therapeutic drugs in counteracting the effects of toxic agents.

Is knowledge of mechanisms of action helpful in understanding neurodegenerative disorders of toxic origin? Again, yes. Searching for the mechanism(s) involved—toxic, viral, nutritional—is essential to the prevention and treatment of disorders such as the progressive degenerative dementias.

Can the search for mechanisms of action help the diagnosis and treatment of genetically induced metabolic faults? Yes, indeed. Phenylketonuria is one example of a very significant behavioral deficiency associated with a genetic failure to produce a specific enzyme phenylalanine (hydroxylase), a condition that results in a block to normal metabolism in a major neurochemical pathway. Very early diagnosis of the fault can enable steps to be taken, usually dietary, to limit brain damage resulting from toxic by-products of the fault.

THE MEASUREMENT OF BEHAVIOR

A quarter of a century ago, a physician employed by a large industrial firm published a challenge in the *Journal of Occupational Medicine* entitled "Functional Testing for Behavioral Toxicity: A Missing Dimension in Experimental Environmental Toxicology" (Ruffin, 1963). The challenge is currently being met by the development of methods for measuring behavior, their standardization, and their applications. Despite the sophistication of our modern methods, questions still arise which indicate that critical issues in the measurement of behavior continue to demand attention. Many of those questions are called to our attention in this volume.

The measurement of behavior was introduced into the process of risk assessment as a means of providing potentially more sensitive measures of toxicity than other properties of living organisms. As Bernard Weiss has asked: "What other discipline is in the unique position of access to a technology for tracing a progression of toxicity from early, subtle effects to clear impairment? What other perspective on toxicology can integrate such a rich configuration of endpoints?"

INTRODUCTION

These rhetorical questions raise many other queries of a more specific nature. A sampling of these critical issues as formulated in this volume includes:

- What are the advantages of measuring behavior in breadth versus depth?
- What should guide the selection of behaviors to be measured?
- Under what conditions should measuring instruments be "tailored" versus taken "off the shelf"?
- What special principles are important in developing a standardized "test battery"?
- What are the current shortcomings in the design of test batteries? What should be done to eliminate those deficiencies?
- Is it possible to develop measuring instruments that can discern functional impairment through the adaptive changes that may camouflage it?
- Are there limiting conditions in the application of behavioral testing in nonlaboratory settings?
- Why is evidence from behavioral tests often not convincing to decision makers?

One of the major developments in behavioral assessment during recent years has been the introduction of microcomputer-based testing. Evaluating the pros and cons of automated testing leads to the conclusions that this general approach to the measurement of behavior, although widely accepted and applied, will be a much debated area for some time to come.

THE LIFE SPAN: WHEN TO MEASURE?

A central theme of many of the most frequently raised questions we have heard is, When during the life span should neurobehavioral toxicologists apply their measurements? The unanimous answer has been, "From the womb to the tomb." The existence of critical issues at the beginning of this broad range was emphasized in 1987 by the appearance of the following statement in the journal *Reproductive Toxicology*: "The most relevant data [for female reproductive risk assessment] come from studies reported in humans, however, these are not generally available" (Sakai, 1987). Clearly here is an area in need of careful study. Risks during in utero development were brought to world attention by the teratogenic horrors of the thalidomide tragedy. Today the number of such critical agents introduced into the uterine environment is recognized more clearly than ever before. Scientists are now aware of the particular significance of susceptibility to toxic agents during the early years of postnatal development, and so we could continue through the life span, perhaps with special consideration of the significant problems associated with an increasingly aging population. Clearly, there is no dearth of critical issues wherever in the life span neurobehavioral toxicologists choose to search.

The fact that chemicals, both natural and synthetic, may trigger long-latency, neurological, and behavioral disorders without early detection emphasizes the importance of longitudinal as well as cross-sectional studies of behavioral toxicants. Examples of this species of critical issues are also to be found among the progressive degenerative dementias.

APPROACHES TO RESEARCH

A number of participants in the workshop pointed out that it is not really possible to divorce questions of experimental design from those related to test selection. The discussion took several directions. One of these stressed the key role played by the objectives for which the research is designed; this led to consideration of critical requirements for the establishment of neurobehavioral test batteries. These are discussed in detail in several of the chapters included in the volume.

The discussion at Canberra took a second direction in which the emphasis was, in a sense, related to the venues in which research in neurobehavioral toxicology takes place. The desirable precision of the *experimental* approach raises questions about extrapolation to conditions of the "real world." Research in the *field* has problems in establishing which among several or many variables is related to the consequent behavioral malfunctions. Research in the *clinic* begins with a disorder and has significant difficulties in establishing its etiology. At one time the comment was made that these three approaches are analogous to the three blind men describing an elephant: one by feeling its tail; another, its leg; and the third, its trunk. Integrating information from the three inputs must certainly be a worthy effort for the next decade.

The importance of using proper control data continues to receive considerable attention. The ideal of using subjects as their own controls—measuring behavior before, during, and after exposure—may be achievable only in the laboratory. Prospective studies in the field are expensive and relatively few have been conducted. Questions then arise about what constitutes an adequate control group with which to compare those exposed occupationally or adventitiously. At Canberra, attention was called to a 1988 publication that clearly illustrates this problem (Gade et al., 1988). Reevaluation of an earlier study found that original behavioral measurements leading to diagnoses of chronic toxic encephalopathy did not refer to population norms for cognitive deficits or to estimates of premorbid levels of functioning. The reevaluation concluded that the poor test performance described in the original report was a consequence of the lower level of general intelligence and education of the exposed subjects, and hence could not be attributed to solvent exposure. It should not be inferred from this one example that exposure to solvents never affects behavior. The recent reevaluation only points to shortcomings in the design of the earlier study. This example has been chosen because it has another important feature. The original diagnosis of solvent-induced dementia precipitated pressures for changes in procedures used by the industries concerned, who may now be skeptical about the value of neurobehavioral toxicology.

Discussion of experimental approaches to the study of neurobehavioral toxicity inevitably leads to consideration of the roles of animal models (Russell and Overstreet, 1984). The use of such models has long characterized the early stages of pharmacological research on particular compounds. Inevitably research using such models raises questions about extrapolation of results from animals to humans. Put more precisely, the basic issue is, For what decisions is a particular animal mode valid? It is not possible to design an experiment to study the effects of potentially toxic substances or even to make systematic observations relevant to such a substance without stating or implying some underlying theoretical model of living organisms and how they interact with their biospheres. The "construct validity" of an animal model is determined by the closeness of fit between the two types of models, the animal and the theoretical. "Predictive validity" requires demonstration of concordance between predictions based on effects of a chemical agent on the animal model and effects of the agent on humans, in essence an ex post facto validation of the model. In considering the wide use of animal models in neurobehavioral toxicology, it seems clear that such models will continue to play critical roles in the future as they have in the past.

REGIONAL REQUIREMENTS

Questions of critical regional requirements are raised whenever neurobehavioral toxicology has an international audience. The questions range from the relatively simple to the exceedingly complex. For example, those participating in an international meeting in Indonesia in 1986 saw an apparatus essential to assaying for potential toxicants in the local environment that had been lying idle for months because of a lack of small replacement parts. At the other extreme are social factors that stand in the way of introducing critical changes in the chemical environment when they are sorely needed. One of the most difficult tasks in efforts to alter existing toxic conditions will continue to be that of social change, of changing human behavior patterns, attitudes, and life-styles.

COMMUNICATION

As the need for solutions to problems of environmental management has become perceived as critical among major social issues, more attention has been turned to questions of communication among a wide variety of groups with different capabilities and different responsibilities. At the level of the present discourse, fuller communication among scientific disciplines involved in environmental toxicology has begun to develop. The Canberra workshop is a splendid example of the extension of such communication at an international level. Paralleling this development is a growing recognition of the importance of multidisciplinary education and training, a critical issue deserving further attention during the decade to come.

Environmental management involves more than the activities of scientists. It involves trade-offs between biological limitations and social cost and benefits. Management is achieved primarily through the imposition of regulations by legislative or administrative bodies within governments. The usual progression involves a policy-making process in which scientific research points to the existence of hazards, scientific data provide a basis for economic analyses of costs and benefits, and finally, both types of information are used by regulators.

Beyond communications at all these levels is the importance, during the next decade, of communicating better with society in general. There are strong social pressures toward conformity; change usually comes slowly. Here are some examples: The city of Los Angeles has a critical need for an improved and more extensive public transport system, a need recognized generally by its inhabitants for many years. City authorities have proposed several routes for such a system. Whenever a proposal involves a particular neighborhood, the "good" local citizens mobilize to block its approval. Very recently, the U.S. Environmental Protection Agency stated ". . . that it may be forced to prohibit all gasoline- and diesel-powered vehicles in smoggy, Southern California within five years unless the federal Clean Air Act is amended to extend the deadline for meeting clean air standards" (*Los Angeles Times*, 1988). Public pressures are arising in several areas of the world to shorten the time taken by responsible governmental agencies to verify the effectiveness and dangers of new drugs for the treatment of human diseases. A basic conflict arises when such pressures prevent adequate evaluations, yet the public is free to sue an agency if adverse effects occur or if a drug is not effective. These are but two of many examples indicating the importance of communication with the general public—a need that, even if given proper attention, will not have disappeared by the year 2000.

CONCLUSION

Risk assessment, as defined by the NRC, involves hazard identification, dose-response relationships, exposure assessment, and risk characterization (NRC, 1983). At its present state of development, neurobehavioral toxicology has critical issues to explore in all of these areas. A 1984 report by the National Academy of Sciences (NRC, 1984) assured us once again that the knowledge and skills of neurobehavioral toxicologists are greatly needed. There is evidence that this is being increasingly recognized by regulators, who must consider neurobehavioral toxicology as ". . . the introduction of a new biological endpoint into the regulatory arena" (Buckholtz and Panem, 1986).

To achieve an integrated body of knowledge and its applications demands a multidisciplinary approach. The breadth of disciplines involved precludes all neurobehavioral toxicologists from being experts in all the relevant disciplines. However, they can be acquainted with the basic concepts and techniques so that interrelations among the specialized areas of expertise can be understood by all. Acquiring this broader perspective can begin with multidisciplinary education and training. It can be continually fostered by meetings such as the Canberra workshop, when new ideas, information, and techniques can be exchanged and examined in an environment of free and positive discussion.

REFERENCES

Berglund, B., T. Lindall, and J. Sundell, eds. 1984. Pp. 284 in *Indoor Air: Recent Advances in the Health Sciences and Technology.* Stockholm: Swedish Council for Building Research.

Buckholtz, N.S., and S. Panem. 1986. Regulation and evolving science: Neurobehavioral toxicology. *Neurobehav. Toxicol. Teratol.* 8:89-96.

Gade, A., E.L. Mortesen, and P. Bruhn. 1988. Chronic painters' syndrome. A reanalysis of psychological test data in a group of diagnosed cases, based on comparisons with matched controls. *Acta Neurol. Jour.* 77:293-306.

Los Angeles Times. 1988. EPA issues threat to ban gas, diesel vehicles in basin. November 29.

National Research Council. 1975. *Evaluating Chemicals in the Environment*. Washington, D.C.: National Academy Press.
National Research Council. 1983. Pp. 17-83 in *Risk Assessment in the Federal Government: Managing the Process*. Washington, D.C.: National Academy Press.
National Research Council. 1984. Reference protocol guidelines for neurobehavioral toxicology tests. Pp. 169-174 in*Toxicity Testing—Strategies to Determine Needs and Priorities*. Washington, D.C.: National Academy Press.
O'Flaherty, E.J. 1986. Dose dependent toxicology. *Comments on Toxicol*. 1:23-34.
Ruffin, J.B. 1963. Functional testing for behavioral toxicology: A missing dimension in experimental environmental toxicology. *J. Occup. Med*. 5:117-121.
Russell, R.W. 1979. Neurotoxins: A systems approach. Pp. 1-7 in I. Chubb and L.B. Geffen, eds. *Neurotoxins: Fundamentals and Clinical Advances*, South Australia: University of Adelaide Press.
Russell, R.W., and D.H. Overstreet. 1984. Animal model in neurobehavioral toxicology. Pp. 23-57 in N.W. Bond, ed. *Animal Models in Psychopathy*. Sydney: Academic Press.
Sakai, C. 1987. Female reproductive risk assessment. *Reproductive Toxicol*. 1:53-54.
Sigerist, H.E. 1958. *The Great Doctors*. New York: Doubleday.

Are Neurotoxins Driving Us Crazy? Planetary Observations on the Causes of Neurodegenerative Diseases of Old Age

Peter S. Spencer

Health planners in developed countries are increasingly concerned with their burgeoning populations of elderly subjects and the consequent rising prevalence of age-associated disorders, notably those involving the nervous system. By the year 2050, current projections for the United States indicate that the proportion of the population aged 65 or over will be almost double (22%) the 1986 level, whereas the prevalence of senile dementia of the Alzheimer type will triple. It is thus entirely appropriate for the elderly of developed countries to be the subjects of intense scientific scrutiny aimed at understanding the causes and methods of prevention of the major neurodegenerative diseases that all too often accompany the second half of life. There are certain other parts of the world, however, notably in the western Pacific region, where such disorders are far more commonly encountered and where prospecting for etiology is more likely to be profitable. Indeed, one would posit that a knowledgeable extraterrestrial investigator, charged with the task of identifying causes of the great neurodegenerative diseases of *Homo sapiens* on planet Earth, would be unlikely to begin by researching elderly populations in Canberra, London, or New York; rather, the hunt for causation would probably commence in places such as Guam or Irian Jaya where, in certain spots, incidence rates for such diseases have exceeded worldwide statistics by more than one to three orders of magnitude.

If the etiologic search can be likened to the proverbial hunt for a needle in a haystack, why not maximize chances of success by focusing investigation on haystacks that contain a hundred such needles? Critics of this view charge that the western Pacific combination of presenile dementia, parkinsonism, and motor neuron disease found in the Mariana islands (Guam and Rota), Irian Jaya (west New Guinea, Indonesia), and the Kii peninsula of Honshu island (Japan), is little more than a medical oddity and distinct from the neurodegenerative disorders that plague the aged in the West. Far from being a curiosity, others observe that the high-incidence foci of neurodegenerative disease in the western Pacific may actually hold the keys that will unlock the door to lookalike disorders worldwide, if not to the secrets of aging itself.

This is a story about a remarkable and terrible affliction that sometimes presents at onset as motor neuron disease (amyotrophic lateral sclerosis) and, in other instances, as parkinsonism, presenile dementia, or various combinations of all three. The disease is known as western Pacific amyotrophic lateral sclerosis (ALS) and parkinsonism-dementia complex (P-D), or ALS/P-D. Each clinical component in isolation closely resembles its namesake worldwide. Intensive study of this disease on Guam has demonstrated that although ALS or P-D may strike several family members and affect several generations of individual families, the disease is neither inherited nor associated with a transmissible (viral-like) agent, and seems instead to be associated with an environmental factor that is disappearing hand in hand with the declining incidence of ALS.

Armed with this information and knowing of an early suggestion linking Guam ALS to neurolathyrism, I began in 1981 to investigate in India, and later in Bangladesh, China, and Ethiopia, a form of motor neuron disease for which a neurotoxic etiology was clearly established, namely, the spastic paraparesis associated with excessive and continuous ingestions of legume *Lathyrus sativus* L. Lathyrism was of special interest because a detailed understanding of this poorly studied condition seemed likely to provide a base of knowledge and experience on which to examine the unfashionable and largely forgotten suggestion that consumption of a potentially neurotoxic food plant (*Cycas circinalis* L.) was also etiologically linked to Guam ALS. My conjecture was that detailed study of the neurotoxic factors in these plants might lead to a precise understanding of the chemical triggers of lathyrism and western Pacific ALS/P-D, information that could then be applied to search for the etiology of related neurodegenerative diseases worldwide. This research has led to the conclusion that chemical factors in the environment are involved in triggering neurodegenerative diseases which remain silent for years or decades before their dramatic clinical consequences are expressed. Intensive research is now underway both to identify the novel molecular and cellular mechanisms that likely underly these newly recognized long-latency neurotoxic diseases, and to determine whether xenobiotics also play a role in triggering neurodegenerative disorders such as Alzheimer's disease and ALS in developing countries.

CLINICAL FEATURES OF WESTERN PACIFIC ALS/P-D

Sporadic ALS, a disorder of middle-aged and elderly people, is more common in males than females (1.6:1), with a mean annual incidence rate in the United States of 1–2/100,000. Much higher prevalence ratios for ALS, and much younger ages (18+) of onset, are reported among the indigenous population (Chamorro) of Guam. Surveys conducted in the early 1950s demonstrated that about 10 percent of deaths among adult Chamorros resulted from ALS, frequencies about 100 times those recorded for the population of the continental United States. Males were more susceptible than females to ALS (2:1) and to P-D (3:1). However, during the past 30 years, the prevalence of Guam ALS has dropped steadily, the sex ratio now approaches unity, and the mean age at onset has risen from the 40s to the 50s.

Those who have witnessed the progressive clinical course of ALS—a disease associated with baseball star Lou Gehrig in the United States and with actor David Niven in Britain and its Commonwealth—can attest to the dramatic consequences of the underlying motor neuron degeneration. Patients usually present with signs of lower-motor-neuron deficits, with weakness, atrophy, and fasciculation of limb muscles, or with involvement of bulbar musculature. Upper-motor-neuron changes signaled by spasticity, hyperactive deep tendon reflexes, and extensor plantar reflexes (Babinski sign) are often present early in the course of the disease. Muscle weakness progresses steadily and becomes more widespread and symmetrical. Eventually, the victim expires from respiratory failure or related causes, usually within three to five years of diagnosis. The underlying damage to the nervous system consists of progressive degenerative changes and loss of nerve cells in the motor cortex (upper motor neurons) and of lower motor neurons in spinal cord (anterior horn cells) and brain stem nuclei. Neuronal compromise is accompanied by loss of axons in corticospinal tracts and in motor nerves, the latter leading to atrophy of denervated muscles.

Parkinsonism with progressive dementia is being recognized increasingly as a variant of Alzheimer's disease both in Europe and in the United States. In Guam, where P-D is phenomenally frequent, patients characteristically show eventful premorbid conditions such as obesity, essential hypertension, late-onset diabetes, hyperuricemia, and significant trauma. One to five years prior to the onset of extrapyramidal dysfunction, psychoneurotic complaints such as dizziness, nervousness, easy fatiguability, loss of appetite or libido, and excessive sleepiness, may mask the insidious onset of organic brain syndrome or parkinsonism. Memory loss, disorientation, and personality changes follow. Early deterioration of fine movements, decreased blinking, bradykinesia, and eventually increasing rigidity with impaired postural reflexes occur. Tremors are said to be less common and sometimes different from those classically associated with Parkinson's disease. As the disease progresses, incontinence of urine and feces, osteoporosis and fractures from falls, anemia, and extensive bedsores develop, and patients finally succumb to intercurrent infections. Upper and lower motor neurons are affected as a rule, and quadriplegia in flexion, irreversible contractures of all joints, extension of head, blepharospasm, and a total mutistic and demented state, develop in the advanced stage. Patients may lie in this tragic and pathetic condition for many years prior to death.

Both ALS and its P-D clinical variant have been phenomenally common among three population groups in the western Pacific: (a) Chamorros of the Mariana islands of Guam and Rota and others who have adopted their life-style, (b) Auyu and Jaqai linguistic groups of the southern lowlands of Irian Jaya, and (c) Japanese residents of northern Kii peninsula of Honshu island. In the 1950s the indigenous Chamorro residents of Guam showed incidence rates for ALS more than 100 times that for the population of the continental United States. Subsequent surveys on Guam revealed that P-D was also encountered with remarkable frequency. Moreover, overlapping forms could be demonstrated, suggesting that the two disorders were closely related. Some patients with slowly evolving motor neuron disease later developed parkinsonism or P-D; in others, the appearance of P-D was followed by amyotrophy and spasticity, whereas in yet others, there was a more or less simultaneous onset of both ALS and P-D. Similarly, among Japanese residents of the Hobara region in the Kii peninsula focus of ALS/P-D, ALS incidence rates for the period 1946 to 1965 exceeded the average for Japan by more than 100 times. In the 1960s, a third focus of ALS and parkinsonism (with or without dementia) was discovered among Auyu and Jaqai linguistic groups residing in the southern lowlands of Irian Jaya (then New Guinea), Indonesia. The crude incidence rate for ALS among these seminomadic people was even greater than that for either the Chamorros of Guam or the Japanese of the Kii peninsula.

Neuropathological studies of Guam ALS and P-D demonstrated in both diseases the presence of Alzheimer-type neurofibrillary tangles with few senile plaques, a critically important observation that unified the two conditions. Comparable neuropathological changes were observed in Kii peninsula ALS/P-D. Additionally, a similar pattern of neuropathology was found in 70 percent of 302 brains of Chamorros who had shown no neurological deficit at the time of death, suggesting that subclinical forms of the disorder pervaded the indigenous population of Guam. According to Dr. Leonard Kurland, the extent of neurofibrillary change that one would expect to see at age 75 in the population of the U.S. mainland would be found at age 45 or 50 in Chamorros who had no overt neurological disease. Recent postmortem studies of Guam ALS/P-D have shown plentiful senile plaques of Alzheimer type; the Davies Alz-50 protein of Alzheimer's disease is also found in Guam ALS/P-D; and the amino acid sequence of the neurofibrillary tangles in both diseases is identical. Taken in concert, therefore, remarkable homologies seem to exist between western Pacific ALS/P-D and Alzheimer's disease.

WHAT IS THE CAUSE OF WESTERN PACIFIC ALS/P-D?

The phenomenally frequent and often familial occurrence of ALS/P-D in three western Pacific pockets has encouraged the search for a common etiology. Because the incidence of motor neuron disease is declining in all three zones, and inherited and viral factors have been virtually ruled out, the search for etiology has focused on nontransmissible environmental factors that are disappearing as the susceptible population groups acculturate to modern ways.

Two major etiologic hypothesis have been advanced, neither of which is mutually exclusive: one invokes metal intoxication promoted by a dietary deficiency of calcium, whereas the other implicates the neurotoxic cycad plant used for food and medicine. Examined here are the evidence supporting these proposals and some of the implications that arise if the favored cycad hypothesis is correct.

The hypothesis of defective mineral metabolism proposes that chronic nutritional deficiency of calcium and magnesium provokes a secondary hyperparathyroidism, leading to increased absorption of potentially toxic metals and the deposition of calcium, aluminum, and other elements in neurons of subjects who then develop ALS/P-D. Intraneuronal accumulation of metal elements is proposed to interfere with slow axonal transport by altering neurofilament production, resulting in excessive neurofilament accumulation and formation of neurofibrillary tangles. One of the major lines of evidence supporting this novel proposal has come from the results of analyses of soil and drinking water in the high-incidence foci of ALS/P-D, where calcium concentrations have been reported to be very low. These data have been challenged recently by an extensive and authoritative study of the rivers of Guam, which found high concentrations of calcium characteristic of borderline hard waters and mean aluminum concentrations similar to those in rivers of the southeastern U.S. mainland. Other recent investigations have revealed adequate calcium concentrations in traditional foodstuffs.

Although these data appear to have dislodged the cornerstone (deficient dietary calcium) of the mineral dysmetabolism hypothesis, there is nonetheless concrete evidence of heavy intraneuronal deposition of calcium and aluminum in Guam ALS/P-D. However, it must be recalled that although aluminum has neurotoxic potential, the associated neuropathologic change of neurofilament accumulation is nonspecific and distinct from the paired helical filaments that characterize the Alzheimer neurofibrillary tangle. Aluminum deposition also occurs in Alzheimer's disease and dialysis dementia, two disparate dementing disorders with distinctive neuropathologic changes. Taken in concert, therefore, the intraneuronal accumulation of aluminum and other metals in ALS/P-D is likely to be a secondary pathological feature of these neurodegenerative disorders. This view is in keeping with the results of experimental studies that have been unable to induce clinical signs of motor-system disease in primates fed a low calcium-magnesium diet, with aluminum lactate added to the drinking water.

Presently favored as the key trigger of western Pacific ALS/P-D is the seed of the cycad (*Cycas* spp.), an established cause of locomotor dysfunction in animals grazing on the plant. Many species contain chemicals with neurotoxic, carcinogenic, and teratogenic properties.

CYCADS: POISONOUS PLANTS USED FOR MEDICINE AND FOOD

The living cycads belong to an ancient line of nonflowering, gymnosperm-like seed plants that arose from the cycad ferns in late Paleozoic times and flourished in the Mesozoic era (200 million years ago). Thus, they are "living fossils" distinct from all other contemporary plants except the equally ancient Ginkgoaceae. Once distributed over the entire planet, cycads became restricted during the last Ice Age to tropical and subtropical climes of both hemispheres. The nine genera in the single surviving order of the Cycadales include the Australian *Macrozamia* (east, central, and west Australia) and *Bowenia* (northeastern

Queensland); *Zamia* of the Americas (southern Florida, West Indies, southern Mexico to the Amazon, and Peru); and *Cycas,* the genus of current interest in relation to the etiology of ALS/P-D. *Cycas circinalis* has the widest distribution, extending over an enormous range from the Mariana islands (including Guam) southeast to Fiji, Tonga, Samoa, and west to New Caledonia, Queensland, New Guinea, the Philippines, Indonesia, coastal Indochina, Bangladesh, Sri Lanka, Madagascar and the Mascarenes, Zanzibar, and part of the eastern African coast. Other relevant *Cycas* species are *C. revoluta* (Japan to Taiwan), *C. media* (northern Australia and Queensland), and *C. cairnsiana* (northern Queensland). Cycads often grow in poor, rocky soils under exposed conditions where other plants are unable to compete, but they also survive in wet forest environs. Hardy by nature, they are resistant to typhoon, flood, and drought, and are among the first plants to regrow after fire has destroyed the vegetation. They are commonly grouped in stands (e.g., Groote Eylandt, Australia), although individual examples survive in relative isolation in the midst of forests of other plants (e.g., southeastern Irian Jaya).

Mistakenly considered by nonbotanists as palmlike, cycads are mostly terrestrial and arborescent, with an unbranched aerial stem (trunklike) covered with persistent leaf bases as in *Cycas,* or shrubby with a subterranean, tuberous stem as in some *Zamia* species. Cycad leaves are borne terminally on the stem in the form of a large crown; when young, the leaves are soft, succulent, and fernlike, but they acquire a stiff, plastic-like quality upon maturation. Reproductively, cycads are either male or female, modified leaves (sporophylls) bearing either pollen or ovules in conspicuous conelike structures (strobiles) resembling pineapples. In *Cycas* spp., only the male develops a strobilus, whereas the female sports elongate, seed-bearing megasporophylls that together form a crown when young, but later droop individually around the stem displaying the mature seed. Cycad seed is often brightly colored (red, brown, green) and has the appearance of an edible fruit, features that have undoubtedly encouraged human contact. The mature seed (commonly mislabeled, nut) has a fleshy outer husk (sarcotesta) covering a thin, stony shell that houses the starchy kernel (female gametophyte and embryo).

Despite the physical beauty of many cycad species, they are all exceedingly poisonous plants. This statement applies especially to the young leaves and immature seed kernel, the pith of the stem, and the roots. The fleshy seed cover is said to lack poisonous properties, whereas the gum that exudes from the seed micropyle or from the broken megasporophyll or leaf stem of *Cycas* spp. appears not to have been assessed for toxic potential. The noxious property of cycads has not deterred many animal species, including *Homo sapiens*, from using them as food. Not only have humans in times past used all parts of cycad plants for their nourishment, they have also employed various components for medicine, recreation (toys such as whistles), and other purposes. Moreover, as if to encourage human contact, some cycads have acquired popular names such as wild pineapple (*Macrozomia*) and wild date (*Encephalartos*). Other names that conceal the toxic potential of these plants stem from the use of their pith for sago: wild sago (*Zamia floridana*); false sago palm (*C. circinalis*); bead palm (*C. rumphii*); sago palm (*C. revoluta*). Not only is the popular English name of the latter identical to the true sago palm (*Metroxylon sagu*), but the Japanese name sotestsu is the subject of a well-known love song, a point that emphasizes the degree to which these highly poisonous plants have been incorporated into human culture and life-style. In general, food, medicinal, and recreational uses of cycads are associated with poverty (and, therefore, denied), ignorance, or lack of modern cultural development, whereas their indoor and outdoor ornamental application in developed countries is primarily a hallmark of the (equally ignorant) affluent. Uses of cycads in the high-risk foci of ALS/P-D in the western Pacific include medicinal purposes (Guam, Irian Jaya, Kii peninsula); food, beverage, and as a confection or chew (Guam); and recreational purposes such as childrens' playthings (Kii peninsula). Individual cases of ALS in these regions have been linked with heavy exposure to *Cycas* seed kernel used as a topical medicine (Irian Jaya, Guam), an oral medicine (Kii peninsula), or a foodstuff (Guam). Obviously, the cardinal principle of the cycad hypothesis is that only certain individuals sustain exposure to cycad chemicals in a sufficient amount to exceed the dose threshold for clear-cut clinical expression of disease. As illustrated below, certain human groups have discovered how to remove the poisonous elements from cycads to a sufficient degree that they are usable as a valuable source of food without recognized adverse health effects. Prominent among these groups is the Australian aboriginal.

Although the practice is now restricted, if not rare, the Australian aboriginal has used cycads as food for at least an estimated 4,500 years. Fire was probably used to clear vegetation and to synchronize cycad growth and seed production. Prior to 1900, the fleshy seed coat of *Macrozamia* spp. was eaten by aboriginals; its flavor was found by some Europeans to resemble that of chestnut. Seed of *C. media* was a significant component of the aboriginal diet in the Northern Territories, including Groote Eylandt where the practice seems to have disappeared. The Tiwi of Melville Island reportedly consumed cycad seeds at least as recently as the 1950s. At least two Australian doctoral theses have discussed in detail the various methods (burial, water leaching, roasting) used by aboriginals to "detoxify" the seed.

The acute toxic effects of cycads—variously exploited in the past to execute criminals (Honduras), poison rats, and kill unwanted children (Celebes) or enemies (Costa Rica)—were discovered unwittingly by some early European explorers of Australia. Although credit for the first bad experience has been given to Cook's party, Beaton argues that he was preceded in his innocent discovery of the poisonous effects of cycad (and of Australia!) by other explorers.

On December 29, 1696, Willem de Vlamingh, in charge of an expedition mounted by the Dutch East India Corporation, anchored his frigate Geelvinck on the east side of Rottnest Island just a few miles from what is now called Perth. Some days later, a party of 86 men went ashore on the mainland and broke up into three reconnaissance parties. In one was an officer who conducted the first known European experiment with cycads:

> The 6th in the morning we split into three Parties, each taking a different route, to try if we could find some Men. After two or three hours we rejoined the company near the River without discovering anything but a few huts and footsteps; so we took a rest. In the meantime they brought me the nut of a certain Fruit *[Macrozamia]* resembling the form of the Drioens, and having the taste of the Dutch great Beans, and those which were younger were like a hazelnut. I ate five or six of them, drunk the water from one of the already mentioned pits; but after about three hours I and five others who had eaten of these Fruits, began to vomit so violently that there was hardly any difference between us and death; so it was with the greatest difficulty that I with the Crew reached the shore...

(Robert, 1972, p. 63).

Some years later, in August 1770, on the east coast of Cape York the present site of Cooktown, Joseph Banks records what Beaton (1977) calls the second Australian experiment in action ethnobotany:

> Palms here were of three different sorts.... The third... was low, seldom ten feet in hight, with small pennated leaves resembling those of some kinds of fern; Cabbage it had none but generaly bore a plentifull crop of nuts about the size of a large chestnut or rounder. By the hulls of these which we found plentifully near the Indian fires we were assurd that these people eat them, and some of our gentlemen tried to do the same, but were deterrd from a second experiment by a hearty fit of vomiting and purging which was the consequence of the first. The hogs however who were still shorter on provisions than we were eat-

ing them heartily and we concluded their constitutions stronger than ours, till after about a week they were all taken extremely ill of indigestions; two died and the rest were savd with difficulty. . . (Beaglehole, 1962).

Later, reflecting on the episode, Banks wrote that some of the crew who only ate one or two of the nuts were ". . . violently affected by them, both upwards and downwards, and our hogs whose constitutions we thought might be as strong as those of the Indians [aboriginals] literally died after having eat them. It is probable however that these people have some method of Preparing them by which their poisonous quality is destroyed. . . " (Beaglehole, 1962).

Many other examples of acute cycad poisoning, sometimes with fatal outcome are recorded among early settlers in Australia and elsewhere. Eventually, however, the migrant Europeans learned how to "detoxify" the cycad. In the 1920s, settlers in western Australia exploited the stem starch for human consumption, as laundry starch, and as a commercial adhesive. Even as recently as the 1950s, cycad starch was found suitable for conversion into glucose. *Macrozamia* starch has also been recommended as food for poultry, pigs, and calves. By contrast, consumption of raw cycad leaves by grazing cattle has been the cause of locomotor dysfunction and death of huge numbers of animals in Australia and elsewhere. Here is an interesting topic for study by the behavioral toxicologist, because the problem of neurocycadism in cattle (zamia staggers) continues to this day.

Lest the American reader be tempted to scoff at the seemingly extraordinary Australian practice of eating patently poisonous plants, read on! Zamia in Florida was eaten by an extinct group of Florida aboriginals (sixteenth century), thereafter by Seminole Indians who relied upon it during the long wars with the United States, and later by slaves and by white settlers. In 1898, Cuzner notes authoritatively in the *Journal of the American Medical Association:*

> When the poor whites on the east coast are greatly in need of money they go to the woods to dig koonti [*Z. floridana*], finding a ready market for the roots. Indeed, it is the sole occupation of many people. The roots are not cultivated, as they grow wild in great abundance. A very fine quality of starch and tapioca is manufactured from them, which may be found at all times in the Key West market. . . . The starch is said to equal the best Bermuda arrowroot and lately its worth as an article of commerce has been fully recognized in Florida. There are a number of factories for its preparation in Southern Florida. A correspondent of the United States Agricultural Department writes: "I ate of koonti pudding, at Miami, and can say that, as it was there prepared, and served, with milk and guava jelly, it was delicious."

At their peak, mills along the Miami river processed 10–15 tons of the tuberous underground stem of *Z. floridana* and *Z. pumila*, most of which was marketed under the name Florida arrowroot for use in infant foods, biscuits, chocolate, and spaghetti. Gifford noted that water used in washing the starch produced "slow poisoning" when drunk by animals. By 1950, nearly 6 million pounds of sago (prepared from *C. revoluta*) was imported into the United States from the Dutch East Indies for use in the preparation of food, syrup, beer, and adhesives, as well as sizing for paper and textiles.

Although the exposure of seemingly large numbers of people in certain developed countries to small amounts of "detoxified" cycad products may hold more than passing interest, the focus here is on populations that have sustained heavy exposure to patently nondetoxified cycad seed. This takes place when the raw seed is used as a topical or an oral medicine, or when time is too short to complete the lengthy procedure of detoxication by drying, water leaching, or fermentation. Such situations occur during famine periods associated with war or following typhoon, when sources of food other than the hardy cycad are destroyed. In 1696, Cleyer emphasized the importance of sotetsu sago (obtained from the stem of *C. revoluta*) as a famine food and, in 1802, Smith reported "the plant is jealously preserved for use by the Japanese army [because] the pith of its stem is [after careful preparation!] sufficient to sustain life for a long time." The nineteenth and early twentieth century literature is replete with examples of the use of cycads for famine food in parts of Asia and Africa. Preparation ("detoxication") methods differ widely not only between societies but also between families. For example, opinions expressed by Chamorro women of Guam and Rota demonstrate a remarkable range of time (2–30 days) considered appropriate to leach out the poisonous principles of *C. circinalis* seed prior to consumption. In the past, Chamorro children were said to fall sick after eating cycad products, a few died if preparation was poor, whereas the majority ingested the material without noticeable adverse health effects.

Few societies would be expected to be able to connect an acute illness with a chronic disease that appeared many years later. Nevertheless, in 1905, Safford wrote that Guamanians were well aware that cycad starch was injurious to health if consumed for any length of time, and Whiting recorded in the 1960s that some residents attributed a variety of ailments, including an incapacitating paralysis (lytico)—many cases of which are diagnosed as ALS—to the handling and consumption of cycad plant material.

Gaudichaud first recorded the use of *Cycas* for food among the aboriginal inhabitants of Guam. By 1900, cycad consumption was much less common, except when maize was scarce or in times of famine after typhoon. However, during the Japanese occupation of Guam (1941–1944), when rice was hard to obtain, cycad seed represented a major source of food for the indigenous residents of Guam. Cycad flour, used to make tortillas (bread), atole (beverage), and soup, is prepared from the kernel; this is removed from the husk, halved, quartered, sliced or crushed, soaked for days or weeks in fresh water that is changed periodically (usually daily) to remove poisonous principles, dried in the sun, and ground into flour. During the wartime occupation, however, the beleaguered Chamorros were sometimes forced through hunger to forego the niceties of cycad preparation and consume products prepared from only briefly washed seed. Additionally, the fresh seed integument was used as a chew "to relieve thirst" and, after drying, as a confection. Cycad seed also enjoyed common use as a Chamorro medicine during World War II: freshly grated cycad seed kernel, or the juice therefrom, was applied to open wounds, leg ulcers, abscesses and boils, and left in place as a poultice for several days. Food and medicine shortages continued after the Americans displaced the Japanese from Guam, and there is a strong possibility that a number of U.S. (and Australian) servicemen active in Guam (and other parts of the Pacific theater) consumed inadequately prepared cycad material during this time. However, as Guamanians have progressively acculturated to the contemporary food and medical practices of the continental United States, the use of cycad for these purposes has declined along with the incidence of ALS. Whereas cycad was the famine food of the Chamorro, today the material is considered a traditional delicacy! Preference for traditional Chamorro food was recently reported to be the only one of 23 risk factors tested that showed a significant relationship with P-D. In the face of this and other data linking cycad to neurodegenerative disease, the governor of Guam strongly discouraged further use of this poisonous plant for anything but decoration.

The link between cycad and ALS was first proposed more than 25 years ago in the writings of Dr. Marjorie Whiting and Dr. F. Raymond Fosberg who were consulted by Dr. Donald Mulder and Dr. Leonard Kurland when they first suspected a toxic nutritional factor operating in Guam ALS. Between 1962 and 1972, six international conferences instigated by the National Institutes of Health in the United States were held to consider the possible relationship between Guam ALS/P-D and the traditional Chamorro practice of employing cycad seed for food. Only the third (Lyon Arboretum, 1964) and sixth (Lyon Arboretum, 1972) conferences were published, but the previously forgotten gems of the other four have become available recently (Lyon Arboretum, 1988). Although the initial focus of these conferences was to elucidate why neurodegenerative disorders were so common in certain of the Mariana islands (a possible link with lathyrism was entertained in the opening remarks), the important discovery of cycasin—a potent, single-shot experimental carcinogen in this and other cycad plants—progressively distracted attention away from the central issue toward the more popular concern of the mechanisms underlying carcinogenesis. By 1972, the cycad hypothesis for western Pacific ALS had fallen into disfavor for two principal reasons: cycads were reported *not* to be used for food in either the Kii peninsula or the Irian Jaya foci of ALS, and nonprimate laboratory animals repeatedly exposed to cycad products failed to develop a paralytic disorder. However, what must remain as a most puzzling oversight was the publication in the third conference of an experimental feeding study in *Macaca mulatta* which resulted in one of the three tested animals developing unilateral arm weakness, with neuropathological evidence of nonreactive degeneration of motor neurons (possibly the first primate model of ALS, this important experiment was neither confirmed nor refuted). Because cycad flour lacked cycasin by the detection methods available at the time, the presence of an unidentified noncycasin neurotoxin in *Cycas* spp. was predicted.

FROM LATHYRISM TO CYCADISM

During the initial cycad conferences, debate on the possible relationship between lathyrism and Guam ALS revolved around the observation of skin and bone abnormalities in Chamorro subjects that recalled those of osteolathyrism. This is a purely experimental disease of skin, bone, and blood vessels caused by the disruption of collagen and induced in animals by a gamma-glutamyl derivative of beta-aminopropionitrile isolated from *Lathyrus odoratus* (sweet pea), a species etiologically unassociated with human (neuro)lathyrism. The neurotoxic species of *Lathyrus* (*sativus, clymenum, cicera*) contain a nonprotein convulsant amino acid, beta-N-oxalylamino-L-alanine (BOAA), which was isolated in the early 1960s; BOAA was assumed to cause human lathyrism, a nonconvulsant spastic disorder of upper motor neurons induced by heavy consumption of neurotoxic *Lathyrus* spp. By drawing an analogy between the superficially similar paralytic effects of lathyrism and cycadism in domestic animals, a search was begun by E. A. Bell for a BOAA-like compound in *Cycas* spp. Bell failed to find BOAA but, at the fifth cycad conference, described the identification of a related convulsant amino acid, beta-N-methylamino-L-alanine (BMAA). However, at the sixth conference in 1972, Bell and his colleagues reported the disappointing news that prolonged (78 days) feeding of rats with subconvulsive doses of BMAA failed to elicit paralysis. His team concluded that BMAA was unlikely to be linked with the etiology of Guam ALS. Thereafter, there was little interest in the possible association between exposure to cycad and the development of human neurological disease despite the existence of clear-cut evidence of the paralytic properties of cycad poisoning in animals (see below).

Almost a decade later, in June 1981, I restated the possible relationship between lathyrism and Guam ALS, and called for an evaluation of the effects of prolonged administration of subconvulsive doses of BOAA and BMAA in a suitable laboratory animal (macaque). The first step, however, was to define the neurology of human lathyrism and produce a satisfactory primate model in which the action of BOAA could be examined. Thus, in succeeding years, the socioeconomic setting as well as the clinical and neurophysiological features of this disease were subjected to detailed scrutiny by colleagues from Europe, Africa, Asia, and the United States.

Lathyrism is a major cause of motor-system disease in endemic regions of Bangladesh, India, and Ethiopia, where prevalence estimates range up to an extraordinary 2.5 percent. In these regions, *L. sativus* is consumed as a component of the staple diet and, after flood or drought when other crops are destroyed, as an insurance crop and famine food. The neurotoxic effects of the chickling pea develop when it constitutes a major part of the diet for a period of weeks or months. Presenting symptoms typically consist of muscle cramping (especially calf muscles), and uncommonly include myoclonus, urgency and frequency of micturition, and nocturnal erection and ejaculation. One or more of these clinical manifestations of central nervous system (CNS) excitation may precede development of leg weakness and then wane or disappear once intake of the toxic diet is reduced or abandoned. The individual seems either to recover or to be left with varying degrees of spastic paraparesis, indicative of permanent dysfunction of selected regions of the corticomotoneuronal system. Those least affected run with difficulty, due to thigh adductor spasm, and walk with a stiff-legged gait; more severely compromised subjects have gastrocnemius spasm and walk on the balls of their feet with a scissoring gait; the most severely disabled have total spastic paraplegia and severe leg weakness, and are forced to crawl on their knees or buttocks (because wheelchairs are unavailable). The few neuropathological studies conducted decades after onset of spastic paraparesis have revealed degeneration of long spinal tracts, notably the corticospinal pathways. Severe loss of Betz cells (upper motor neurons) may occur, most noticeably in the upper part of the precentral sulcus and in the paracentral lobule. Anterior horn cells are not lost, and in studying

Spanish peasants who developed lathyrism during World War II, we were unable to confirm reports from Israel that long-standing subjects with the upper-motor-neuron deficits of lathyrism eventually develop additional disease of anterior horn cells, a picture that would resemble ALS. Thus, the clinical picture of lathyrism is dominated by spastic paraparesis; the disease is distinct from ALS, and neither parkinsonism nor dementia is a recorded feature.

The next challenge was to develop a model of lathyrism in well-nourished primates and compare the effects of repeated, subconvulsive doses of BOAA. These studies demonstrated that myoclonus, hindlimb extensor posturing, and neurophysiological evidence of corticomotoneuronal dysfunction appeared in cynomolgus monkeys fed for 3–10 months with an exhaustively analyzed diet of chickling pea that had been supplemented beyond the minimum nutritional requirements for that species. Similar changes were brought on more rapidly (weeks) when animals were fed either *L. sativus* plus BOAA or BOAA alone. Cessation of dosing led to disappearance of characteristic signs of BOAA-induced neurobehavioral dysfunction, indicating the successful modeling in the primate of the early reversible stage of human lathyrism. International efforts are now underway to remove BOAA selectively from *L. sativus* so that this hardy crop may be used safely by human and animal populations in Asia and Africa.

Once the primate studies had confirmed BOAA as the culpable agent of lathyrism, my attention turned to the relationship between the acute neurotoxic properties of BOAA and BMAA, and then to the primate toxicity of the latter. Initial studies were performed with mouse spinal cord explants. Application of micromolar concentrations of BOAA resulted in the appearance of discrete, postsynaptic CNS edematous vacuolation in the circumventricular organs of postnatal rodents administered convulsive doses of BOAA. Armed with these data, in September 1983 I approached Dr. Peter Nunn, a member of Bell's original team, which had discontinued work on the *Cycas* amino acid more than a decade earlier. Nunn kindly provided samples of BMAA which, at higher concentrations than BOAA, proved to induce comparable stereospecific neuronal pathology in mouse CNS explants. Thus, by 1984 it was predictable from the results of these tissue culture experiments that BMAA would likely produce some type of motor-system abnormality when fed to primates. These studies with Nunn and other colleagues began in the summer of 1985. Our predictions were proved correct within a few weeks of dosing the pilot animal, although the motor-system disorder that developed was markedly different from lathyrism and clinically recalled the manifestations seen in humans with ALS/P-D. In the majority of animals, the forelimbs were affected first, with wrist drop, clumsiness, and difficulty in picking up small objects. Muscle weakness and loss of muscle bulk followed. Many animals displayed unilateral or bilateral extensor hindlimb posturing, with or without leg crossing (a feature of spasticity seen in primate lathyrism), stooped posture, unkempt coat, and tremor and weakness of the extremities. Both resting and action tremor were noted in the same animal. More prolonged intoxication led to periods of immobility with an expressionless face and blank stare, crouched posture, and a bradykinetic, shuffling, bipedal gait performed with legs flexed and rump close to the ground. Two of these animals treated with an oral antiparkinsonian drug showed selective recovery of marked facial movement and spontaneous activity. Additional features of BMAA intoxication included brief "wet-dog" shaking and limb/torso scratching, reduction or loss of aggressive behavior, disinterest in the environment, changes in normal diurnal patterns of vigilance, urinary incontinence, altered vocalization, slowed mastication, and whole-body tremor.

Electrophysiological studies demonstrated changes in the entire motor pathway, and neuropathological examination showed a hierarchy of regional susceptibility: motor cortex (most affected), spinal cord (less affected), and substantia nigra (mostly unaffected). Striking changes were found in giant Betz cells which, with smaller pyramidal cells in the cortex, underwent central chromatolysis, neurofilament accumulation, and chronic cell degeneration similar to that seen in ALS. Similar, though less marked, abnormalities were found in motor neurons of Rexed laminae VI–IX of the spinal cord. Clusters of glial cells suggestive of neuronophagia were present in one animal that had received BMAA for 17 weeks. This animal also displayed abnormal neuritic swellings, containing twisted filamentous structures embedded in an amylaceous core, in the pars compacta of the substantia nigra. Otherwise the basal ganglia, hippocampus, and cerebellum of BMAA-treated animals were similar to controls.

These data, reported for the first time in the summer of 1987, demonstrated an intriguing parallelism between chronic BMAA intoxication and human ALS/P-D. However, because the experimental disorder lacked certain important features of the human disease, notably nigral degeneration and plentiful paired helical filaments, clearcut loss of motor neurons, and muscle denervation, it was inappropriate to refer to the primate disorder as a model of ALS/P-D. However, it is noteworthy that the disorder appeared rapidly and that the neuropathology was studied weeks to months (rather than years to decades) after initial exposure to BMAA and commencement of the pathological process. Furthermore, primates were not exposed to whole cycad seed, to other environmental factors (discussed earlier) proposed as causally related to ALS/P-D, or to the combined effects of toxic damage and age-related attrition of vulnerable neurons (see below).

Although BMAA is able to induce convulsions in rodents and a motor-system degenerative disease in primates, the role of this compound in the production of seizures in humans and paralytic disease in grazing animals has yet to be established. In the past, these effects have been associated with hepatotoxic and neurotoxic properties of methylazoxymethanol (MAM), the aglycone of the cycad seed glycoside cycasin. The effects of acute cycad poisoning in humans and grazing animals are strikingly similar. Signs of acute toxicity in humans commence 12–40 hours after ingestion of incompletely detoxified cycad components. Nausea and vomiting develop suddenly, hepatomegaly and convulsions then appear, and the subject loses consciousness and usually dies. The disease in grazing animals (cycadism) is also marked by hepatotoxicity, enterotoxicity, and death. Cycadism has been responsible for the loss of thousands of head of cattle in Australia alone. Rapid twitching of the eyelids, nostrils, lips, and jaw muscles, with periodic tremors of the body, are reported in sheep, and muscle fasciculation has been noted in poisoned heifers. Animals that survive acute cycad poisoning develop some weeks later a locomotor disorder associated with weakness and wasting of hindlimbs. Initially, there is a staggering, weaving gait, with crossing of the legs, incoordination, and "ataxia." More severe forms are characterized by posterior motor weakness, dragging of extended hindlegs and, occasionally, a stringhalt-like action of the hocks. Function of bladder, anus, and tail is said to be unimpaired. The few pathological reports of this condition in cattle described degeneration of long, presumably motor tracts in the lumbar region, with changes in the fasciculus gracilis and dorsal spinocerebellar tracts in the cervical area. Information on the status of motor neurons that innervate the weakened and atrophied limbs of animals with cycadism is unavailable. Studies of the brain, spinal cord, peripheral nerves, and muscles of animals with long-standing cycadism (available in Australia and Japan) are urgently needed to determine whether changes are related to those seen in humans with western Pacific ALS/P-D.

LINKING CYCADS TO ALS IN IRIAN JAYA AND JAPAN

The second major reason for discarding the cycad hypothesis in the early 1970s was the reported absence of cycad use for food by the

Japanese of the Kii peninsula of Honshu island, or by the Auyu and Jaqai of southeastern Irian Jaya, both of whom suffered from a high incidence of ALS/P-D. Thus, in 1987, after familiarizing myself with the literature on the distribution and usage of cycads in Asia and Oceania, I set out with Valerie Palmer to determine whether the reported absence of cycad use for food in these regions was correct. Indeed it was, but other facts came to light which demonstrated heavy exposure to cycad seed of individuals who subsequently developed ALS.

Nobel Laureate D. Carleton Gajdusek had shown that cases of ALS and parkinsonism occur with remarkable frequency among Auyu and Jaqai people of the remote southern inland plain of Irian Jaya where head hunting was practiced as recently as the 1940s. The Auyu and the Jaqai (their former head-hunting neighbors) now appear to live peaceably in organized riverine villages. These were constructed by Dutch missionaries who brought the Stone Age hunter-gatherers out of their forest dwellings in the first half of this century. The villages are reachable only by helicopter or canoe. My transportation was restricted to the latter because my budget also had to cover the expenses of two Indonesian physicians, a nurse, a translator, and a policeman equipped with a machine gun for protection in the event of civil unrest! Although the physical environs were not conducive to the comfortable pursuit of research, and no cycad trees were sighted in the long hours spent canoeing through the forest rivers, it turned out to be a relatively simple matter to discover the sought-after link between ALS and cycad seed. Questioning revealed that *Cycas* seed, obtained from solitary trees whose exact location in the forest was known to the Auyu, was considered an ideal medicine for topical treatments (in individuals of all ages) of various skin lesions, including open sores. For this purpose, scrapings of the kernel of a raw seed are crushed, the resulting pulp is immersed in the poisonous milky exudate, and the sodden mass is applied directly to the lesion on a leaf, which is then strapped into position. The poultice is replaced daily with freshly prepared pulp until the skin is healed. Although this medical use of cycad seed kernel was acknowledged by many, all vigorously denied employing any part of the plant for food because it was considered poisonous. Instead, the people relied for food on sago obtained from the stem of the true sago palm (*Metroxylon* spp.). Thus, in the absence of shops, these people obtained their food and medicine from the forest. On a single occasion at age 15, one 29-year-old male with ALS of recent onset told how he had applied the preparation for a month to an open sore (5–10 cm diameter) on the ankle. An elder brother with no clinical disease said he had used a cycad poultice for only two weeks to promote healing of a deep cut of one foot. Their mother, who had taught them the procedure, was "paralyzed" before death at age 50 and, according to the ALS patient, suffered from the same disease. Immediately, the association between cycad exposure and familial ALS became clear.

Establishing a link between cycad and ALS in the Kii peninsula focus of neurodegenerative disease was more difficult. My initial suggestion of the need to study this question was greeted with surprise by a prestigious group of Japanese neuropathologists that had honored me with a guest lectureship in Osaka. Nevertheless, with the help and interest of Dr. Masayuki Ohta and Valerie Palmer, it was possible between 1987 and 1988 to establish the link. Intensive questioning of residents of Kii peninsula confirmed earlier reports that neither the seed nor the stem of *C. revoluta* had ever been used for food in this region of Japan. Indeed, sotetsu trees in Kii are used only as ornamentals for temples, schools, parks, public buildings, and adorning the gardens of the affluent. However, continued study revealed that pharmacies in the high-risk ALS area stocked untreated sotetsu seed and occasionally filled prescriptions written by practitioners of folk medicine (kitoshi). The seed was imported selectively into the region from one of the southern Ryukyu islands. Japanese tests recommend daily use of an oral aqueous steep prepared from 3–10 g of the potentially poisonous seed for the treatment of ailments such as diarrhea, dysmenorrhea, tuberculosis, neuralgia, and to "strengthen the body." Subsequently, the latter was discovered to be the reported reasoning of a concerned grandmother who administered the cycad potion repeatedly to a young girl five years of age. Additionally, the grandmother periodically provided her with a basket of young, brightly colored, and highly poisonous orange sotetsu seeds which were used by the girl as marbles, to make necklaces, and to fashion whistles. She developed into an apparently healthy teenager and did well in long-distance running races until, at the age of 18, she began to complain of backache and spasm of the calf muscles. She was diagnosed with ALS at age 20 and died some years later. Thus, in this case, as in the Auyu man from Irian Jaya, a pathological process that was not to become clinically evident for many years had silently begun at the time of the toxicologically uneventful cycad treatment. We appeared to be dealing with a previously unrecognized type of long-latency neurotoxicity involving a slow-acting toxin.

These challenging but fruitful field investigations focused on the youngest available ALS cases; if cycad exposure had occurred, it seemed likely that it would have been greatest in subjects who developed the disease at an early age. Recent events would be more memorable, and parents (who would probably still be living) would be able to record events in the earliest years of the victim's life. Using this approach, therefore, I was able to establish unequivocally that human exposure to untreated cycad occurred in all three high-risk regions for ALS/P-D. However, the common thread was not food use of "detoxified" cycad components—a practice that some cultures (e.g., Australian aboriginals) apparently had perfected—rather, it was exposure to the *untreated* cycad seed kernel as medicine. In Guam and in Irian Jaya, this took the form of topical exposure to open wounds (tropical ulcers), whereas in the Kii peninsula, oral medicinal use had been incorporated into the local culture. This is not denying the possible additional role of improperly prepared cycad food, especially since a recent epidemiological study linked traditional Chamorro food with P-D.

These studies also revealed that cycad use was declining in all three regions, thus accounting for the declining incidence of ALS. In the Kii peninsula, cycad prescriptions are now issued by a few elderly kitoshi and the practice is dying out. In Irian Jaya, the recent introduction of Dutch and Indonesian education, coupled with the limited availability of relatively modern methods of medical treatment, is reducing Auyu dependence on cycad seed for medicine. Finally, in Guam, the medical practices of the Chamorro folk doctor (surahana) have been curtailed by law, and food products prepared from cycad seed are viewed as delicacies. Thus, the declining patterns of utilization of cycad seed in all three communities at risk for ALS/P-D fulfill the criterion of a disappearing environmental factor required as an etiological link to ALS/P-D.

TIMING OF CYCAD EXPOSURE

If cycad is the principal trigger for western Pacific ALS/P-D, when does the critical exposure occur, and do the timing and degree of intoxication influence the nature of the resulting clinical compromise? Although the answers to these important questions are unknown, some clues link motor neuron disease in the western Pacific loci with cycad exposure at a remarkably early age. One is the presence of multinucleated and misplaced neurons in the cerebellum and vestibular nuclei of some Japanese and Guamanian subjects with ALS/P-D. This suggests exposure during the later phases of brain development (up to the age of 1 year) to an agent that arrests the developmental mitotic and migratory responses of neurons; one such substance known from experimental rodent studies is the neuroteratogen MAM. Administration of MAM to newborn rats results in cerebellar microplasia associated with misplaced and multinucleated neurons, an experimental observation not previously connected with the human disease. A second, more direct link, is the appearance of ALS in Japanese and Irian Jaya subjects within 10–15 years after exposure to cycad seed in the first or second decade of life, as discussed earlier. This is consistent with the observation that some Chamorro subjects developed ALS 1– 34 years after leaving Guam; based on age of migration, the minimum duration of exposure to the Guam environment to have acquired disease susceptibility was the first 18 years of life. Additionally, according to longitudinal data for the incidence of neurodegenerative disease in the Chamorro, the peak for ALS among males in 1955 followed approximately 10 years after the period estimated for maximum reliance on cycad for food and medicine, whereas the peak for P-D occurred in 5–10 years. Taken in concert, therefore, these data indicate that ALS/P-D is a long-latency disorder that may be acquired years or decades prior to clinical expression.

Changes in the demographics of ALS and P-D that have occurred as cycad use has declined provide clues as to why some patients develop ALS, and others—who tend to be older—develop P-D. The incidence of ALS among male Chamorros was twice that of P-D in the early 1950s, whereas 20 years later the relative proportion of such cases was inverted. Although comparable data are unavailable for Kii peninsula ALS/P-D, a similar proportional decrease of ALS (relative to parkinsonism) appears to be occurring in the epicenter of the Irian Jaya disease focus. Additionally, teenage cases of ALS are no longer seen in any of the high-incidence disease areas and, on Guam, the mean ages for onset of ALS and P-D (seen in older subjects) have increased over the past 30 years. These several pieces of data are consonant with the proposal that degree of intoxication is a critical factor dictating both the age of onset and the clinical characteristics of the resultant disease. Specifically, heavy exposure may precipitate ALS by lethally damaging motor neurons. The heavily exposed subjects also sustain some damage to the apparently less susceptible nigrostriatal pathway—a silent change demonstrable neuropathologically and by fluorodopa positron emission tomography, but because the lesion is insufficient to overcome the large functional reserve of this pathway (thereby permitting the clinical expression of parkinsonism), the victim seemingly suffers only from motor neuron disease (ALS). Other subjects who are less heavily intoxicated (or have a lower degree of susceptibility) may survive for many years with motor neuron compromise and eventually develop parkinsonism as a consequence of the additive effects of toxic damage and attrition of the age-susceptible nigrostriatal neurons. A further possibility is that dementia represents the late effects of the lowest clinically significant level of cycad exposure, whereas the majority of elderly Chamorro subjects with subclinical neurofibrillary tangles represent those with extremely low exposure or a relatively high tolerance. In summary, therefore, this working hypothesis states that the various forms of western Pacific neurodegenerative disease represent individual points on a three-dimensional dose-response curve for plant toxicity in which time represents the third dimension.

RELEVANCE FOR VETERANS OF THE PACIFIC WAR

If our working hypothesis is correct, individuals who receive small doses of cycad toxins develop a dementing illness in old age that could be readily mistaken for Alzheimer's disease. Because many U.S., Australian, and Japanese veterans were subjected to shortages of food and medicine in the Pacific theater of World War II, it is possible that some resorted to cycads during this period. This point is of special relevance for U.S. personnel serving on Guam during the immediate postwar period. However, surveys conducted in the 1970s revealed that neither the 2 million U.S. Armed Forces veterans who passed through Guam or the northern Marianas since 1945, nor some 10,000 U.S. construction workers in Guam from 1945 to 1954, had a measurable increased risk for ALS/P-D. Unfortunately, these studies were conducted on many subjects who had yet to reach the present mean age for onset of (male) Guam ALS (52 years) or P-D (60 years). In view of this consideration and several anecdotal reports of U.S. veterans who developed ALS or Alzheimer's disease after serving on Guam or New Guinea, it seems important to reexamine this issue.

TOXIC COMPONENTS OF CYCADS

Two toxic agents have been identified in *Cycas* seed: cycasin(s) (the active form of which is MAM) and BMAA: MAM is a potent hepatoxin, carcinogen, and teratogen, which is also able to interrupt cerebellar development in mice; BMAA is an excitant neurotoxin that produces a motor-system disorder in primates after repeated subconvulsive dosing. Although BMAA is present in low concentrations relative to that of cycasin in cycad seed kernel, what constitutes a significant dose of BMAA has yet to be established. Dr. Glen Kisby has recently developed a method to quantify BMAA in plant and animal tissue after derivitization of amino acids with fluorenylmethyl chloroformate (FMOC), separation by high-performance liquid chromatography, and detection of FMOC-BMAA by fluorescence. Recent studies from our laboratory show that BMAA (1) increases in concentration with time after the seed kernel of *C. circinalis* has been crushed, (2) appears when serum is incubated in vitro with serum alanine, and (3) is found in serum of primates after intraarterial injection of MAM. The potent methylating action of MAM therefore has the potential to increase the concentration of BMAA in both plant and animal tissue. Other studies show that cycasin in cycad leachate diminishes in concentration prior to BMAA. Thus, individuals ingesting seed kernel that has been soaked in water for only a few days would receive doses of both MAM and BMAA. However, larger doses of both compounds would probably be absorbed/generated in subjects with an open wound treated topically with a cycad poultice; this practice provides 24-hour/day subcutaneous exposure for weeks, with daily replacement of fresh plant material.

The precise locus of initial action of BMAA in the nervous system has been determined with a combination of tissue culture and animal studies in which synthesized *L. sativus* neurotoxin (BOAA) has served as a comparison compound. These studies showed that micromolar concentrations of both compounds rapidly induced postsynaptic dendritic neuronal changes comparable to those seen with potent glutamate analogues (excitotoxins) which cause sustained depolarization of neuronal membranes and thereby trigger convulsions; BMAA was less potent and slower acting than equimolar concentrations of BOAA.

The differential neuronotoxic (excitotoxic) action of these compounds was delineated by pretreatment of the tissue explant with drugs acting selectively at glutamate receptor subtypes. Glutamate receptors, presumably located on the surfaces of dendrites and neurons, have been classified according to their responses to three potent agonists: N-methyl-D-aspartate (NMDA), quisqualate, and kainate. Whereas the neuronotoxic action of BMAA in cortical explants is dose dependently attenuated by 2-amino-7-phosphonoheptanoic acid, a selective antagonist for the NMDA receptor, BOAA neuronotoxicity was similarly impaired by pretreatment with piperidine dicarboxylic acid (PDA), a nonspecific antagonist that blocks the action of the plant-derived excitotoxic amino acids kainate and quisqualate. These data, collected by Dr. Stephen Ross, suggest that the neurotoxic action of BMAA is mediated by the NMDA receptor complex.

Dr. Ross made comparable observations in young CD-1 mice by measuring the duration of hyperexcitability induced by BMAA or BOAA administered by intracerebroventricular injection (i.c.v.). The BMAA induced a transient hyperexcitable state followed by long-lasting whole-body shaking and wobbling. Pretreatment of mice i.c.v. with the NMDA antagonist AP7 provided complete protection against this BMAA-induced behavioral change; AP7 also showed a nonsignificant trend for protection against the early, transient hyperexcitable state caused by injection of BMAA. By contrast, PDA had no effect on BMAA-induced hyperexcitability, although this drug was dose dependently active in attenuating the seizuregenic responses triggered by administration of BOAA i.c.v. These results confirmed the differential acute neurotoxic actions of BOAA and BMAA, as well as suggesting again that the neurotoxic action of BMAA is mediated via the NMDA receptor complex.

These neuropharmacological studies have been supplemented and extended by measuring electrophysiologically the time course and patterns of BOAA- and BMAA-induced depolarization of synaptically interconnected mouse hippocampal cells grown in primary cell culture. Voltage-clamp recordings made using the whole-cell configuration of the patch-clamp technique have shown that similar waveforms of currents activated by BOAA also displace specific ligands for the quisqualate site in mouse brain tissue. In comparable electrophysiological studies, pressure injection onto the cell surface of either NMDA or the BMAA elicited similar patterns of depolarization. Under the same conditions, both the NMDA and the BMAA currents were blocked by AP7. Taken in concert, therefore, these studies suggest that whereas BOAA preferentially (and reversibly) binds to the quisqualate receptor, BMAA acts in its parent form at a site associated with the NMDA complex.

Although the different neuronal receptors targeted by BOAA and BMAA may be etiologically linked to the distinct patterns of neuronal vulnerability in human lathyrism (cortical motor neuron) and western Pacific ALS (upper and lower motor neurons, substantia nigra, and hippocampus), the motor-system disorders show another clinical distinction that must be cardinally important to an understanding of their pathogenesis. Whereas lathyrism is a largely *self-limiting* disorder that typically appears subacutely in individuals who consume excessive amounts of BOAA-containing chickling pea for several weeks or months, ALS/P-D appears to be a long-latency disease that is triggered years or decades prior to the clinical appearance of a *progressive* disease. Because the onset of western Pacific ALS in teenagers cannot be explained by concurrence of toxic damage and age-related attrition of the same cellular population, there must be another explanation for this tardive phenomenon. Thus, it seems likely that cycads contain factors ("slow toxins") which are able to penetrate neurons and establish irreversible changes that trigger their progressive downfall. Because changes of this type are unlikely to be mediated via the cell surface, it is important to determine whether BMAA or other cycad toxins are able to enter selected neurons (perhaps by way of the structural entity that is critical for cell survival). One important question under investigation is whether this event occurs at the level of the cell nucleus.

SUMMARY

Three of the most devastating degenerative disorders of the human nervous system—ALS, parkinsonism, and progressive presenile dementia—have been present in combined form and remarkably high incidence at three foci in the western Pacific region. In all three areas, ALS is associated with oral or percutaneous exposure to the untreated seed kernel of *Cycas* plants. Consumption of *Cycas* or other cycads precipitates a poorly defined locomotor disorder in grazing animals (cycadism). Although the culpable neurotoxic agent has yet to be identified, at least two compounds with differential neurotoxic properties have been previously isolated from *Cycas* seed: BMAA and MAM, the latter being capable of generating BMAA in serum. Methylazoxymethanol is an experimental hepatotoxin, carcinogen, and neuroteratogen: it arrests rodent cerebellar neuronal migration and mitosis during development, and multinucleated and displaced neurons in some cases of ALS/P-D may represent a biological marker of early exposure to this compound. Beta-N-Methylamino-L-alanine is an excitant neurotoxin in rats and a neurotoxin in mouse cord and cortex explants; neuropharmacological and neurophysiological studies indicate that the depolarizing, excitotoxic, and neuronotoxic actions of BMAA are mediated by the NMDA-receptor complex. Although BMAA produces an interesting and potentially important constellation of motor neuron, extrapyramidal, and behavioral dysfunction in cynomolgus monkeys repeatedly fed subconvulsive doses of the pure compound, insufficient neuropathological changes have been generated to merit description of the primate response as an animal model of western Pacific ALS/P-D. The role of BMAA, MAM, and other factors in the etiology of this disease is being studied.

Epidemiology and other data strongly suggest that human exposure to cycad toxins may precede by years or decades the onset of clinical ALS or P-D, the nature of the clinical disease and the age at which it appears possibly being linked with cycad doses and host susceptibility. The mechanism underlying the proposed long-latency adverse effects of cycad exposure is believed to represent the single most important question in elucidating the pathogenesis of this neurodegenerative disorder.

An understanding of the chemical factors that trigger western Pacific ALS/P-D is expected to lead to the identification of comparable chemical factors in other environments. Research of this type should not be restricted to chemical components in plants used for medicine or food but should also take into account the wider potential for chemical factors that might have a role in triggering some forms of motor neuron disease, parkinsonism, and senile dementia of the Alzheimer type. By intensive exploration of the chemical exposure cupboards of young-onset patients with neurodegenerative diseases, it may be possible to obtain pertinent leads that can be tested in the laboratory. However, because exposure and disease onset may be separated by prolonged periods of time, and the nature of the putative factors is unknown, such an effort will require every weapon in the combined armamentaria of the epidemiologist and the neurotoxicologist. Experience in the western Pacific loci teaches that individual patients and their families, rather than large statistically selected populations, are the preferred targets of initial study, and intensely probing interviews are more likely to generate clues than prescribed questionnaires which miss the key questions and fail to record important responses. This approach, combined with intensive laboratory studies, is surely the most rapid way to find out if neurotoxins are driving us crazy in developed countries, just as they are tragically the apparent cause of dementia in certain communities of the western Pacific.

REFERENCES

Beaglehole, J. C. ed. 1962. The Endeavour Journal of Joseph Banks 1768–1771. Sydney: Angus and Robertson.

Beaton, J. M. 1977. Dangerous Harvest: Investigations in the Late Prehistoric Occupation of Upland Southeast Central Queensland. Dissertation. Australian National University. Canberra. (unpublished)

Cuzner, A.T. 1898. Arrowroot, cassava, and koonti. J. Am. Med. Assoc. 1:366–369.

Lyon Arboretum. 1964. Proceedings of the Third Conference on the Toxicity of Cycads. Fed. Proc. 23:1337–1387.

Lyon Arboretum. 1972. Sixth International Cycad Conference. Fed. Proc. 31:1465–1538.

Lyon Arboretum. 1988. Transcripts of Four Conferences on the Toxicity of Cycads. New York: Third World Medical Research Foundation.

Robert, William C. H. 1972. The explorations of Willem DeVlamingh, 1696–1697. In Contributions to a Bibliography of Australia and the South Sea Islands. Amsterdam: Philo Press.

Safford, W. 1905. The Useful Plants of Guam. Contributions from the U.S. National Herbarium. Washington, D.C.: Smithsonian Institution.

PART I
Assessment of Neurobehavioral Tests Now in Use

Methods in Behavioral Toxicology:
Current Test Batteries and Need for Development

Helena Hanninen

Behavioral toxicology is an applied branch of psychology with a very practical aim: to detect neurotoxic effects for the planning of prevention among exposed populations.

The earliest research in clinical and epidemiological behavioral toxicology was done within occupational medicine by psychologists working close to it or by medical doctors with some familiarity with psychological methods. Most of the research in the area is still done in this context. It provides the best possibilities for collaboration with the other disciplines working toward the same goal. In this context, however, the connections of behavioral toxicology with other branches of psychology—applied, as well as theoretical—have been loose. Reports are seldom published in psychological journals but rather in those of occupational medicine or neurotoxicology because of the practical relevance of study results within these disciplines. As a consequence, research on behavioral toxicology has been little known to those working in other fields of psychology. Another consequence is a limitation of scientific feedback from colleagues in other branches of psychology.

However, psychologists doing research on behavioral toxicology do use the knowledge and methods developed in other fields of psychology. Methods of behavioral toxicology originate from different branches of applied and academic psychology: clinical psychology, neuropsychology, aptitude testing, psychophysiological psychology, and in recent years, cognitive psychology. Tests were borrowed from these fields, mostly without trying to incorporate the corresponding theoretical framework into behavioral toxicology. The theoretical framework has been regarded as less important than the efficiency of the tests in detecting neurotoxic effects and their applicability in epidemiological research.

TEST BATTERIES NOW IN USE

Epidemiological and Field Studies

The methods now in use in epidemiologically oriented behavioral toxicology compose an interesting variety of psychological tests. Eight studies published in 1985–1987 serve as examples (Table 1). The variety of tests used reflects the history of behavioral toxicology and, to a certain extent, the advances of psychological testing during that period.

One of the first comprehensive test batteries for the detection of neurotoxic effects was designed at the Institute of Occupational Health in Finland, in connection with a study on carbon disulfide (Hanninen, 1971). It was composed of the best tests on hand at that time. First a very large selection of tests was used; there were tests of clinical psychology and neuropsychology, as well as aptitude tests standardized at the institute. Then, after preliminary data analyses, those tests of different areas of functioning were selected which seemed to be sensitive and suitable for use in clinical practice as well as in epidemiological research. Over the years the battery has been modified, and in each of the Finnish studies a somewhat different selection of tests has been used, depending on the exposure agent in question and the overall study design. Study 1 (Mantere et al., 1984) in Table 1 is one example. Descriptions of the tests were published in a separate booklet (Hanninen and Lindstrom, 1979), and several of them have spread to wider use in behavioral toxicology: the Santa Ana dexterity test, the Bourdon-Wiersma Vigilance test, the Benton test for visual memory, and selected subtasks of the Wechsler Adult Intelligence Scales (WAIS). Tests of WAIS and WMS (Wechsler Memory Scale) as well as the Benton Visual Retention test have been much used also because they are well documented and have been standardized in several countries. Other neuropsychological tests have been adapted to behavioral toxicology, too.

Studies 2 and 3 are other examples of test batteries composed mainly of much-used neuropsychological tests. The tests are not the same, however. The battery used in study 2 (Baker et al., 1984) was composed of subtasks of the Wechsler Scales; verbal memory was emphasized. Santa Ana was used for psychomotor ability. The British battery (Cherry et al., 1984) used in study 3 (Cherry et al., 1985) has three psychomotor tests and three for visual-motor performance. Of the latter, the Visual Searching task was devised by Goldstein as an indicator of brain damage, and Trail Making is from the Halstead Reitan Battery. The two most applied tests of the WAIS, Digit Symbol and Block Design, are also included in this battery. Memory is assessed by one test only. The British National Adult Reading Test (NART) is included as a measure of premorbid ability.

The battery used in study 4 (Jeyaratnam et al., 1986) concentrated even more on visual, motor, and visual-motor functions; each was measured with several tests. Additionally there was Digit Span for memory performance.

Study 5 (Maizlish et al., 1985) is an example of studies that apply methods developed within experimental psychology. There were no common elements with the five previous test batteries, but the battery covers principally the same broad functional domains as these do. Each of the tests contains a long series of measurements (about 15 minutes each), and gives detailed information about the function in question.

Many of the recent test batteries in behavioral toxicology include single computer-assisted tests adapted from other branches of psychology. The Neurobehavioral Evaluating system (NES) developed by Baker and Letz (Baker and Letz, 1986) and used in studies 6 (Maizlish et al., 1987) and 7 (Fidler et al., 1987) is totally computerized. Some of its tests can be described as computerized versions of certain conventional tests that have shown sensibility in toxicity testing; some were chosen or designed as promising new tests for that purpose. Being an extensive battery, the NES provides an opportunity to choose relevant methods for different kinds of studies. In addition to the tests used in studies 6 and 7, the regular sequence of NES tests also includes a verbal Associate Learning task and a vocabulary test for obtaining a measure of initial intelligence level.

The last study in Table 1 (Williamson and Teo, 1986) is another example of new methodological approaches in behavioral toxicology. Its tests were selected on the basis of information-processing theory, which also provides the framework for interpreting the results.

BEHAVIORAL TOXICOLOGY IN CLINICAL PRACTICE

Behavioral toxicology aims at examining and identifying toxic effects on the functional capacity of the central nervous system (CNS).

TABLE 1 Eight Batteries for Behavioral Toxicity Testing

	Study 1 Mantere et al. (1984)	Study 2 Baker et al. (1984)	Study 3 Cherry et al. (1985)	Study 4 Jeyaratnam et al. (1986)
1.				Flicker Fusion
2.	Santa Ana Dexterity	Santa Ana	Dotting Grooved Pegboard Simple Reaction Time	Santa Ana Simple Reaction Time
3.	Bourdon-Wiersma Vigilance	Digit Symbol (WAIS)	Digit Symbol (WAIS) Visual Search Trail Making	Bourdon-Wiersma Digit Symbol (WAIS) Visual Search Trail Making
4.	Block Design (WAIS) Picture Completion (WAIS)	Block Design (WAIS)	Block Design (WAIS)	
5.	Visual Reproduction (WMS)	Visual Reproduction (WMS) Digit Symbol Recall		
6.	Digit Span (WAIS, WMS) Logical Memory (WMS)	Digit Span (WAIS, WMS) Logical Memory (WMS) Associate Learning (WMS) Mental Control (WMS)	Animal Names (Bushke)	Digit Span (WAIS)
7.	Similarities (WAIS)	Similarities (WAIS) Vocabulary (WAIS)	Reading Test (NART)	
8.				

	Study 5 Maizlish et al. (1985)	Study 6 Maizlish et al. (1987)	Study 7 Fidler et al. (1987)	Study 8 Williamson and Teo (1986)
1.				Flicker Fusion
2.	Tapping; Fitts Law Task	NES Finger Tapping NES Hand-Eye	NES Hand-Eye	Hand Steadiness Simple Reaction Time Visual Pursuit
3.	Stroop Test	NES Continuous Performance NES Symbol Digit	NES Continuous Performance NES Symbol Digit	Sustained Attention
4.	Mental Rotation	NES Pattern Comparison		
5.	Memory Scanning Memory Span Continuous Recognition Memory	NES Pattern Memory	NES Pattern Memory	Sensory Store Memory Paired Associates
6.	Memory Scanning Memory Span Continuous Recognition Memory		NES Digit Span NES Memory Scanning	Sensory Store Memory Paired Associates
7.				
8.		NES Mood Scale	NES Mood Scale	

NOTE: Areas tested: 1 = sensory function; 2 = motor ability; 3 = attention and sensory-motor speed; 4 = visual intelligence/spatial reasoning; 5 = visual memory; 6 = verbal memory; 7 = verbal intelligence/vocabulary; 8 = mood.

The subject matter links it with neuropsychology. The methodological choices described above reflect this link, but not consistently. The neuropsychological orientation is particularly relevant when neurobehavioral methods are used for clinical assessment of individuals afflicted by neurotoxic exposures.

In Scandinavia and in some other European countries, psychological assessment is a rather regular part of diagnostic procedure when a neurotoxic occupational disease is suspected. Moreover, at least in Finland, it is rare for referral of such a patient to be based on psychological examination done in a neurological clinic or a public health center. Neuropsychologists in other countries have been less aware of the neurotoxic syndrome, but the situation seems to be changing rapidly.

Neurobehavioral dysfunctions seen in intoxication patients provide useful information that directs the choice of methods when the early effects are studied, and vice versa: results of the epidemiological studies have guided the selection of tests in clinical test batteries. Table 2 lists the tests included in three diagnostic test batteries: the one used by Bolla-Wilson and Bleecker in the Johns Hopkins School of Medicine (Bolla-Wilson and Bleeker, 1987); the Pittsburgh Occupational Exposures Test Battery, or POET (Ryan et al., 1987), which is meant for use at clinics as well as in research; and the battery now in use at the Finnish Institute of Occupational Health (IOH).

The first battery represents traditional neuropsychological methodology; the two others were designed especially for neurotoxic effects. Differences among the three batteries are not striking. All of them sample roughly the same broad functional domains, with one or more tests. The tests selected are not the same, however. Only four WAIS tests appear in all three batteries. Seven tests are included in two of them. Additionally, each battery contains tests not present in the two others. There are also differences in emphasis. The Finnish battery attempts a detailed description of the motor performance and its disturbances, whereas there is only one ordinary psychomotor test in the two other batteries. On the other hand, the set of memory tests included in POET allows a much more elaborate description of the memory function than the Finnish battery does, and the battery used by Bolla-Willson and Bleecker is the only one that assesses verbal functions that may be affected by toxic exposure.

BASES OF TEST SELECTION

Published studies in behavioral toxicology seldom give reasons for selection of test methods: reports emphasize results, not methodol-

TABLE 2 Three Clinical Test Batteries for Neurotoxic Syndromes

	Bolla-Wilson and Bleecker[a]	Ryan (POET)	IOH, Helsinki
1.			
2.	Simple Reaction Time	Grooved Pegboard	Finger Tapping Santa Ana Mira Staircases
3.	Digit Symbol (WAIS-R)	Digit Symbol (WAIS-R)	Digit Symbol (WAIS) Stroop Test Bourdon-Wiersma[b]
4.	Block Design (WAIS-R)	Block Design (WAIS-R) Picture Completion (WAIS-R) Boston Embedded Figures	Block Design (WAIS) Picture Completion (WAIS-R) Symmetry Drawing[b] Benton Test

5.	Visual Reproduction (WMS)
	Symbol-Digit Learning
		Visual Reproduction (WMS)
		  —immediate and delayed
		Symbol-Digit Learning
		  —immediate and delayed
		Digit Symbol Retention
			Digit Symbol Retention
			Visual Reproduction (WMS)[b]

6.	Serial Digit Learning Verbal-Verbal Associates Rey Auditory Verbal Learning Logical Memory (WMS)	Digit Span (WAIS-R) Verbal Paired Associates Recurring Words	Digit Span (WAIS-R) Associative Learning (WMS) Logical Memory (WMS)[b]
7.	Similarities (WAIS-R) Vocabulary (WAIS-R) Verbal Fluency (FAS)	Similarities (WAIS-R) Information	Similarities (WAIS-R) Synonyms
8.			POMS Rorschach[b]

NOTE: Areas tested: 1 = sensory function; 2 = motor ability; 3 = attention and sensory-motor speed; 4 = visual intelligence/spatial reasoning; 5 = visual memory; 6 = verbal memory; 7 = verbal intelligence/vocabulary; 8 = mood.

[a] Only tests used in follow-up are listed.
[b] Optional tests.

ogy. This does not mean that there was no rationale for choosing the particular tests used. Very often, however, the rationale can only be guessed.

In the choice of tests, the first questions considered are the following: What behavioral/neuropsychological functions are likely to be affected? Which tests are available for examining those functions? The answer to the second question yields, in principle, a wide selection of relevant tests. To choose among them, further questions must be considered. Important criteria are (1) previous empirical evidence of the sensitivity of the tests with regard to the effect looked for; (2) other previous documentation of the test: its psychometric properties, general population norms, documentation of its construct validity, etc.; (3) suitability and acceptability of the tests for the subjects to be tested; and (4) costs and benefits of the tests: costs in terms of time, expertise, and equipment needed; benefits in terms of information provided by the method.

The existing differences in test selection (Tables 1 and 2) reflect different answers to these questions or a different importance given to them.

Most test series include measures of a wide area of behavioral functions. This indicates that the authors either expected the effects to be widespread and diffuse, or did not have specific expectations concerning the functions to be affected. It is also possible that they had such expectations, but wanted to sample a broader area of function to be better able to circumscribe the effect. Some differences in the emphasis given to different functional areas were also found: some batteries emphasized psychomotor tasks; others, cognitive tests, or memory tests in particular. These differences seemed not to be related to the exposure agent under study, but rather to reflect different concepts of the nature of the neurotoxic syndromes in general.

The tests chosen to measure different behavioral areas varied more than the general structure of the batteries. Some classic neuropsychological tests seem to be rather well-established general tools in behavioral toxicology, but there were also batteries that had very few common elements with the others, or none at all. One explanation is that the authors represented different branches of psychological research and preferred tests that best corresponded to their education and experience. Moreover, the studies were conducted in different countries where different tests may have been available and well documented with general population norms, etc.

It seems also that the previously established sensitivity of the test in regard to toxic effects was not considered equally important by all of researchers. Rather, the authors can be classified as conservative or progressive. The former prefer the "old" tests because there is previous documentation about their sensitivity and the possibility of comparing their results with earlier ones. At the same time, the users of these tests may be troubled by their weaknesses: questionable reliability, low acceptability by the subjects, laborious scoring, etc. The "progressive" authors, on the other hand, strive to introduce recent theoretical or methodological progress in different branches of psychology into the field of behavioral toxicology. Their problem is that in using new promising and potentially sensitive methods, they lose the possibility of comparing their results with early research. Also, in case of negative results, they have difficulty in deciding whether the exposure in question actually had no behavioral effect or whether the promising methods indeed were not sensitive enough to display them.

Consideration of the population to be investigated is also necessary when selecting a test. The suitability of a test depends partly on the general educational level and culture of the subjects, and on the cultural and educational homogeneity of the subject group. In culturally and linguistically very heterogeneous populations the methodological choices are limited mainly to sensory and motor tests. Moreover, the practical circumstances of the study outlined by the study design, the testing facilities (particularly in field studies), and economic matters may strongly restrict methodological choices, making a wise test selection a real challenge.

THE NEUROBEHAVIORAL CORE TEST BATTERY

The Neurobehavioral Core Test Battery (NCTB) was designed by an expert group of the World Health Organization (WHO). It is intended for use in health hazard evaluations and field studies when the testing time is limited and the use of sophisticated equipment is not possible (Johnson, 1987). The seven tests included in the NCTB can also be recommended as common tests in more comprehensive test batteries, to allow comparison of results obtained by different research groups.

The NCTB includes Pursuit Aiming for motor steadiness, Simple Reaction time for attention and response speed, Digit Symbol for perceptual motor speed, Santa Ana for manual dexterity, the Benton Visual Retention test for visual perception and memory, and the Digit Span test for auditory memory. The Profile of Mood States (McNair et al., 1981) is included as a measure of affect. When tests were chosen for the NCTB, special consideration was given to their applicability for examining working populations all over the world where neurotoxic hazards exist. The tests had to be minimally dependent on the culture and education of the subjects.

It is evident that such a short series of tests cannot sample all the possible effects of all possible neurotoxicants. Moreover, it may not be sensitive enough to discover subtle effects of low-level exposure. Nevertheless, it covers the functions that are most likely to be affected and can be expected to be able to detect neurotoxic effects in more hazardous exposure situations.

FUTURE DIRECTIONS: SOME CRUCIAL ISSUES

Neuropsychological Description/Definition of Neurotoxic Syndrome(s)

Selection of relevant methods would be easier if we knew more about the CNS dysfunction being studied. Is there a general neurotoxic syndrome, affecting many different areas of behavioral function? If so, then the effect is most likely to be seen in those functions that are most vulnerable in CNS dysfunctions in general, or in functions for which the most accurate and reliable tests exist. If many performances are only slightly impaired, mere chance can decide which of the tests yields a result that reaches the desired level of statistical significance.

If there is a general diffuse neurotoxic syndrome (at the behavioral level), is it similar to that seen in other diffuse brain conditions, or does it have any unique, or at least semispecific, features? In spite of the advances of neurotoxicology during the last decades, not too much is known today about the mechanisms that mediate between the toxic effect on the brain and behavioral dysfunction. Several effects on brain tissue have been found, these effects being somewhat different for different exposure agents. Nevertheless, the operation of different mechanisms may produce similar effects on behavioral output. The toxic syndrome differs from other diffuse brain syndromes, and the syndromes caused by different toxicants differ from each other, mainly as far as the effect on the brain is selective and some areas of the brain are affected more than others. Very little is known about this today.

There appear to be both a general diffuse behavioral dysfunction caused by all neurotoxic exposures, or most of them, and more specific impairments that may be agent dependent and related to the different sensitivity of various brain structures to the agent in question. It can be hoped that advances in neurotoxicology research will yield more information about the mechanisms and sites of action of various neurotoxicants. In addition, better acquaintance with the advances of neuropsychology might help to put the recent and future pieces of information together, and to formulate fruitful hypotheses concerning the neuropsychological nature of the toxic syndrome(s), to guide further research.

Even without knowing the pathological processes in the brain that are behind the syndrome(s), their more accurate descriptions should be attempted. This can be done, for example, by investigating crucial behavioral disturbances with theoretically well-based test combinations that allow a more detailed analysis of these disturbances. This approach was used in the study by Williamson and Teo (1986; study 8 in Table 1), and in a study on solvent-induced memory impairment by Stollery and Flindt (1988).

Natural History of the Neurotoxic Effects

The time course of behavioral effects is also poorly known. The severity of an effect is apparently a function of both the intensity and the duration of exposure, as demonstrated by Mutti and coworkers (1984) in their study on styrene. However, it can also be assumed that the time course and the dose-effect curve differ for various effects. The different exposure history of the subjects may, in fact, explain many of the discrepancies in empirical results. Longer duration or a higher level of exposure can be expected to cause not only more severe but also more widespread behavioral effects.

At the other end of the exposure history, accentuation of behavioral impairment depends on the time elapsed since cessation of exposure. Clinical observations on patients with occupational intoxication and the few follow-up studies reported so far indicate different reversibility of the affected functions: intellectual functions and visual memory tend to recover, whereas deficits in verbal memory and motor performance are more often irreversible or progressive (Lindstrom et al., 1982; Orbaek and Lindgren, 1988). Different recovery curves for varying dysfunctions suggest a different brain pathology behind them.

Dilemma Between Conservative and Progressive Approaches

There are several challenges for new and more sophisticated methods in this area of research: the need to increase the sensitivity of tests, the need to get a better understanding of the phenomena being studied, and the need to increase the efficiency of research, particularly in large-scale field studies, by using timesaving computer-based methods.

The problem is that the application of many new methods will increase the methodological diversity of the field, complicate the comparison of results, and slow down the accumulation of a common body of knowledge. Another serious problem is that new methods with unknown sensitivity to neurotoxic effects should not be used as the only behavioral methods in studies evaluating the behavioral toxicity in exposed populations. For both practical and ethical reasons there is a very limited possibility of conducting purely methodological research on exposed populations. Thus, it can be expected and even recommended that the methods which have been most useful in previous studies will in the future also be used in behavioral toxicity studies, complemented—when possible—with promising new methods. However, experimentation with new methodological approaches in clinical situations and in field studies, whenever possible, is vitally important too.

The construction of good new tests is an arduous procedure. Even more work is usually needed to gather different types of validity data for new methods. Empirical research done with a test in different populations and addressing different questions increases the validity of the interpretation of its results. Recognizing this fact, the authors of the NES battery organized international collaboration to gather versatile data on the reliability and validity of the NES tests. A similar procedure is going on concerning the WHO battery, even though its tests have already been widely used. This is, I think, the right way to proceed with new methodological approaches.

Further Questions of Test Selection

Simple or Complex Tasks?

Some behavioral toxicologists prefer testing simple elemental functions (sensory, motor, and cognitive) instead of complex ones, partly because this strategy may yield a more precise description of the dysfunction. Another reason for measuring narrow elemental functions is that they offer fewer possibilities for compensating an impaired function by use of other, unaffected ones. Simple reaction time is one example of a well-circumscribed elemental function sensitive to neurotoxic exposures.

However, two other arguments speak for the use of more complex tasks. Although requiring complex interaction of several brain structures, such tasks may be sensitive to effects not detected by tests focusing on narrow functions. In addition, if many elemental functions are impaired, then performance may be more grossly impaired in a test requiring several of them than in tests measuring single functions.

The optimal strategy would be to use tests of varying complexity, but in a real-life research situation this may be impossible, because a too long test battery tends to distort the study design by raising the costs and increasing the risk of dropouts, thereby decreasing the statistical power of the data analyses.

Reliability of Tests and Stability of Functions

High reliability (i.e., minimizing measurement error) is one of the basic qualifications of a good test. Reliability has several components: objectivity of the measurement, internal consistency of the test, and stability of the test result over time. The latter requirement is somewhat problematic. Changes in the test score over time do not reflect only the error variance, but also the instability of the measured function (i.e., its sensitivity to situational factors). Certainly, there is no sense in measuring very unstable functions. On the other hand, stressing the importance of test-retest reliability too much may lead to a selection of insensitive tests.

According to the results of a recent solvent study (Hanninen et al., 1986), those tests that displayed the best dose-response relationship were least stable in the exposed group. It seems that variability of the performance level over time, and even during a single exposure session, is itself a characteristic of subjects handicapped by a neurotoxic effect.

Changes in Personality, Mood, and Affect

The behavioral effects of neurotoxic substances include changes in mood, affect reaction, and personality. These changes can be present alone or along with performance decrements (Cranmer and Goldberg, 1986). Even in the latter case they often are the most disturbing and handicapping effects from the point of view of afflicted subjects and their closest environments. Compared to the work done to find good methods for the sensory, motor, and cognitive dysfunctions, research on the changes in affect has been badly neglected.

Test batteries described in Tables 1 and 2 did not contain measures of affect or personality, except for the POMS (Profile of Mood States) included in the NES and the WHO battery. However, most studies contained symptom enquiry with questions on mood states, emotional distress, fatigability, etc. In some studies, as well as clinical examinations, various personality tests have been applied (Forzi et al., 1975; Lindstrom, 1984; Maroni et al., 1977).

Valid assessment of emotional changes is difficult. Moods and emotions have a large interindividual and intraindividual variation, their quantification is problematic, and confounding factors are difficult to rule out. One problem is the difficulty in distinguishing primary toxic effects from more secondary reactions due to diminished functional capacity or to anxiety about being exposed to a toxic substance. Though the role played by exposure in emotional changes and other psychological symptoms is difficult to evaluate, their assessment should not be neglected when toxic effects are studied. Improvement and refinement of methods to be used for this purpose are major challenges for behavioral toxicologists.

Other Areas for Method Development

Deficient memory and learning ability are common complaints among subjects exposed to neurotoxic substances. In some studies the tests of learning and memory were among the best detectors (Eskelinen et al., 1986; Hanninen et al., 1976), but in several others they yielded more marginal or negative results. Even patients with marked subjective memory deficits sometimes have normal or near-normal scores in memory tests. Improvement of the sensitivity of these tests is another major challenge.

Sensitivity of some memory tests can be improved by improving their psychometric qualities. For example, the Digit Span of the Revised WAIS yields more accurate scores than the old Digit Span test, and sensitivity of the WMS Associate Learning can be improved by increasing the number of difficult items. Yet improving old tests may not be enough.

Memory is not one single function but has several components. Selection of the best tests to detect neurotoxic impairment of memory and learning requires a detailed description of the deficit in question, i.e., the application of paradigms of contemporary research on memory and its impairments. Further development is similarly needed in regard to assessment of attention and its disturbance in neurotoxic conditions. In this area, the main difficulty may be finding or developing tests that yield a valid measurement of attention (and not of sensory or motor speed or something else) and are still short and simple enough to be applicable in epidemiologic research.

Another area needing special consideration is the category of deficits in the executive functions: goal formulation, planning, carrying out activities, etc. (Lezak, 1984). Though it may be difficult to find good methods to apply these disturbances as research tools, their assessment should not be neglected when examining patients with suspected neurotoxic diseases.

SUMMARY

The variety of tests in use has been illustrated by eight studies published in the last four years. In addition, three clinical test batteries have been described. These examples were selected to give a picture of the primary methodological approaches, and not to present a comprehensive review.

The most usual approach has been an eclectic use of tests previously employed in other areas of psychology, mainly in neuropsychological research and practice. Many of these tests have displayed good sensitivity to neurotoxic effects. The variety of choices is large, but some of the tests have been used by many research groups in several countries.

Some of the recent studies have striven toward the application of newer methodological and theoretical approaches to behavioral toxicology. Certain tests have been computerized to increase the objectivity and cost-effectiveness of the testing, and to obtain a more detailed measure of the performance and its deficits. The theoretical paradigms of contemporary psychology have been applied to acquire a better and more elaborated understanding of the neuropsychological dysfunction being studied.

However, the new methods often require equipment or expertise that is not always available when research on neurotoxicity is urgently needed, or they are still lacking sufficient validation with regard to their sensitivity to neurotoxic substances. The use of older tests with documented validity is therefore to be recommended in many field study situations. At the same time, however, it is important to work on new methodological paradigms aiming particularly at a better neuropsychological description of the neurotoxic syndrome and its agent-specific variants. Future research should also be directed to the time course of neurotoxic effects during continuing exposure and after its cessation.

Special consideration should be given to the toxic-induced changes of mood, affect, and personality, as well as to the deficits of attention, learning, and memory. These are common concomitants of a neurotoxic affliction for which practical, reliable, and sensitive methods are still lacking.

REFERENCES

Baker, E. L., and R. Letz. 1986. Neurobehavioral testing in monitoring hazardous workplace explosures. J. O. M. 28:987–990.

Baker, E. L., R. G. Feldman, R. A. White, J. P. Harley, C. A. Niles, G. E. Dinse, and C. B. Berkeley. 1984. Occupational lead neurotoxicity: A behavioral and electrophysiological evaluation. Study design and one year results. Br. J. Ind. Med. 41:a352–361.

Bolla-Wilson, K., and M. L. Bleeker. 1987. Neuropsychological impairment following inorganic arsenic exposure. J. O. M. 29:500–503.

Cherry, N., H. Hutchins, T. Pace, and H. A. Waldron. 1985. Neurobehavioral effects of repeated occupational exposure to toluene and paint solvents. Br. J. Ind. Med. 42:291–300.

Cherry, N., H. Venables, and H. A. Waldron. 1984. Description of the tests in the London School of Hygiene test battery. Scand. J. Work Environ. Health 10(Suppl 1):18–19.

Cranmer, J. M., and L. Goldberg. eds. 1986. Proceedings of the workshop on neurobehavioral effects of solvents. Neurotoxicology 7:45–56.

Eskelinen, L., M. Luisto, L. Tenkanen, and O. Mattei. 1986. Neuropsychological methods in the differentation of organic solvent intoxication from certain neurological conditions. J. Clin. Exp. Neuropsychology 8:239–256.

Fidler, A. T., E. L. Baker, and R. E. Letz. 1987. Neurobehavioral effects of occupational exposure to organic solvents among construction painters. Br. J. Ind. Med. 44:292–308.

Forzi, M., M. G. Cassitto, R. Gilioli, G. Armeli, and V. Foa. 1975. Personlichskeit-fehlentwicklungen in Arbeitern bei der electolytischen Chlor-Alkali-Gewinnung. Pps. 70–73 in Proceedings of the 2nd Industrial and Environmental Neurology Congress, E. Klimkova-Deutschova and E. Lukas, eds. Prague: Universitas Karlova.

Hanninen, H. 1971. Psychological picture of manifest and latent carbon disulphide poisoning. Br. J. Ind. Med. 28:374–381.

Hanninen, H., L. Eskelinen, K. Husman, and M. Nurminen. 1976. Behavioral effects of long-term exposure to a mixture of organic solvents. Scand. J. Work Environ. Health 2:240–255.

Hanninen, H., and K. Lindstrom. 1979. Behavioral test battery for toxico-psychological studies used at the Institute of Occupational Health in Helsinki. Institute of Occupational Health Reviews 2:1–58.

Hanninen, H., E. Tuominen, K. Rantala, and R. Luukkonen. 1986. Psychological tests in the detection of early neurotoxic effects: Construction and preliminary validation of a screening test battery. Tyoterveyslaitoksen tutkimuksia 4:68–71 (Finnish with English summary).

Jeyaratnam, J., K. W. Boey, C. N. Ong, C. B. Chia, and W. O. Phoon. 1986. Neuropsychological studies on lead workers in Singapore. Br. J. Ind. Med. 43:626–629.

Johnson, B., ed. 1987. Prevention of Neurotoxic Illness in Working Populations. New York: John Wiley & Sons. 257 pp.

Lezak, M. 1984. Neuropsychological assessment in behavioral toxicology—Developing techniques and interpretative issues. Scand J. Work Environ. Health 10(Suppl. 1):25–29.

Lindstrom, K., M. Antti-Poika, S. Tola, and A. Hyytiainen. 1982. Psychological prognosis of diagnosed chronic organic solvent intoxication. Neurobehav. Toxicol. Teratology 4:581–588.

Lindstrom, K. 1984. The Rorschach test in behavioral toxicology. Scand. J. Work Environ. Health 10(Suppl. 1):20–23.

Maizlish, N. A., G. D. Langolf, L. W. Whitehead, L. J. Fine, J. W. Albers, J. Goldberg, and P. Smith. 1985. Behavioural evaluation of workers exposed to mixtures of organic solvents. Br. J. Ind. Med. 42:579–590.

Maizlish, N., M. Schenker, C. Weisskopf, J. Seiber, and S. Samuels. 1987. A behavioral evaluation of pest control workers with short-term, low-level exposure to the oganophosphate diazinon. Am. J. Ind. Med. 12:153–172.

Mantere, P., H. Hanninen, S. Hernberg, and R. Luukkonen. 1984. A prospective follow-up study on psychological effects in workers exposed to low level of lead. Scand. J. Work Environ. Health 10:43–55.

Maroni, M., C. Bulgheroni, M. G. Cassitto, F. Merluzzi, R. Gilioli, and V. A. Foa. 1977. A clinical, neurophysiological and behavioral study of female workers exposed to 1,1,1-trichlorethane. Scand. J. Work Environ. Health 3:16–22.

McNair, D. M., M. Lorr, and L. F. Droppelman. eds. 1981. Manual for the Profile of Mood States. California: Education and Industrial Testing Services. 29 pp.

Mutti, A., A. Mazzucchi, P. Rustichelli, G. Frigeri, G. Artini, and I. Franchini. 1984. Exposure-effect and exposure-response relationships between occupational exposure to styrene and neuropsychological functions. Am. J. Ind. Med. 5:275–281.

Orbaek, P., and M. Lindgren. 1988. Prospective clinical and psychometric investigation of patients with chronic toxic enchephalopathy induced by solvents. Scand. J. Work Environ. Health 14:37–44.

Ryan, C. M., L. A. Morrow, E. J. Bromet, D. K. Parkinson. 1987. Assessment of neuropsychological dysfunction in the workplace: Normative data from the Pittsburgh Occupational Exposures Test Battery. J. Clin. Exp. Neuropsychology 9:665–679.

Stollery, B. T., and M. L. Flindt. 1988. Memory sequelae of solvent intoxication. Scand. J. Work Environ. Health 14:45–48.

Williamson, A. M., and R. K. Teo. 1986. Neurobehavioural effects of occupational exposure to lead. Br. J. Ind. Med. 43:374–380.

The Current Status of Test Development in Neurobehavioral Toxicology

Ann M. Williamson

There has been a proliferation in the number of tests used in neurobehavioral toxicology in recent years and an increase in the number of groups producing test batteries. Nevertheless, the area still remains a difficult one despite the increased interest in it, and many questions still remain unanswered.

Neurobehavioral methods have been used to examine an increasing number of toxicants, but the impact of the findings from such studies has differed considerably across countries. Evidence from neurobehavioral testing has been highly influential in lowering acceptable health standards for lead and solvent exposure, for example, in Scandinavian countries, but it has had little effect in many other countries (e.g., Australia and Britain). Political forces no doubt play a large part in these differences, but the fact remains that evidence from neurobehavioral tests is simply not convincing to many decision makers in the latter countries.

The reasons for this are the problems encountered in neurobehavioral testing, namely, problems of selection of controls; accounting for confounding variables such as alcohol, drug use, education, age, and socioeconomic status; selection of sensitive and comprehensive tests; and quantifying exposure. It is apparent that these problems are often regarded as sufficient evidence to reject an overwhelming weight of evidence that would otherwise be seen as convincing.

For occupational lead exposure, for example, a review of the literature since 1980 demonstrates that of the 14 or so papers published on neurobehavioral effects, all but 2 show statistically significant impairments in some tests in lead workers compared to controls. In all studies, lead exposure levels were within the subclinical range (i.e., below 3.8 µmol/L). Despite this, critics of the field tend to "throw the baby out with the bathwater" and concentrate their attention on the undeniable flaws in most of the studies without looking at the literature as a whole. It therefore becomes essential to ensure that neurobehavioral testing for toxic effects is as rigorous as possible if it is to have a significant impact on decision makers.

RESEARCH DESIGN AND TEST SELECTION

Although the major factor of interest here is the choice of tests, it is not really possible to divorce questions of experimental design from those relating to test selection. A number of workers have reviewed the study design problem for neurobehavioral toxicology (Gamberale, 1985; Valciukas and Lilis, 1980).

Virtually all studies in this area are cross sectional, in which exposed workers are compared to nonexposed workers on their performance of a battery of neurobehavioral tests. Test selection is a problem in this type of design because many tests are sensitive to extraneous differences between exposed and nonexposed workers such that they contribute to, and potentially confound, the finding of differences in test performance. For example, in a study by Parkinson et al. (1986), when the effects of age, education, and income were removed statistically, significant differences between lead-exposed workers and controls on some neurobehavioral tests disappeared.

Problems of confounding in this design are usually dealt with by matching exposed and control groups or by statistical means. Appropriately chosen tests however can produce the same effect through the selection of tests that are not vulnerable to the effects of confounding factors such as age, education, or ethnic background, at least for working populations. This has not been done in any study to date. It is common though for researchers to investigate the effect of putative confounding variables prior to using various statistical techniques to minimize confounding (Hogstedt et al., 1983; Valciukas and Lilis, 1982).

For the few prospective cohort studies done in this area, the problem of test selection lies in selecting tests that are not susceptible to the effects of practice. Because workers are tested on more than one occasion, it is essential that test-retest results are not muddied by the fact that workers will improve from one test session to another simply because they have seen the test before. The results of the single prospective study performed on occupational lead exposure (Mantere et al., 1984), for example, were less convincing because of strong training effects on the performance of a number of tests.

REASONS FOR TEST SELECTION

In most studies a set or battery of psychological or behavioral tests is used. Typically, the tests are chosen for one or more of the four following reasons:

1. The test is a well-standardized psychological test for which the distribution of scores in the population is already established, e.g., subtests from the Wechsler Adult Intelligence Scale (WAIS; Wechsler, 1955).

2. The tests will measure some aspects of functioning that are known to be influenced by the toxic substance based on clinical evidence; e.g., knowing that inorganic mercury exposure produces unintentional hand tremor would result in a tremor test or some other motor test being included in the test battery.

3. Each test corresponds roughly to a particular psychological function, e.g., including the Santa Ana test for motor functions, the Digit Symbol test for attention, and Digit Span test for memory functions.

4. Tests may be chosen on some theoretical grounds such that each test is juxtaposed against another in some logical manner (Williamson et al., 1982).

In some studies however, no rationale is provided for the choice of tests in a particular battery.

Various test batteries are currently employed for a range of neurotoxic substances, and there are a number of reviews that describe them (Anger, 1984, 1986; Gullion and Eckerman, 1986; Johnson and Anger, 1983). Some of these batteries, such as that devised at the Finnish Institute of Occupational Health (Hanninen and Lindstrom, 1979), have been used in multiple studies of a range of toxic hazards. Consequently, some estimate can be made about their sensitivity. Many batteries, however, are simply put together for a single purpose, and if no rationale is provided for test selection, the reliability, validity, and sensitivity are largely unknown, particularly if no sound rationale exists from previous work.

RECENT APPROACHES TO TEST BATTERY DESIGN

There have been three new approaches to test battery design in the last few years, each of which focuses on a different aspect of battery design.

Neurobehavioral Core Test Battery

The Neurobehavioral Core Test Battery (NCTB) battery was developed at a meeting of a group of experts from a range of disciplines as a joint initiative of the World Health Organization (WHO) and the National Institute for Occupational Safety and Health (NIOSH). The aim of the meeting was to devise a test battery that could be used to screen for neurotoxic effects with particular reference to its use in developing countries. Although the rationale for test selection was not much different from that used in many other studies, this initiative constitutes a significant leap forward because it is an attempt to set up norms for each test that are applicable to comparisons between and within cultural boundaries. It is argued that by applying the test battery in a range of countries it should be possible to estimate the influence of cultural differences on test performance. Moreover, on a more practical level, such cross-cultural comparison should allow for better interpretations between studies performed in different countries.

The Computerized Battery Approach

The computerized battery approach capitalizes on the recent boom in personal computer development by designing a battery that is administered by computer. Two advantages of computerized testing are that test administration is standardized and reproducible, requiring minimal involvement by the test administrator, and that data handling and scoring are made easier so that the results can be reported immediately.

Two main research groups have developed computer-administered test batteries. Probably the most well known is the battery developed by Baker and Letz at Harvard (Baker et al., 1985). After some standardization procedures, the final test battery includes three memory tests from the Wechsler Memory Scale (WMS) or Wechsler Adult Intelligence Scale (WAIS); two tests classified as measuring verbal concept formation both from the WAIS; four visuomotor tests two of which are from the WAIS; and a mood scale. Three of these tests are from the NCTB.

This battery has been subjected to some validation (Baker et al., 1983), although the description of the investigation of test reliability is not clear (Fidler et al., 1987). Some estimates can be made as to the sensitivity of this battery because it has been used to show impairments in both lead-exposed (Baker et al., 1984) and solvent-exposed workers (Fidler et al., 1987).

The second well-known computer-assisted battery was developed at the University of North Carolina by Foree et al. (1984). This battery (also known as the Microtox battery) was based on the work of Carroll (1980), who proposed a theory of 10 factors to represent the range of cognitive abilities. Carroll holds that all test performance is partitionable into smaller building blocks which he called elementary cognitive tasks (ECTs). The battery consists of three sensory tests, two psychomotor tests, two attention tests, eight memory tests, and one test that is classified as "other."

Investigations of validity and reliability were performed by Carroll in the development of the theory. The sensitivity of the battery has been evaluated to some extent in studies of the effects of carbon monoxide and alcohol (Foree et al., 1984).

Model or Theory-Based Tests and Test Batteries

A few test batteries have some theory of psychological function as their basis. The aim of this approach is to facilitate interpretation of results. Smith and Langolf (1981), for example, take the view that tests selected for a battery should be ability-specific and have some underlying theoretical structure. This, they argue, allows interpretation to be made in terms of the processing stages or systems that produce performance on a test; furthermore, scores can be given for individual stages in processing. The argument advanced by Smith and Langolf appears to rest mostly on their use of the Sternberg memory scanning test (Sternberg 1966, 1975), which has a well-developed theoretical basis. There has been considerable debate, however, on the adequacy of Sternberg's theory to account for all aspects of performance on this test. Gullion and Eckerman (1986) state that this debate is sufficient to make unwarranted any inferences about the intactness of underlying cognitive processes on the bases of performance on the memory scanning task.

Rather than choosing particular tests that have well-developed theoretical structures, two research groups have employed theory-based test batteries. The first is the Microtox test battery described above. In this battery, Carroll's elementary cognitive task theory (Carroll, 1980), is used to guide test selection. The second is a battery devised by Williamson and colleagues (Williamson and Teo, 1986; Williamson et al., 1982, 1987), in which the choice of tests is based on information-processing theory (Wickens, 1984, 1987). This battery includes a sensory test, three psychomotor tests, a sustained attention test, and four memory tests: a sensory store memory test, two short-term or working memory tests, and a long-term memory test. Validation has been carried out on this test battery, and some limited studies of reliability have been done, but these results have not been published. The sensitivity of this battery has also been evaluated in detecting effects of inorganic mercury (Williamson et al., 1982), inorganic lead (Williamson and Teo, 1986), and prolonged exposure to the underwater environment (Williamson et al., 1987).

Model or theory-based test batteries have the advantage of providing a comprehensive coverage of neurobehavioral functions that is often missing from other approaches to battery design. Screening test batteries in particular should be designed on this basis. For screening purposes, battery design should proceed as if nothing is known about the neurobehavioral effects of the toxin in question because clinical symptoms may be very misleading. For example, a commonly reported symptom in toxically exposed workers is fatigue (Fidler et al., 1987; Hanninen et al., 1979; Valciukas et al., 1979). However, a researcher would have great difficulty selecting an appropriate test to investigate this symptom further unless a holistic or theoretical approach was taken in designing the test battery. Clinically manifested fatigue can have mental or physical origins so which should be tested for? In addition, if, for example, the fatigue is due to problems in maintaining mental effort, it would be impossible to determine which of the functional areas or stages of processing is responsible. Unless all possibilities are tested no clear conclusions can be made regarding the action of the supposed toxin.

Another difficulty with designing a battery without some global structure is that interpretation of test results can be made only for single tests. For example, a typical study of the effects of lead exposure in which the battery included tests of a range of neurobehavioral functions may have found impairments in reaction time, learning, and memory functions in lead workers, compared to nonexposed controls, but no apparent impairment of other functions. In this case the conclusion would be made that lead exposure affects motor, learning, and memory functions. If, however, the tests could be related on the basis of a cohesive theory, the interpretation might be very different. For example, in the study of lead exposure by Williamson and Teo (1986) the clustering of performance impairments seen in lead workers compared to controls suggested that sensory motor, learning, sensory store, and short-term but not long-term memory functions were affected. By using the information-processing principles on which the tests were selected, however, because vision was involved in the performance of each test and vision was impaired, it is just as likely that lead is affecting only the sensory function measured (in this case, vision) and that problems relating to the adequacy of stimulus input would explain the impairments in the other functions. This possibility is being pursued.

INFLUENCE ON TESTING SELECTION OF DIFFERENT OBJECTIVES FOR TESTING

The position taken above regarding the utility of theory-based test batteries is also relevant to the question of the most important objectives for neurobehavioral testing. It is commonly agreed that there are two levels of testing (Hanninen, 1981). The primary level focuses on screening for neurobehavioral insult by particular substances; the secondary level, on determination of the functional and, if possible, the neurological sites and mechanisms of the toxic action. Test selection is affected by the level or reason for testing. For screening batteries it is argued that tests should be quick and easy to administer and should concentrate on the known effects of the toxin, whereas for the second level of testing it is maintained that tests can be more complex and time-consuming.

Although this dichotomy of levels has real merit on practical grounds, there has been a tendency to focus too much on the "getting the job done" approach (Gullion and Eckerman, 1986) at the expense of "understanding the phenomenon." There has been a tendency in doing so to use tests that are fast and expedient rather than comprehensive and informative about the toxic effect on the system. This does not necessarily mean understanding the phenomenon but, rather, being aware of the breadth of the problem. A test battery designed around an information-processing model, for example, will provide a proper screening tool that reduces the possibility of Type 1 errors occurring simply because the affected function was not tested adequately or at all. As Wickens (1987) states about information-processing theory,

> From the perspective of human factors, the importance of the distinction between processing stages results because knowing that a particular environmental stressor, chemical toxicant or system characteristic influences one processing stage and not another has important implications for system redesign or reconfiguration. For example, knowing that a given stressor influences response processes and not encoding should lead the designer to focus on the improvement in control, rather than the display interface.

In the same way, neurobehavioral toxicology needs to be able to distinguish effects on function by appropriate choice of tests at the screening level of testing. Questionnaires are simply not adequate for initial screening. Rather, screening should first involve a very careful analysis of the functional effects of the putative toxin. Once this is done and the boundaries of the toxic effect have been established, a second battery can be developed which could consist of a screening questionnaire for subjective symptoms and a short, economical objective test battery (Hanninen, 1981).

EXPECTED DIRECTIONS IN TEST DEVELOPMENT

The development of tests in this area will almost certainly turn to concentrating much more on establishing valid, reliable, and sensitive test batteries. What tests are employed in these batteries will depend on the reason for using neurobehavioral tests in the first place.

The main reason for using neurobehavioral tests is either to identify whether the substance is toxic at a particular level of exposure or to determine the nature of the effect on the nervous system, or both. If the rationale for testing is the former, the types of tests needed are the same as those currently in use. As discussed above, development in this area should be toward simple tests that can be broken down into basic functional elements and have a place in a comprehensive, theoretical framework in order to improve interpretation of the toxicity question. If the latter reason constitutes the rationale for testing, development would be in the direction of finding appropriate tests that are analogues of underlying electrophysiological and biochemical processes. This type of development, however, has considerably further to go than the first and will be discussed again in the next section.

A third reason for testing that is likely to emerge is to demonstrate the need to respond to the toxicity problem. As discussed in the beginning of this chapter, a major problem in this area is establishing, to the satisfaction of the wider community, that impairments in neurobehavioral functioning of the type typically shown do constitute both a health effect and a possible compromise to safety (e.g., the slowed reaction time that occurs in solvent-exposed workers may cause more accidents). If this is the aim of the investigation, the direction of test development would be toward tests that mirror "real-life" functioning in much the same way as tests were developed to investigate the effects of alcohol on driving (e.g., Moskowitz, 1973).

A likely offshoot of improvements in the validity, reliability, and sensitivity of test batteries, as well as the significant efforts currently being made toward setting up norms for neurobehavioral tests (e.g., the WHO/NIOSH test program), is that tests will become much more useful as clinical tools for diagnosing individual responses to toxic substances. This development has important implications for prevention of toxic effects if early or "subclinical" effects can be detected in particular workers. Last, an important area for investigation in test development is in designing tests that will examine the strength of adaptive capacities to overcome the effects of toxic insult to the nervous system. This is an extremely important question because there are strong possibilities that "behavior's global nature also may allow compensatory mechanisms to thwart the early detection of an irreversible pathological process" (Weiss, 1983).

This was demonstrated in a study by Albers et al. (1987) in which a previously mercury-exposed cohort who had shown no ill-effects related to their exposure during their working lives were compared after retirement with a group of matched controls. The previously exposed group showed statistically significant impairments in psychomotor performance that were also related to the extent of their lifetime exposure. These findings were interpreted in terms of changes with age in the capacity of the nervous system to adapt or compensate for deficiencies or injury.

One possible facet of functioning that could provide a window on such concealed compensatory changes involve the strategies that individuals use in tackling tasks or solving problems. For example, in a study of professional (abalone) divers (Williamson et al., 1987), analysis of the results of the memory scanning task showed that divers responded to stimuli faster than matched controls but made significantly more errors in doing so. The divers were clearly sacrificing accuracy for speed. It was not surprising that these divers showed risk-taking tendencies, given the nature of their work. This tendency obviously had implications for the results of other tests in the battery, and for any testing that might be carried out in the future, which would not have been revealed if a less informative test had been used.

Tests need to be developed that will focus on the way that the problem is solved, not just on the speed (although this measure may reflect a slowing due to the adoption of less familiar or less efficient strategies) or the number of errors made. The approach of using tests that can be broken down into elementary cognitive tasks (Carroll, 1980) would provide a good beginning for this new direction.

REMAINING BARRIERS TO TEST DEVELOPMENT

Most of the remaining barriers to test development are due to the state of knowledge in neurobiology and neuropsychology. To proceed much further with test development, more must be known, for example, about the biological correlates of existing tests, and tests need to be devised for which this is known. In addition, psychological theories or models of behavior also need to be refined further in terms of tests that measure fundamental functions. Gullion and Eckerman (1986) argue very strongly that the present status of psychological theory is too weak to be useful in neurobehavioral toxicology. They maintain that the theories have not been tested adequately for each type of validity or for reliability and, furthermore, that such testing needs to be carried out in the field, on populations similar to those that are likely to be exposed to toxic hazards, not just on healthy college students.

It should be noted, however, that Gullion and Eckerman base their argument on criticism of a theory that attempts to describe only one aspect of psychological functioning, namely, short-term memory. Focusing only on individual tests could in itself be seen as a barrier to the development of useful tests for neurobehavioral toxicology. It is important that efforts in theory-based test development concentrate on holistic theories of behavior or information processing, not just on particular aspects.

Another potential barrier to the effective development of tests is the widespread use of computer-aided testing. There is little doubt that computers are extremely useful in making the task of data collection much quicker and easier, and they can increase the scope of the data collected (Fidler et al., 1987; Gullion and Eckerman, 1986). Their use can present problems, however, in that because of the ease of administration, untrained and inexperienced testers can be used. Test administration involves much more than simply showing the subject what to do. A great deal of insight can be gained about subjects' performance by watching how they perform. This would be of particular importance in the early stages of screening for the effect of a toxic hazard and can easily be dispensed with if the computer is doing all the work. In addition, although standardizing test administration is important, the conditions under which testing is carried out are also important. This point can be overlooked if the tester is not intimately involved in the test process and again, particularly, if the tester is untrained.

Problems can also be encountered due to the subject's lack of familiarity with computers. This is especially likely to be a problem for the type of workers who will be exposed to toxic hazards. Unless this is closely monitored, test results may well be confounded by extraneous factors due to the computer itself.

There is a significant possibility that the development of tests will be forced to take a back seat in favor of research focusing on the effects of the many toxic substances that have not yet been studied. This may well be the major barrier to test development. To policymakers and the providers of research grants, test development may look like "contemplation of the navel." Neurobehavioral toxicology must make test development at least as important as the analysis of toxic effects if it is to make further progress.

RECOMMENDATIONS FOR FURTHER RESEARCH OR DEVELOPMENT

One of the major problems in neurobehavioral toxicology is designing studies which eliminate, or at least minimize, the effects of extraneous variables that confound any demonstration of impairment in exposed compared to nonexposed workers. The use of prospective or cohort designs, which is the most satisfactory solution to this problem, has implications for the development of test batteries. Further research is needed to devise tests that are resistant to learning or practice effects.

Research is also needed in the development of tests that can detect the effect of adaptive changes which may camouflage functional impairment. From the point of view of prevention, this particular area holds real promise for the future. If this approach becomes successful it would then be possible to conceive of early detection criteria for long-term and even delayed effects. This would also provide a much more comprehensive picture of the breadth of the effects of a particular toxic hazard.

Finally, research should concentrate on development of tests that can be used for screening of individual workers. The initiative taken by the WHO/NIOSH where the NCTB is being applied worldwide is a step in this direction. It should, however, now be followed up by careful standardization in terms of a full investigation of the validity (particularly construct and concurrent validity) and reliability of the tests and the test battery. Once this is done, neurobehavioral toxicology will have available a set of well-defined tests that have established norms. Not only will testing of individuals be a realistic possibility, but the effectiveness of group testing will also be much improved. This is particularly important in attempting to overcome the vexing problem of comparison of results from different laboratories and different parts of the world.

REFERENCES

Albers, J., D. Escheverria, P. Donofrio, L. Fine, L. Kallenbach, G. Langolf, and R. Wolfe. 1987. Persistent adverse health effects of occupational mercury exposure. Paper presented at International Congress on Occupational Health, Sydney, Australia.

Anger, W. K. 1984. Neurobehavioural testing of chemicals: Impact on recommended standards. Neurobehavioural Toxicology and Teratology 6:147–153.

Anger, W. K. 1986. Workplace exposures. In Neurobehavioural Toxicology, Z. Annau, ed. Baltimore: Johns Hopkins Press.

Baker, E. L., R. G. Feldman, and R. F. White. 1983. Monitoring neurotoxins in industry—development of a neurobehavioural test battery. Journal of Occupational Medicine 25:125–130.

Baker, E. L., R. G. Feldman, and R. F. White. 1984. Occupational lead neurotoxicity—a behavioural and electrophysiological evaluation. Study design and year one results. British Journal of Industrial Medicine 41:352–361.

Baker, E. L., R. Letz, and A. Fidler. 1985. A computer administrated neurobehavioural evaluation system for occupational and environmental epidemiology. Journal of Occupational Medicine 27:206–212.

Carroll, J.B. 1980. Individual Difference Relations in Psychometric and Experimental Cognitive Tasks. L. L. Thurstone Laboratory Report No. 163. Chapel Hill, N.C.: University of North Carolina.

Fidler, A., E. L. Baker, and R. Letz. 1987. Neurobehavioural effects of occupational exposure to organic solvents among construction painters. British Journal of Industrial Medicine 44:292–308.

Foree, D., D. Eckerman, and S. L. Elliott. 1984. MTS: An adaptable microprocessor-based testing system. Behavioural Research Methods, Instrumentation Computing 16:223–229.

Gamberale, F. 1985. Use of behavioural performance tests in the assessment of solvent toxicity. Scandinavian Journal of Work Environment and Health 11(Suppl 1):65–74.

Gullion, C.M., and D. Eckerman. 1986. Field testing for neurobehavioural toxicology: Methods and methodological issues. In Neurobehavioural Toxicology, Z. Annau, ed. Baltimore: Johns Hopkins Press.

Hanninen, H. 1981. Behavioural methods in the assessment of impairments in central nervous function. In Biological Monitoring and Surveillance of Workers Exposed to Chemicals, A. Aito, V. Riihimaki, and H. Vainio, eds. Washington: Hemisphere Publishing Corporation.

Hanninen, H., and K. Lindstrom. 1979. Behavioural Test Battery for Toxicopsychological Studies: Used at the Institute of Occupational Health in Helsinki, second revised edition. Helsinki: Institute of Occupational Health.

Hanninen, H., P. Mantere, and S. Hernberg. 1979. Subjective symptoms in low-level exposure to lead. Neurotoxicology 1:333–347.

Hanninen, H. 1981. Behavioural methods in the assessment of impairments in central nervous function. In Aito, A et al. (Eds.), Biological Monitoring and Surveillance of Workers Exposed to Chemicals, Washington: Hemisphere.

Hogstedt, C., M. Hane, and A. Agrell. 1983. Neuropsychological test results and symptoms among workers with well-defined long-term exposure to lead. British Journal of Industrial Medicine 40:99–105.

Johnson, B. L., and W. K. Anger. 1983. Behavioural toxicology. In Environmental and Occupational Medicine, W. R. Rom, ed. Boston: Little, Brown.

Mantere, P., H. Hanninen, and S. Hernberg. 1984. A prospective follow-up study on psychological effects in workers exposed to low levels of lead. Scandinavian Journal of Work Environment and Health 10:43–50.

Moskowitz, H. 1973. Laboratory studies of the effects of alcohol on some variables related to driving. Journal of Safety Research 5:185–199.

Parkinson, D. K., C. Ryan, E. J. Bromet, and M. M. Connell. 1986. A psychiatric epidemiologic study of occupational lead exposure. American Journal of Epidemiology 123:261–269.

Smith, P. J., and G. D. Langolf. 1981. The use of Sternberg's memory-scanning paradigm in assessing effects of chemical exposure. Human Factors 23:701–708.

Sternberg, S. 1966. High speed scanning in memory. Science 153:652–654.

Sternberg, S. 1975. Memory scanning: New findings and current controversies. Quarterly Journal of Experimental Psychology 27:1–32.

Valciukas, J. A., and R. Lilis. 1980. Psychometric techniques in environmental research. Environmental Research 21:275–297.

Valciukas, J. A., and R. Lilis. 1982. A composite index of lead effects. International Archives of Occupational and Environmental Health 51:1–14.

Valciukas, J. A., R. Lilis, and H. A. Anderson. 1979. The neurotoxicity of polybrominated biphenyls: Results of a medical field survey. Annals of the New York Academy of Sciences 320:337–367.

Wechsler, D. 1955. Wechsler Adult Intelligence Scale. New York: Psychological Corporation.

Weiss, B. 1983. Behavioural toxicology and environmental health science: Opportunity and challenge for psychology. American Psychologist 91:1174–1186.

Wickens, C. D. 1984. Engineering Psychology and Human Performance. Columbus: Charles Merrill.

Wickens, C. D. 1987. Information processing, decision-making and cognition. In G. Salvendy, ed. Handbook of Human Factors. New York: John Wiley & Sons.

Williamson, A. M., and R. K. C. Teo. 1986. Neurobehavioural effects of occupational exposure to lead. British Journal of Industrial Medicine 43:374–380.

Williamson, A.M., R.K.C. Teo, and J. W. Sanderson. 1982. Occupational mercury exposure and its consequences for behaviour. International Archives of Occupational and Environmental Health 50:273–289.

Williamson, A. M., B. Clarke, and C. E. Edmonds. 1987. The neurobehavioural effects of professional abalone diving. British Journal of Industrial Medicine 44:459–466.

Human Neurobehavioral Toxicology Testing

W. Kent Anger

Twenty-five years ago, Joseph Ruffin, a staff physician with Kaiser Steel Corporation, published a call for "Functional Testing for Behavioral Toxicity: A Missing Dimension in Experimental Environmental Toxicology" in the *Journal of Occupational Medicine* (Ruffin, 1963). Since that time, the field of behavioral toxicology or, more broadly, neurotoxicology has shown a rapid growth and become one of the first independent specialty fields under the general rubric of toxicology. One result of this growth is that sufficient research has accumulated to allow the development of screening programs for behavioral toxicity.

There are two reasons for developing standardized tests or test batteries to screen for (i.e., identify) effects of neurotoxic chemicals: (1) premarket testing or related regulatory needs and (2) development of a neurotoxicity data base. Although the former reason is a relatively recent development impacting this field, the latter has a longer history. Scientists within the field have encouraged standardization of tests (Buck et al., 1977; Dews, 1975; Morgan and Repko, 1974) and the use of reference chemicals as positive controls (Buelke-Sam, 1980; Laties, 1973). Though never explicitly stated, the reason is to allow us to relate findings in one laboratory to findings in other laboratories or to relate findings in one country to those in other countries. The ultimate purpose is to develop enough information on a range of chemicals that general principles of neurotoxicity can be gleaned or commonalities identified. Only then can the chemical-by-chemical approach to testing, now typical in this field, be replaced by a more expeditious approach. To identify the hoped-for commonalities, a data base must be assembled from research on diverse chemicals studied with common methods. This suggests the need to select standard methods to be used in research to identify neurotoxic effects. It is also consistent with the regulatory needs for pre- and postmarket screening.

The selection of tests for a neurotoxicity screening battery has occupied the field of neurotoxicology for several years, especially since the Environmental Protection Agency (EPA) asked the field to select neurobehavioral screening tests in 1979. In that year, the EPA sponsored a conference in San Antonio with the purpose of identifying behavioral tests that could be used to evaluate new or untested chemicals for behavioral/nervous system effects, a primary regulatory need of the Toxic Substances Control Act (TSCA) (Geller et al., 1979). At the end of the meeting, Weiss and Laties (1983) of the University of Rochester summed up their opinion for EPA: "This collection of papers provides the most emphatic statement so far of how essential it is for the Environmental Protection Agency to shun test [selection] standardization. . . . A behavioral analog of the Ames test. . . is an impossible dream." These comments exemplify the strong trend of opinion opposing the development of screening test batteries in the psychology community. This is especially notable among those researchers studying neurotoxic effects in adult animals. Those in the field of behavioral teratology have been more tractable on this topic, generally adopting test batteries common to several laboratories. The National Center for Toxicological Research (NCTR) collaborative laboratory study established the replicability of one such battery, as described by Buelke-Sam et al. (1985). Those scientists conducting human neurobehavioral worksite research have also tackled the problem of battery development with enthusiasm in recent years. Several investigators in the United States are in the process of developing or validating human neurotoxicity test batteries (Otto and Eckerman, 1985).

RATIONALES FOR DEVELOPING TEST BATTERIES

Two rationales can be formulated for selecting tests into a reasonably comprehensive battery of the sort needed to assess the range of behavioral changes produced by the variety of chemicals found in the United States. The first and most comprehensive rationale would be to assess all major nervous system functions to identify all potential problems that might be produced by chemical exposures. This would require a taxonomy of nervous system functions. Although such taxonomies have been developed by Carroll (1980) for cognitive functions and by Fleischman and Quaintance (1984) for a range of performance tasks, there is no widely accepted taxonomy that attempts to specify all nervous system functions. Of course, any such list would be too broad to accommodate in a test battery of reasonable length. The second rationale would be to select tests that would measure neurotoxic effects typically found following occupational or environmental exposures. This rationale for developing a battery of tests is likely to be an evolutionary one. That is, as certain effects replace others as the most frequently reported problems, the makeup of the battery would be altered.

Neither of these approaches has been followed exclusively. Eckerman and others (Gullion and Eckerman, 1986) developed one test battery based on eight cognitive factors identified by Carroll (1980). However, the battery has not been used in field evaluations. On the other hand, several approaches to the development of human neurotoxicity test batteries have been followed, and the resulting test batteries are undergoing field trials. The more prominent approaches are discussed next.

Finland's Institute of Occupational Health Approach

Historically, the first approach to worksite neurotoxicity testing was developed at Finland's Institute of Occupational Health (FIOH) in the 1950s. Investigators at FIOH developed a test battery that is well adapted to studying the main concerns in Finnish industry, particularly exposure to a limited number of solvents. Their tests (Table 1), which have been streamlined through factor analysis over the years, reflect various psychological domains that can also be seen in the table (Hanninen and

TABLE 1 Finland's Institute for Occupational Health (FIOH) Test Battery

Test	Domain
Benton Visual Retention	Visual perception
Bourdon-Wiersma	Visual perception
Symmetry Drawing	Visual perception
Mira Test	Motor performance
Reaction Time	Motor performance
Santa Ana	Motor performance
Wechsler Memory Scale (portions)	Cognitive/memory
Wechsler Adult Intelligence Scale (portions)	Cognitive

SOURCE: Hanninen and Lindstrom (1979).

Lindstrom, 1979). The battery is now used routinely in neurotoxicity evaluations of worker groups in Finland, including prospective studies involving new workers.

Problem-Based Approach

The problem-based approach to worksite testing has its origins in the wide variety of neurobehavioral problems and neurotoxic chemicals found in the United States, where field investigators have typically adopted a unique battery of tests for each particular situation. The tests have been selected based on two factors: (1) the type of symptoms reported by the exposed group to be tested and (2) the established neurotoxic effects of the chemical under study or structurally related chemicals. This general approach has led to the use of literally hundreds of different tests in various worksite studies conducted over the years (Johnson and Anger, 1983) and has also been characteristic of National Institute for Occupational Safety and Health (NIOSH) research (Anger, 1985).

Approach Recommended by the World Health Organization

A third approach represents a melding of the first two approaches, and was well articulated in the World Health Organization (WHO) meeting held in Cincinnati during May 1983. At that meeting, a small group of established researchers in neurotoxicology recommended a core set of tests (the Neurobehavioral Core Test Battery, or NCTB) that could be used as a basic screen to identify a broad range of neurotoxic effects, particularly for use in developing countries. Tests selected into the core set (1) had been used successfully in worksite studies (i.e., they had identified group differences produced by chemical exposures), (2) were portable, (3) required minimal training to administer, and (4) were expected to be valid and reliable in most cultures. Most of the core tests (Table 2) were well-known "paper and pencil" tests specifically chosen to avoid mechanical or other instrumentation problems (a special concern in developing countries). For the one test requiring a source of electricity (the reaction time test), the instrument selected can be operated by using batteries as well as 110- or 220-volt current.

Another series of tests was identified as supplemental to the core battery. Their use was to be dependent on the chemical involved, the type of personnel available to conduct the tests, and the setting in which the tests were to be administered. The core set of tests was

TABLE 2 World Health Organization Neurobehavioral Test Battery

Test	Functional Domain
Santa Ana	Manual dexterity
Aiming	Motor steadiness
Simple Reaction Time	Attention/response speed
Digit Symbol	Perceptual-motor speed
Benton Visual Retention Test	Visual perception/memory
Digit Span	Auditory memory
Profile of Mood States (POMS)	Affect

SOURCE: Johnson (1987).

intended to generate more uniform, more consistent data from a broad variety of occupations and neurotoxic exposure conditions. The supplemental tests were intended to provide more in-depth information based on the known effects of the chemical under study and symptoms reported by the exposed workers (Johnson, 1987).

Neurobehavioral Evaluation System Approach

The Neurobehavioral Evaluation System (NES) implemented several neurobehavioral tests that had been used successfully in clinical settings or in previous field studies on IBM-PC and Portable Compaq computers. Some 17 tests were available on the NES as of 1986 (Letz and Baker). The tests are listed in the Table 3, along with the functions assessed. Tests are frequently added to this battery (the most recent version includes three additional tests not found in the table),[1] which is following an evolutionary course dictated by current interest in the field. The NES includes variants of five of the seven WHO-NCTB tests (noted by asterisks in Table 3). As with the problem-based model, developers of the NES recommend that the user "select tasks which are appropriate for specific exposure situations" (Baker et al., 1985).

Each of the four approaches described below is pragmatically based and used past research findings in selecting tests. Three approaches (FIOH, WHO-NCTB, NES) involve a limited battery of tests, and each battery is sensitive to important psychological functions. However, none of the batteries aspires to assess the broad range of human functions proposed above as one basis for test selection. Further, it is not clear if the functions assessed by these batteries are representa-

TABLE 3. Computer-Administered Neurobehavioral Evaluation System (NES)

Domain	Test	Function
Psychomotor Performance	Symbol-Digit[a]	Coding speed
	Hand-Eye Coordination	Coordination
	Simple Reaction Time[a]	Visuomotor speed
	Continuous Performance Test	Attention/speed
	Finger Tapping	Motor speed
Perceptual ability	Pattern Comparison	Visual perception
Memory and learning	Digit Span[a]	Short-term memory/attention
	Paired-Associate Learning	Visual learning
	Paired-Associate Recall	Intermediate memory
	Visual Retention[a]	Visual memory
	Pattern Memory	Visual memory
	Memory Scanning	Memory processing
	Serial Digit Learning	Learning/memory
Cognitive	Vocabulary	Verbal ability
	Horizontal Addition	Calculation
	Switching Attention	Mental flexibility
Affect	Mood Test[a]	Mood

[a]Variant of WHO Core Test.
SOURCE: Letz and Baker (1986).

tive of the range of functions that might potentially be affected by neurotoxic chemicals, the other basis for test selection noted above.

To assess how well these batteries would detect the health effects typically caused by neurotoxic chemicals at low concentrations (i.e., target organ effects), the target organ health effects identified in the research literature are described, followed by a comparison of the potential of the test batteries to identify those effects. Because the batteries do not assess the broad range of human functions, it is important to consider how effectively they assess health effects frequently caused by neurotoxic chemicals. There is no direct evidence on this point. Several threads of evidence do, however, provide the data to begin such an assessment. These threads, discussed in order below, demonstrate that (1) a large number of chemicals produce effects on the nervous system, and 65 of these chemicals have exposure populations in excess of one million, and (2) many nervous system-related health effects occur at lower exposure concentrations than most other effects for certain chemicals.

TARGET ORGAN EFFECTS

Neurotoxic Chemicals

There are 60,000 (Reiter, 1980) to 100,000 (NIOSH, 1983) chemicals in commerce today. How many affect the nervous system? Anger and Johnson (1985) have reviewed major secondary reference sources in this field (American Conference of Governmental Industrial Hygienists, 1980, 1982; Clayton and Clayton, 1981; Damstra, 1978; Gosselin et al., 1976; Lazerev and Levina, 1976; Norton, 1975, revised 1980; Spencer and Schaumburg, 1980; Weiss, 1978) to identify chemicals for which there is evidence of nervous system effects. They identified just over 750 chemicals or chemical groups for which evidence of direct or indirect nervous system effects exists. It is clear that there are a large number of industrial chemicals that affect the nervous system adversely.

Exposure Populations

There is also evidence that people are exposed to those chemicals. In U.S. workplaces, the National Occupational Hazard Survey (NOHS) identified 200 chemicals, each of which had an estimated one million or more persons exposed. These estimates are based on statistical extrapolation from extensive sampling data and have been published by NIOSH (1977). The number of workers exposed to each chemical is very likely inflated due to the exposure identification strategy of that survey. (The strategy was to list chemicals in work environments whether or not there was indication of actual use or exposure.) It is also obvious that many of the people in that survey were exposed to multiple chemicals and were thus counted for more than one chemical. (A more recent national survey conducted by NIOSH, the National Occupational Environmental Survey [NOES], has not yet been published, but its results are expected to modify this picture considerably.)

Cross-referencing the 750 chemicals noted above that affect the nervous system (Anger and Johnson, 1985) with the 200 chemicals to which one million or more people are exposed, by NIOSH (1977) estimate, indicates that 65 of the 200, or about one third, are also found in the list of 750 (Anger, 1986). From these data, it is possible to conclude that a large number of U.S. workers work with chemicals that are known to affect the nervous system. These and other factors have led NIOSH to identify neurotoxic disorders as one of the 10 leading occupational problems in the United States (Centers for Disease Control, 1983, 1986) and to focus efforts on their prevention.

Behaviors Affected by Many Chemicals

The review that identified 750 chemicals which affect the nervous system also provided an indication of the universe of nervous system-related effects that are known to be produced by industrial chemicals. The various behavioral deficits induced by industrial chemicals were categorized into some 120 nervous system-related effects (Anger and Johnson 1985). Behavioral effects that have been reported as caused by 25 or more of the 750 chemicals are listed in Table 4, along with the number of chemicals with which each has been linked (in the last column). Some of the effects are vague; others are specific. Despite the lack of parallelism, which is quite predictable given the diversity in the source reports, Table 4 contains the 35 behavioral effects most frequently recognized and reported in the reference literature as occurring following exposure to industrial chemicals (Anger, 1986).

This collection of effects provides one measuring stick against which to judge the available test batteries. However, it is not clear if these neurotoxic effects are realistic concerns in the industrial environment. This would be the case if they occur at relatively low exposure levels. That is, are they target organ effects or health effects that occur at the lowest concentrations relative to other effects for a given chemical, rather than curiosities that occur only at high exposure concentrations. One line of evidence suggesting that they are target organ effects is the fact that these are cited as the basis for recommending workplace standards by one federal agency and one independent professional group.

NIOSH Recommendations

Over the years, NIOSH has produced 91 criteria documents on chemicals/chemical groups/physical agents (counting only once those documents that have been revised), under its mandate in the Occupational Safety and Health Act of 1970. These documents provide a review of the literature and recommend exposure maxima. The basis for the recommendation is explicitly stated in each document. A review of those documents indicates that nervous system effects have been an explicitly stated basis in about 36 of them, or approximately 40 percent. Generally, the nervous system effect was identified as one that occurred at very low concentrations and can be described as a target organ effect. The chemical/physical agents are listed in Table 5.

TABLE 4 Neurobehavioral Effects Reported Following Chemical Exposures for 25 or More Chemicals

Effect	Of 750 Chemicals[a]
Motor	
Activity changes	32
Ataxia	89
Convulsions	183
Incoordination/unsteadiness/clumsiness	62
Paralysis	75
Pupil size changes	31
Reflex abnormalities	54
Tremor/twitching	177
Weakness	179
Sensory	
Auditory disorders	37
Equilibrium changes	135
Olfactory disorders	37
Pain disorders	64
Pain, feelings of	47
Tactile disorders	77
Vision disorders	121
Cognitive	
Confusion	34
Memory problems	33
Speech impairment	28
General	
Anorexia	158
Autonomic dysfunction	26
Cholinesterase inhibition	64
Depression of the central nervous system	131
Fatigue	87
Narcosis/stupor	125
Peripheral neuropathy	67
Affect/personality	
Apathy/languor/lassitude/lethargy/listlessness	30
Delirium	26
Depression	40
Excitability	58
Hallucinations	25
Irritability	39
Nervousness/tension	29
Restlessness	31
Sleep disturbances	119

NOTE: Adapted from Anger (1984, 1986), Anger and Johnson (1985).
[a]Numbers below denote number of chemicals for which effect has been reported.

TABLE 5 Agents/Classes Cited by NIOSH Criteria Documents as Producing Nervous System Effects at Low Concentrations

Chemical/Physical Agent	Document No.
Acrylamide	(77-112)
Alkanes	(77-151)
Anesthetic gases, waste	(77-140)
Carbaryl	(77-107)
Carbon disulfide	(77-156)
Carbon monoxide	(HHS 73-11000)
Carbon tetrachloride	(76-133)
Chloroform	(75-114)
Cresol	(78-133)
Dinitro-o-cresol	(78-131)
Ethylene dibromide	(77-221)
Fluorocarbon polymers, decomposition products	(77-193)
Formaldehyde	(77-126)
Hydrogen cyanide and salts	(77-108)
Hydrogen sulfide	(77-158)
Ketones	(78-173)
Lead, inorganic/revised	(78-158)
Malathion	(76-205)
Mercury, inorganic	(HHS 73-11024)
Methyl alcohol	(76-148)
Methyl parathion	(77-106)
Methylene chloride	(76-138)
Nitriles	(78-212)
Noise	(HHS 73-11001)
Parathion	(76-190)
Petroleum solvents, refined	(77-192)
Styrene	(83-119)
Tetrachloroethane (perchloroethane)	(76-185)
1,1,2,2-Tetrachloroethane	(77-121)
Thiols (n-alkane monothiols, cyclohexanethiol, benzenethiol)	(78-213)
Toluene	(HHS 73-11023)
1,1,1-Trichloroethane (methylchloroform)	(76-184)
Trichloroethylene	(HHS 73-11025)
Tungsten and cemented tungsten products	(77-127)
Xylene	(75-168)
Zinc oxide	(76-104)

NOTE: NIOSH Criteria Documents for each chemical listed in the table are available by document number (in parentheses) from NIOSH Publications, 5555 Ridge Ave., Cincinnati, OH 45213.

SOURCE: NIOSH Criteria Documents.

Recommendations of the American Conference of Governmental Industrial Hygienists

In the United States, the earliest national sources of recommended limits for exposures to industrial chemicals were the publications of the American Conference of Governmental Industrial Hygienists (ACGIH). They have published reviews of a far larger number of chemicals than has NIOSH or any other group, and their recommendations are discussed at some length because their stated intent is to be comprehensive in reviewing toxic chemicals in commerce in the United States (ACGIH, 1982).

The ACGIH recommendations are provided by their Threshold Limit Value (TLV) committee, composed of voting practitioners from academia and government. Persons from NIOSH and industry serve as nonvoting consultants. The ACGIH publishes annually a list of the chemicals most frequently encountered in industry for which there is documented evidence of untoward symptoms or occupational disease. For each chemical listed, exposure maxima (TLVs) are recommended (ACGIH, 1982), based on the relevant literature and the personal experience of members and consultants.

The ACGIH (1982) recommended TLVs for 588 chemicals. To support these recommendations, the ACGIH also published a book of documentation (ACGIH, 1980) which describes the basis for each recommended TLV. Anger (1984) abstracted 36 organ systems, health effects, or other bases cited in the documentation as the most relevant information leading to the ACGIH recommendations. These presumably reflect those effects produced by low exposure concentrations. In considering only those categories labeled nervous system, unpleasant taste/odor, and eye (other than irritation), a total of 167 chemicals (roughly one-quarter of the total 588) listed through 1982 have TLVs based on these direct neurologic or behavioral effects (Anger, 1984). This and the NIOSH criteria documents suggest that the nervous system is an important target organ for industrial chemicals in use today.

To return to the 35 effects in Table 4 (produced by 25 or more of the 750 chemicals), most are also found cited under at least two chemicals in the ACGIH documentation. Thus, most of the effects in Table 4 may be presumed to occur following low-concentration exposures. An additional six effects, which were cited at least twice by ACGIH but less than 25 times in the list of 750, are also listed in Table 4. Those effects cited by ACGIH as caused by only one chemical are not included in this table.

POTENTIAL OF TEST BATTERIES TO ASSESS TARGET ORGAN HEALTH EFFECTS

The data in Table 4 provide a sample of health effects occurring frequently following exposures to toxic industrial chemicals. How effectively would the test batteries noted above identify or screen for these frequently occurring effects? The 35 health effects produced by 25 or more chemicals have been repeated in Table 6 in a single column. Table 6 also identifies the tests in the WHO, FIOH, and NES batteries that would presumably assess the various effects, if the sensitivity of each test and the sample size are adequate.

As can be seen in Table 6, many of the motor changes would be identified by the three test batteries, and the Profile of Mood States (POMS) or mood tests found in two of the batteries would detect changes in affect. Sensory changes would be poorly identified, as would ataxia and weakness, two effects that occur frequently after toxic chemical exposure. The forte of the batteries, cognitive testing (particularly memory), is aimed at central nervous system (CNS) functions. This is based on the assumption that such effects are not reported very frequently because they are only rarely assessed with any degree of sophistication. This assumption is somewhat substantiated by past worksite research with carbon disulfide, mercury, lead, and methyl chloride. That research has identified subtle CNS deficits in worker groups exposed to concentrations that did not produce peripheral effects or other signs of frank poisoning (Anger and Johnson, 1985). Human laboratory research on acute exposure effects of many solvents also supports this assumption (Dick and Johnson, 1986).

The WHO-NCTB, FIOH, and NES test batteries are reasonable, defensible, research-based approaches to the assessment of neurotoxic chemicals. Further, each battery tests for well-established health-related effects that have been accepted by the public health community in the past. They also include tests aimed at assessing the more subtle CNS deficits that occur at lower exposure concentrations than do the more frank poisoning effects that have been the focus of attention in the past.

There are clearly limitations to these batteries, however. The WHO-NCTB and FIOH batteries use a test of motor performance (the Santa Ana) that has previously demonstrated differences between a group exposed to toxic chemicals and an unexposed group, but the NES hand-eye coordination test has not yet identified chemical effects. Also, only the FIOH battery and the WHO-NCTB have more than one test of coordination. The WHO-NCTB is slightly superior to both the NES and the FIOH batteries in its use of the POMS test. The FIOH battery has no test for affect, and the NES uses a mood test that

TABLE 6 Neurobehavioral Effects Reported Following Chemical Exposures

	Tests		
Neurobehavioral Effects	WHO	NES	FIOH
Motor			
Activity changes			
Ataxia			
Convulsions			
Incoordination/ unsteadiness/clumsiness	Santa Ana	Hand-Eye Coordination	Santa Ana
Paralysis	Santa Ana	Hand-Eye Coordination	Santa Ana
Pupil size changes			
Reflex abnormalities			
Tremor/twitching	Santa Ana	Hand-Eye Coordination	Santa Ana
Weakness			
Sensory			
Auditory disorders			
Equilibrium changes			
Olfaction disorders			
Pain disorders			
Pain, feelings of			
Tactile disorders	Santa Ana		Santa Ana
Vision disorders	Benton	Pattern	Benton
Cognitive			
Confusion			
Memory problems	Benton	Pattern	Benton
Speech impairment			
Affect/personality			
Apathy/languor/lassitude/ lethargy/listlessness	POMS	Mood Test	
Delirium			
Depression	POMS		
Excitability	POMS		
Hallucinations			
Irritability	POMS	Mood Test	
Nervousness/tension	POMS	Mood Test	
Restlessness	POMS		
Sleep disturbances			
General			
Anorexia			
Autonomic dysfunction			
Cholinesterase inhibition			
Depression of the central nervous system			
Fatigue	POMS		
Narcosis/stupor			
Peripheral neuropathy	Santa Ana	Hand-Eye Coordination	Santa Ana
Pathology			
Psychic disturbances			

is based on the 65-item POMS, but it employs only 25 of those items to identify five of the six factors (all but vigor) on the POMS. The reliability and validity of the resulting test have not been assessed thoroughly, and an item-by-item analysis of the mood test indicates that it does not appear to assess some types of affect that are assessed by the POMS (and occur frequently following chemical exposures). Overall, this assessment would suggest that the WHO battery is slightly superior to the FIOH and the NES batteries, based on the criteria proposed above. However, the NES has the advantage of being administered by computer, which reduces administration costs substantially. On the negative side, some of the most frequently occurring neurotoxic effects, particularly some forms of peripheral neuropathy and affective symptoms, weakness, ataxia, and sensory effects, would be missed by all of the batteries. Of course, it is quite possible that CNS changes are correlated with some of these effects and these tests would thus perform their function of detecting health effects. A "screening" battery is developed for detection, not characterization.

The established test batteries must undergo constant scrutiny. An evolutionary course of test battery development is essential when the battery is based on established health effects, because these effects may change as chemicals in use change. Further, the immense range and diversity of behavior noted above suggest that simple test batteries are inadequate for comprehensive screening. There is the danger of not detecting those effects for which the batteries lack tests. One hopes that the currently recommended test batteries noted above are designed for the problems of the future by assessing more subtle CNS effects rarely tested adequately in the past.

DEVELOPMENT EFFORTS UNDER WAY

The test batteries described above are not without their problems. One major problem lies in interpreting the result of these batteries. The neurobehavioral tests in the various test batteries, with only a few exceptions, are not clinical instruments. They do not have established norms based on extensive population testing. Rather, they are performance tests that provide objective measures ideally suited to test-retest (before-after) assessments. Of course, baseline (preexposure) performance data on workers exposed to chemicals almost never exist. Performance data on the same tests administered to unexposed people must be used for comparison. That is, performance effects are defined, not by established norms, but by comparison data from people believed to be healthy—a referent or control group.

An appropriate control group should consist of people who are not only unexposed to toxic chemicals, but are also similar to the exposed subjects in terms of age, education, job activities or movements, and socioeconomic variables. This is extremely difficult to achieve, and field researchers are virtually always concerned with the accuracy of the controls as a basis for judging the performance of the exposed group. The most pervasive problem in judging the validity of group differences has been age differences between exposed and comparison groups. Therefore, the World Health Organization has recommended the development of normative data from unexposed or control subjects in five age ranges for the NCTB. This recommendation may be extended to the other major neurotoxicity screening batteries. The five arbitrarily selected age ranges are: 16–25, 26–35, 36–45, 46–55, and 56–65 years. Ideally, the subjects would be employees with occupations relatively typical of those found throughout the country (or at least not atypical) and with a fairly homogeneous educational background. NIOSH is conducting such a study using the NES and the NCTB.

It is the World Health Organization's intention to carry out an assessment of the NCTB in eight nations (WHO, 1987). Because the NCTB tests were developed in western European-derivative countries (primarily Finland and the United States), the assessment is aimed at comparing the results from the U.S. and European countries with results from culturally diverse people of different countries. As of this writing, some 15 research groups had applied to participate in the assessment, although some cultural groups are not represented. If test performance in various countries/cultures is within certain ranges, the WHO-NCTB battery can be used to assess poisoning incidents or other neurotoxic exposures worldwide, and the results can be generalized to people throughout the world. Thus, data from all countries could be used to assess safe exposure concentrations of specific chemicals. This is intended to accelerate the development of a data base of chemical effects through the use of common tests. It will, in turn, advance the process of identifying chemical classes or mechanisms associated with neurotoxicity and thus lead to the ultimate goal of prediction of neurotoxicity, rather than identification through post hoc discovery of adverse health effects in humans.

NOTES

1. Current information on this battery and the software to implement it are available from Dr. Richard Letz, Environmental Sciences Laboratory, Mt. Sinai School of Medicine, 10 East 102nd Street, New York, NY 10029.

ACKNOWLEDGMENT AND DISCLAIMER

Appreciation is extended to Mrs. Pat Amendola and Pam Schumacher for preparation of the typescript. Mention of company or product names does not imply endorsement by NIOSH. This article is reproduced with permission from *Toxicology and Industrial Health* (Anger, 1989).

REFERENCES

American Conference of Governmental Industrial Hygienists. 1980. Documentation of the Threshold Limit Values, fourth edition, 1980 (and supplemental documentation for 1981 and 1982). Cincinnati, Ohio: ACGIH Publications Office.

American Conference of Governmental Industrial Hygienists. 1982. Threshold Limit Values for Chemical Substances and Physical Agents in the Workroom Environment with Intended Changes for 1982. Cincinnati, Ohio: ACGIH Publications Office.

Anger, W. K. 1984. Neurobehavioral testing of chemicals: Impact on recommended standards. Neurobehav. Toxicol. Teratol. 6:147–153.

Anger, W. K. 1985. Neurobehavioral tests used in NIOSH-supported worksite studies, 1973–1983. Neurobehav. Toxicol. Teratol. 7:359–368.

Anger, W. K. 1986. Workplace exposures. Pp. 331–347 in Neurobehavioral Toxicology, Z. Annau, ed. Baltimore: Johns Hopkins University Press.

Anger, W. K. 1989. Human neurobevioral toxicology testing: current perspectives. Toxicology and Industrial Health 5(2):165–180.

Anger, W. K., and B. L. Johnson. 1985. Chemicals affecting behavior. Pp. 51–148 in Neurotoxicity of Industrial and Commercial Chemicals, J. L. O'Donoghue, ed. Boca Raton, Fla.: CRC Press.

Baker, E. L., R. E. Letz, A. T. Fidler, S. Shalat, D. Plantamura, and M. Lyndon. 1985. A computer-based neurobehavioral evaluation system for occupational and environmental epidemiology: Methodology and validation studies. Neurobehav. Toxicol. Teratol. 7:369–377.

Buck, W. B., D. L. Hopper, W. L. Cunningham, and G. G. Karas. 1977. Current experimental considerations and future perspectives in behavioral toxicology. Pp. 2.1–2.10 in Behavioral Toxicology: An Emerging Discipline, H. Zenick and L. Reiter, eds. US EPA Pub. No. EPA-600/9-77-042. Research Triangle Park, N.C.

Buelke-Sam, J. 1980. Standardization is not an ugly word. Neurobehav. Toxicol. 2:289–290.

Buelke-Sam, J., C. A. Kimmel, and J. Adams. 1985. Design considerations in screening for behavioral teratogens: Results of the collaborative behavioral teratology study. Neurobehav. Toxicol. Teratol. 7:537–789.

Carroll, J. B. 1980. Individual difference relations in psychometric and experimental cognitive tasks. L. L. Thurstone Psychometric Laboratory Report No. 163. Chapel Hill, N.C.: University of North Carolina.

Centers for Disease Control. 1983. Leading work-related diseases and injuries in the United States. Morbidity and Mortality Weekly Report 32(2):24–26, 32.

Centers for Disease Control. 1986. Leading work-related diseases and injuries—United States. Neurotoxic disorders. Morbidity and Mortality Weekly Report 35(8):113–116, 121–123.

Clayton, G. D., and F. E. Clayton, eds. 1981. Patty's Industrial Hygiene and Toxicology, third revised edition. New York: John Wiley & Sons.

Damstra, T. 1978. Environmental chemicals and nervous system dysfunction. Yale J. Biol. Med. 51:457–468.
Dews, P. 1975. An overview of behavioral toxicology. Pp. 439–445 in Behavioral Toxicology, B. Weiss and V. Laties, eds. New York: Plenum.
Dick, R. B., and B. L. Johnson. 1986. Human Experimental Studies. Pp. 348–387 in Neurobehavioral Toxicology, Z. Annau, ed. Baltimore: Johns Hopkins University Press.
Fleishman, E. A., and M. K. Quaintance. 1984. Taxonomies of Human Performance. New York: Academic Press.
Geller, I., W. C. Stebbins, and M. J. Wayner, eds. 1979. Test methods for definition of effects of toxic substances on behavior and neuromotor functions. Neurobehav. Toxicol. 1(Suppl. 1):1–225.
Gosselin, R. E., H. C. Hodge, R. P. Smith, and M. N. Gleason. 1976. Clinical Toxicology of Commercial Products, fourth edition. Baltimore: Williams and Wilkins.
Gullion, C. M., and D. A. Eckerman. 1986. Field testing. Pp. 288–330 in Neurobehavioral Toxicology, Z. Annau, ed. Baltimore: Johns Hopkins University Press.
Hanninen, H., and K. Lindstrom. 1979. Behavioral Test Battery for Toxicopsychological Studies Used at the Institute of Occupational Health in Helsinki. Helsinki: Institute of Occupational Health.
Johnson, B. L., and W. K. Anger. 1983. Behavioral toxicology. Pp. 329–350 in Environmental and Occupational Medicine, W. N. Rom, ed. Boston: Little, Brown.
Johnson, B. L., ed. 1987. Prevention of Neurotoxic Illness in Working Populations. New York: John Wiley & Sons.
Laties, V. G. 1973. On the use of reference substances in behavioral toxicology. Pp. 83–88 in Adverse Effects of Environmental Chemicals and Psychotropic Drugs, M. Horvath, ed. New York: Elsevier.
Lazerev, N.V., and E. N. Levina. 1976. Harmful Substances in Industry. Leningrad: Khimiya Press, (translated by Literature Research, Annandale, Va.; available from NIOSH Library, 4676 Columbia Pkwy., Cincinnati, OH 45226).
Letz, R., and E. L. Baker. 1986. Computer-administered neurobehavioral testing in occupational health. Seminars in Occup. Med. 1:197–203.
Morgan, B. B., and J. Repko. 1974. Methodological problems in behavioral toxicology research. Pp. 478–479 in Behavioral Toxicology, C. Xintaras, B. L. Johnson, and I. DeGroot, eds. US DHEW (NIOSH) Pub. No. 74–126. Washington, D.C.
National Institute for Occupational Safety and Health. 1977. National Occupational Hazard Survey. 1972–1974. USDHEW (NIOSH) Pub. No. 73-114. Cincinnati, Ohio: NIOSH Publications Office.
National Institute for Occupational Safety and Health. 1983. Registry of Toxic Effects of Chemical Substances, 1981–3. USDHHS (NIOSH) Pub. No. 83-107. Cincinnati, Ohio: NIOSH Publications Office.
Norton, S. 1975; revised 1980. Toxicology of the central nervous system. Pp. 179–205 in Toxicology, C. D. Klassen, M. D. Amdun, and J. Doull, eds. New York: Macmillan.
Otto, D., and D. Eckerman, eds. 1985. Workshop on neurotoxicity testing in human populations. Neurobehav. Toxicol. Teratol. 7:283–418.
Reiter, L. W. 1980. Neurotoxicology, meet the real world. Neurobehav. Toxicol. 2:73.
Ruffin, J. B. 1963. Functional testing for behavioral toxicity: A missing dimension in experimental environmental toxicology. J. Occup. Med. 5:117–121.
Spencer, P. S., and H. H. Schaumburg, eds. 1980. Experimental and Clinical Neurotoxicology. Baltimore: Williams and Wilkins.
Weiss, B. 1978. The behavioral toxicology of metals. Fed. Proc. 37:22–27.
Weiss, B., and V. G. Laties. 1983. Assays for behavioral toxicity: A strategy for the Environmental Protection Agency. Neurobehav. Toxicol. 1:213–215.
World Health Organization. 1987. Unpublished Report: Field Evaluation of WHO Neurobehavioral Core Test Battery. Geneva: WHO Publications.

Neurobehavioral Tests:
Problems, Potential, and Prospects

J. Graham Beaumont

There seems to be general agreement that any monitoring of the effects of environmental and occupational exposure to neurotoxins should include behavioral measures. An important element in the effects of known toxins is the response of the nervous system, including peripheral sensory and motor components and higher central effects upon the function of the forebrain. This response has clear behavioral aspects following gross acute exposure and significant chronic exposure to a range of neurotoxins. There are considered to be more subtle behavioral effects of less severe acute exposure or of sustained exposure to lower levels of the relevant substances.

The assessment of behavioral effects is considered to be the primary approach to the systematic monitoring of neurotoxic exposure, and where mass screening is considered for large populations at risk, it may be the only practicable approach, at least for initial selection. It is obvious that automated screening by the use of computer-based assessment could contribute significantly to the development of appropriate techniques.

The essential context for the adoption of acceptable assessment techniques is that the potential behavioral changes should have been identified and reliable measures of these changes should be available, which have been demonstrated to be valid, and for which appropriate normative data are available.

It may also be desirable that the test be stable under conditions of repeated testing. Particularly when relatively subtle changes, with a low base-rate in the population (as may be typical of mass screening), are to be detected, it is essential that the validity (and therefore the reliability) be exceptionally high.

None of this is in conflict with the preceding chapters, indeed there is remarkable agreement as to the current state of the field, the methodological principles that apply, and the standards that should be adopted. Areas in which there is some potential disagreement are as to whether the currently available tests are sufficient for their purposes, and whether the introduction of new test instruments is to be encouraged. This chapter therefore concentrates principally upon those issues.

TESTS CURRENTLY IN USE

The last three chapters have covered the history and description of current tests, with particularly helpful tabulations by Hanninen and Anger, and it would be redundant to repeat much of this material.

It is worth, however, drawing attention to the version of the World Health Organization's Neurobehavioral Core Test Battery (WHO-NCTB) in a computer-based form developed by the Institute of Occupational Health at the University of Milan. The battery is much as in its original form except that the Santa Ana Rotation Test of the original battery has, for pragmatic reasons, been replaced by a test which assesses rather different cognitive functions, and the modality of the Digit Span task has been changed in a way that is known to alter the cognitive functions involved (Beaumont, 1985).

A preliminary study of the psychometric characteristics of this implementation of the NCTB has been reported (Camerino, 1987). This indicates that there are some serious questions concerning the validity of these tests in terms of their suitability for the assessment purposes under consideration.

A group of 30 volunteers—young, relatively well-educated adults—were retested at weekly intervals on certain of the tests [excluding Benton Visual Retention Test (VRT) and Aiming Pursuit], and estimates of the reliability and validity of the measures were made from the results. It is a little unclear what reliability should be expected from an instrument that assesses mood "over the past week," given at weekly intervals: the range of values of r from -0.24 to +0.87 on the various individual scales is probably not remarkable. The reliabilities on the cognitive tasks are more acceptable, being in the range 0.62 to 0.89, if reaction time (RT) variability and Digit Learning are excluded.

The reliabilities of the Digit Learning test at 0.40 and 0.19 (for occasions 1–2, 2–3, respectively) are clearly quite inadequate and suggest that the test should be abandoned as part of this assessment.

Correlations were also calculated with paper-and-pencil versions, as a crude measure of construct validity. Correlations were modest ranging from 0.55 (Serial Digit) to 0.79 (Benton VRT). These values are not atypical of values that might be expected on psychometric tests of this type.

However, these results do raise certain doubts about the psychometric suitability of these tests to the purposes for which they are being employed. For purposes of debate, assume that the validity of the measures is on average about 0.75. This is probably rather generous: reliability limits the upper extent of validity, and reliabilities are in some cases below this level. In addition, the sample employed was likely to provide relatively high levels of reliability and validity. At this level of validity, if we are trying to identify pathological effects which are present in 50 percent of those tested, the best that the test can theoretically achieve is 77 percent correct classification of the test subjects. In practice, a much more unfavorable base-rate of the condition is likely to apply in the test population. If the incidence to be detected falls to 1 in 10, the theoretical maximum achievement of the test will be 90 percent overall correct classification, but of those affected only 50 percent will be correctly identified. Of those achieving "positive" results on the test, half will be misclassified because they are false positives. As the base-rate or the validity falls, these statistics become even more unacceptable. It should be clear that in psychometric terms, these tests as implemented in the Milan study are insufficiently powerful to allow any valid assessment of the neurobehavioral functions under study.

This must be a serious concern because (quite reasonably) the NCTB has been adopted in a number of centers around the world. Swedish studies conducted at the National Board of Safety and Health (Iregren, 1986) have used these tasks among a battery of others administered in both traditional and computer-based formats, as well as some other automated modes. The computer-based tests include Memory Reproduction (letter and digit sequences, rather like Digit Span), Simple and Choice Reaction Time, and Color Word Vigilance, and others are under development. Studies conducted with the full range of tests have demonstrated some significant interesting findings between criterion groups selected for contrast on relevant variables, mostly relating to solvent exposure. Some reliability data are reported by Iregren in this volume. The methodological rigor of this approach is to be applauded, and the data show some reliabilities for certain of the assessments significantly higher than the Milan data for their battery. Nevertheless, with assessments of higher cognitive functions and of affect, the psychometric adequacy of the instruments remains a problem.

A battery that shares some provenance with the WHO-NCTB, although it strictly just predates it, is Baker and Letz's Neurobehavioral Evaluation System (NES). The NES has been adopted by a major study being carried out by the Institute of Occupational Health in Birmingham, United Kingdom (Spurgeon and Harrington, 1987). This study will use the Clinical Interview Schedule together with the Hogstedt Symptom Questionnaire, Stress and Arousal Checklist, Cognitive Failures Questionnaire, and Prospective Memory Test, in addition to the NES tests. At present only preliminary pilot data are available.

Of course there have been a large number of other studies published in the literature which have employed a wide variety of tests. A survey of the literature of the effects of lead on intelligence reveals the WISC-R to be the most popular test in a traditional format to have been employed in this research (Yule and Rutter, 1985). A great variety of more specific tests of individual functions have also been employed (Anger, 1985).

Further contributions concerning the use of computer-based assessment in this domain are to be found in Braconnier (1985). A useful collection of papers concerned more generally with the issues raised by computer-based assessment appeared in *Applied Psychology* (e.g., see Huba, 1987).

The preceding chapters seem to be in agreement that (1) there are problems evident in the construction of various batteries, (2) most of the tests currently in use are relatively inadequate, and (3) there is poverty in the current psychological descriptions of neurotoxic syndromes.

SOME SPECIFIC POINTS

Some specific points made in the preceding chapters are highlighted here.

Methods in Behavioral Toxicology (Hanninen)

The problems concerning the definition and description of the neurotoxic deficit are well taken: this is clearly of crucial significance for any advance in the field and emphasizes the need for more fundamental research into the cognitive processes affected.

The suggestion that the in-depth study of individual patients might be profitable is also a valuable one. There are now many good single-case experimental designs that might be appropriately deployed in this area, and they should be considered in order to further clarify the description of the relative deficits.

The dilemma that Hanninen discusses between the "conservative" and "progressive" approaches is a real one and, to some extent, is fundamental to much of the discussion that follows. It is of central importance to decide whether to make the best of the rather poor tests that are currently in use, or whether to adopt a more radical reevaluation of current tests and the potential new instruments that might be created.

Current Status of Test Development (Williamson)

Williamson sensibly highlights the potential for tests that relate explicitly to psychological theory (although the distinction between those that relate to "cognitive structure" and those that are "theory-based" may not be so easy to sustain). If it becomes possible to elaborate our understanding of the psychological processes (and, perhaps, as a contribution to that understanding), there is obvious merit in the use of such tests.

The "potential barrier" of computer-based testing must be taken seriously. There is clearly no value in developing computer-based tests if they confer few advantages, introduce extraneous sources of error, and hinder the wide application of tests. There may be benefits from the application of computers that outweigh these disadvantages—at least in parts of the world where they can practicably be used—but it is important to be clear about the advantages in any given case.

The need for more basic research is again emphasized, and the proposal that the adaptive nature of some of the changes which take place be considered may be a particularly useful insight.

Human Neurobehavioral Tests (Anger)

Anger's useful and authoritative view is clear and correct about the potential contribution that the test batteries may make in this field. It is necessary, however, to ensure that this potential is realized in practice. It is certainly possible that the relevant changes could be detected. It is much less certain that current batteries are capable of detecting the changes (and some reason to believe that they are not).

The case also has to be argued more clearly for the value of cross-cultural data collection. It is naturally important, indeed essential, that appropriate local norms be available. However, given that there are inevitably differences among cultures in education, cognitive processes, cultural experience, exposure to testing and test materials, and even (some believe) in intelligence, test performance will differ in different cultures and subcultural contexts. In this situation, differences underlying test performance and exposure to toxins will undoubtedly be confounded.

The results will be difficult, perhaps impossible, to interpret, and little will have been gained by international comparisons. The idea of a worldwide pool of test results may be superficially attractive, yet not based in the psychometric realities of the situation.

THE ADEQUACY OF CURRENT TESTS

The problems inherent in current assessment batteries appear to be twofold. First, the tests employed have been selected on the basis of their previous use in experimental studies of the effects of exposure to neurotoxins. It is natural that, when a test has been shown to distinguish between a criterion group of exposed individuals and a control group, this test should be considered suitable for inclusion in an assessment battery. This is, however, not necessarily the case. Only if the test can be shown to have sufficient psychometric power for the role of general screening can it be considered useful in this way. It is important throughout to maintain a careful distinction between tests that are useful for group experiments and those that may be used for individual screening.

Second, there is a temptation to select tests that are generally considered to be capable of indicating central nervous system (CNS) dysfunction. Here the temptation has been to take tests that are believed capable of revealing the effects of dementia, cerebral disease, or gross trauma, and to adopt them for detection of the effects of neurotoxins. This procedure is open to two misconceptions: that the effects of neurotoxins will be the same (in cognitive terms) as the effects of dementia, cerebral disease, or trauma, and that there are tests capable of simply discriminating among these other disorders. There seems to be little basis for accepting either of these proposals. It is unlikely that CNS poisoning is similar in its effects to other cerebral pathology, any more than the similarity between, say, dementia and trauma. The history of neuropsychology is littered with failed attempts to identify, by means of a single measure or small group of measures, general cerebral pathology. In particular, if the effects are relatively diffuse, the problem is especially difficult. An example is the difficulty of distinguishing, by cognitive measures alone, dementia of the Alzheimer type in the elderly—at least in its early stages—from either functional psychiatric illness or acute systemic illnesses. Much the same problem must apply to the effects of neurotoxins.

It is therefore not surprising that the battery of tests now generally employed is not of strong validity and is probably inadequate for the general detection of the behavioral effects of neurotoxins. There is simply insufficient power in the basic psychological instruments being employed.

The critical problem is the psychometric power of the tests, and the critical question is, Is the WHO-NCTB (including related batteries such as the NES) adequate to the task? It is important at this point to be clear as to what the task is—either to conduct group experiments or to undertake individual screening.

If the task is to investigate the differences between criterion groups, then the NCTB may be adequate to the task. Its psychometric power is still weak, and there might well be better tools available. It is probably, as a psychometric instrument, best described as "premature." Nevertheless, the fact that it is available, and already quite widely adopted, is of some importance, and it is clearly capable of discriminating between carefully selected groups under favorable conditions. Its use is certainly justified in this context, although efforts should be made to dramatically increase the size of the standardization samples available and to improve the basic reliability of the tests. In the context of such studies using the NCTB, it might be that computers are an impediment and that administration in the standard form is to be preferred.

However, if the aim is to carry out screening for exposed and affected individuals, the NCTB is likely to be quite inadequate on psychometric grounds. As discussed above, the available data suggest that the battery is not reliable enough to permit sufficiently accurate classification of affected and nonaffected individuals.

This implies that if screening is a goal of the research (or if significant improvements are to be made in the sensitivity of the tests for detecting differences between criterion groups), then the whole basis of the assessments currently employed needs to be reexamined. Better fundamental research is needed to generate a psychological description of the deficits and better models of the effects which can be related to that description. In achieving this it may well be advantageous to make better use of new developments in psychometrics and in the explicit models of cognitive performance. It is at this point that computers might well be introduced. One way in which this might be done is described below.

SOME NEW DEVELOPMENTS IN COMPUTER-BASED ASSESSMENT

It seems worth inquiring whether there are alternative approaches that could potentially provide more satisfactory solutions to the assessment of cognitive performance. There seem to be at least two potentially fruitful avenues of exploration. One is rather better charted: the use of adaptive testing systems, although it is not considered further here. The other is through the explicit incorporation of cognitive models into intelligent assessment systems. Such systems would not radically overthrow the traditional psychometric approaches, but would complement and extend such approaches so that the advantages of both could contribute to the power inherent in the assessment procedure.

If an intelligent and powerful assessment system is to be developed, it must incorporate appropriate psychometric models of the reference domain as well as a psychological (cognitive) model of that domain. The solution may well come from a progressive integration of psychometric theory, together with selection of those methods with greatest utility on the basis of empirical study. There is, after all, no reason why more than one psychometric model should not be operated concurrently, and the respective processes cross-referenced, as long as the assumptions of each are properly respected.

Cognitive Componential Models

Functional models, increasingly explicit in the cognitive domain, might allow an assessment system to possess an internal representation of the function that is under examination. One of the fruits of the growth of cognitive information-processing approaches into the dominant zeitgeist of contemporary psychology has been the production of explicit functional models. Some of these models are now presented in a sufficiently well-articulated form to make them useful in the description of functional status. Such descriptions can, in turn, be used in the identification of dysfunctional elements in performance and in the design and monitoring of instructional and remedial schemes.

Perhaps the most well known of these models relate to reading ability. Here the interaction between the developmental study of normal reading ability and neuropsychological investigation of the dysfunctions to be observed in brain-injured patients has stimulated the production of general models of reading competency. Over the past few years the analyses of developmental dyslexia and of adult acquired dyslexia have converged into a common view of the processes that may be defective in reading failure.

The point about this and similar models is that each component is capable of identification by manipulations in an explicit experimental paradigm. The evidence is derived from studies on normal subjects by which the processing components can be inferred and from study of clinical patients in whom the failure of one component of the system can be identified.

A number of models in a variety of domains (spelling, arithmetic functions, algebra, reasoning, number-series identification, map interpretation) illustrate how human abilities can be analyzed in terms of componential subprocesses. The relationships among the subprocesses are described in the model. The components in each model, both functional elements and channels of information transfer, can be assessed by experimental paradigms that are amenable to automated implementation. A system which incorporated an explicit model about the function under investigation should be capable of intelligently describing the nature and level of that function in the psychological domain within which the model has been created.

Inferential Systems

It remains to be shown how explicit cognitive functional models, in association with adaptive testing systems technology, might be incorporated into a practical and intelligent assessment system. The way in which this might be achieved is through the use of an intelligent knowledge-based systems approach.

The differences between traditional psychometrics and "expert systems" are not as fundamental as might be supposed. Although expert systems as commonly expressed within a rule-based programming environment appear very different from a psychometric test instrument, they have several fundamental constructs in common. The parallels become more clear if the elements of each procedure are considered. The objectives of the expert system are the test items of the conventional test; the values, the responses; the questions and user interface are equivalent to the administration procedures; the rules are represented in the scoring norms; the inference engine is matched by the psychometric model being employed. The goal that the expert system is set is, of course, the test result of the conventional test instrument.

It is possible to establish the validity of these parallels. The author has a demonstration system, created under a popular expert system "shell," that administers the Mill Hill Vocabulary Test in a form indistinguishable from a number of computer-based implementations of that test which have been realized by procedural programming systems that simply simulate the conventional administration of the test. It may well not be the most efficient way to achieve this result, and the use of the expert system shell may be to some degree artificial, but it nonetheless provides evidence for the parallels that are being proposed between these kinds of systems.

Given these parallels, it is a short step to suggest that a cognitive componential model might be explicitly incorporated within a knowledge-based system to permit intelligent assessment of the cognitive function modeled. This would simply require that the model be sufficiently well articulated to be expressed in terms of the contents of a rule base. A variety of procedures will, of course, also be defined which permit data to be established pertinent to the rule-based inferences that are to be made. These procedures may be prior values held within the system, they may be the responses to questions put to the test subject or to the test examiner, or they may be the results of ancillary procedures (including independent subprocedures defined within a procedural programming environment). The procedures may operate at the level of individual test "items" or may refer to a higher level of "subtest" investigation.

These subprocedural levels may reflect the structures that have already been developed within the adaptive testing context. The statistical procedures that have been derived for use within adaptive testing systems may also operate at this level of the organization of the system. Traditional psychometric (statistical) techniques may be applied at this level, within the lower level subprocedures, or at the level of the implementation of the rule base. The statistical procedures may operate within the definition of the rules derived from the cognitive model, or else be applied in parallel with the cognitive model, so that estimates derived from each inferential process may be compared and combined in generating the overall test outcome (see Huba, 1987).

This is, after all, no more than a formalization of what an expert human examiner does in performing an assessment. Elements of the assessment procedure are composed into the battery of tests to be applied, according to some model (often implicit) that the examiner maintains of the functions to be assessed. The individual tests are then administered, often with some degree of selection and modification of the battery, depending upon earlier test results. Statistical estimates derived from the test are obtained and interpreted in line with hypotheses generated from the functional model that the test examiner holds. A psychological description (the "report") is generated which is relevant to the assessment question being investigated.

The potential advantages of the kind of scheme envisaged above are that the internal cognitive model is explicit and can be more rigorously applied (and improved); the investigation of data relevant to the inferences being tested is systematic and should therefore be more efficient; and intelligence, in the form of the inferencing procedures, is automatically and consistently applied to the problem. In addition, the behavioral description generated from the system is inevitably formulated in terms of the cognitive model being maintained: it is a psychological and not a statistical description. It must therefore be relevant to the application for which the test is being employed and be more useful in response to questions about diagnosis, management, treatment, selection, or adjustment.

IMPLICATIONS FOR NEUROBEHAVIORAL ASSESSMENT

The adoption of techniques such as these implies that a number of conditions should be met before the development of assessment procedures in this area can advance.

The first is a better understanding of the psychological functions affected by neurotoxins. A vague formulation in terms of effects upon psychomotor performance, slowing of response, impaired eye-hand coordination, diminished concentration, recent memory, and affective state, is insufficient. A better model is needed of the physiological vectors that are generating these effects, as well as a better-elaborated description of the effects in behavioral terms.

Second, these functions should be summarized in the form of a psychological description of the general dysfunctional state that follows from exposure to neurotoxins. This needs to be sufficiently detailed to allow a clear account of the psychological processes implicated in these functions to be deduced.

Third, these processes should be formulated in terms of a cognitive componential model of the relevant functions, in a sufficiently coherent form to allow decomposition of the observed performance and analysis of the functional status of the subject.

Fourth, this should be translated into an assessment system in terms of the individual component elements of performance. These should be assessed separately by testing routines (either criterion- or norm-referenced) that will allow an intelligent computer-based diagnostic analysis of performance.

This analysis will probably be dismissed by those whose immediate concern is for an instrument that can be used now to address the very real problems of assessing current levels of neurotoxic exposure. However, if a valid assessment is needed, one is forced to conclude that no adequate instrument is currently available. Radical improvements must be made in our understanding of the target behavioral effects, which must be based on more extensive fundamental research. These can then be translated into effective assessment instruments. Such an approach has already been shown to yield dividends in other neuropsychological areas [particularly in assessing reading disorders (Seymour, 1987) and errors in arithmetic processing], and could well be profitably applied to the neurobehavioral testing of exposure to toxins.

A final, and unrelated thought: No one seems to have taken seriously the possibility of the individual baseline testing of workers potentially open to exposure. Such an approach could completely transform what could be achieved by psychological assessment. Many of our psychometric difficulties would be eliminated at a stroke if data on each worker before exposure were available. Compare the advances made in neuropsychology during World War II, largely because of the availability of psychological data collected upon induction.

If a battery were administered on recruitment (and perhaps every 5 or 10 years subsequently), it would be possible to establish the effects upon cognitive functioning in an individual worker with relative ease and to dramatically improve our understanding of the relevant processes in general. Even if legislation to introduce this is unattainable, the introduction of such a system by a number of major employers would at least make a worthwhile contribution. The suggestion is no doubt naive, but of such immense potential value that it deserves to be discussed.

REFERENCES

Anger, W. K. 1985. Neurobehavioral tests used in NIOSH-supported worksite studies, 1973–1983. Neurobehavioral Toxicology and Teratology 7:359–368.

Beaumont, J. G. 1985. The effect of microcomputer presentation and response medium on digit span performance. International Journal of Man-Machine Studies 22:11–18.

Branconnier, R. J. 1985. Dementia in human populations exposed to neurotoxic agents: A portable microcomputerized screening device. Neurobehavioral Toxicology and Teratology 7:379–386.

Camerino, D. 1987. Presentation, Description and Preliminary Evaluation of the Automated Form of WHO-NCTB. Institute of Occupational Health, University of Milan.

Huba, G. J. 1987. On probabilistic computer-based test interpretations and other expert systems. Applied Psychology 36:357–374.

Iregren, A. 1986. Effects of Industrial Solvent Interactions: Studies of Behavioral Effects in Man. Arbete och Halsa, (ISSN 0346-7821) Solna, Sweden.

Seymour, P. H. K. 1987. Individual cognitive analysis of competent and impaired reading. British Journal of Psychology 78:483–506.

Spurgeon, A., and M. Harrington. 1987. The Neuropsychological Effects of Long-Term Exposure to Organic Solvents. Institute of Occupational Health, University of Birmingham, U.K.

Yule, W., and M. Rutter. 1985. Effects of lead on children's behavior and cognitive performance: A review. In Dietary and Environmental Lead: Human Health Effects, K. R. Mahaffey, ed. Amsterdam: Elsevier.

PART II
Assessment of Animal Models:
What Has Worked and What Is Needed

Exposure to Neurotoxins Throughout the Life Span: Animal Models for Linking Neurochemical Effects to Behavioral Consequences

Hanna Michalek and Annita Pintor

Exposure to toxic substances is a potential threat throughout the human life span. Therefore, it is essential that risk assessment of potentially toxic substances be carried out at critical periods over the entire range. Although much valuable information may be obtained from epidemiological and clinical studies, it must be supplemented by research using animal models and the advantages of experimental methods. This is particularly apparent when information is needed about the modes and sites of action of toxic substances that constitute the substrates of adverse behavioral effects. The discussion that follows uses the laboratory rat as its animal model and a class of compounds affecting the cholinergic neurotransmitter system, organophosphorus anticholinesterases, as examples of potentially neurotoxic substances to which the possibility of human exposure is widespread. Examples have been chosen to illustrate the basic characteristics of research designs, their implementation, and the analysis and interpretation of results. The discussion begins with consideration of normally occurring changes in neurochemical events during early development and later aging.

NEUROCHEMICAL CHANGES DURING DEVELOPMENT AND AGING

Chemical substances entering the body, toxic or nontoxic, produce their effects by altering biochemical events already underway. Many of the events affected, particularly those in the nervous system, are involved in the behavior of an organism as an integrated whole. In any individual the nature of these events is determined by interactions between genetic and environmental factors, interactions that are characterized by changes throughout the life span. In general, most neuronal functions in the neonatal organism are incompletely developed. The evidence is that processes involved in the synthesis, storage, release, and inactivation of neurotransmitter substances are less well developed in early life than in the adult. The blood-brain barrier is generally not as effective in the immature, developing brain, allowing penetration of chemicals that are manifested only by peripheral effects at later ages. Neurochemical changes during early ontogenesis have been shown to parallel behavioral development. Furthermore, declines in behavioral functions with normal or pathological aging suggest that the developmental trends in the central nervous system (CNS) are reversed during aging.

Because of its key roles in behavior, the cholinergic system provides examples of how neurochemical changes during the life span influence behavioral effects of neurotoxic agents. A considerable number of compounds exist that have specific cholinergic effects, some widely used throughout the world as pesticides (Koelle, 1975). Examples from one class of these, organophosphate (OP) compounds, serve our present purposes of assessing neurotoxicity throughout the life span. Human intoxication by OP may result from occupational exposure (agricultural or industrial), adventitious contact indoors or outdoors, or consumption of contaminated food or water. The fact that the risk of exposure may be greater in the indoor than the outdoor environment (Reinert, 1984) places all members of the family—from pregnant women and young children to the elderly—in jeopardy. Epidemiological data indicate that as many as 500,000 people in the world are exposed annually to these compounds at levels requiring clinical attention and that about 5,000 poisonings are fatal (Russell and Overstreet, 1987). Animal models are essential for experimental analyses of the mechanisms by which these compounds produce their effects, for determination of threshold limit values beyond which exposures are unacceptable, for creating therapeutic procedures to treat adverse symptoms, and for monitoring public health programs designed to protect against misadventures.

THE CHOLINERGIC SYSTEM IN BEHAVIOR

"Most impressive is the singular fact that ACh (acetylcholine) is the only substance that can influence every physiological or behavioral response thus far examined" (Myers, 1974). This statement takes into consideration the roles of ACh as the transmitter at neuromuscular junctions and in various pathways in the CNS (Butcher and Woolf, 1986). Normal functioning of the cholinergic system may be impaired when an individual is exposed to OP compounds. Upon entering the body through any of several routes (inspiration, ingestion, injection, transdermally), an OP is first carried to its site of action, the "pharmacokinetic" phase of its journey. Significant molecular modifications may occur during the transit, e.g., the relatively inactive compound parathion is converted to its active metabolite, paraoxon, predominantly in the liver.

The "pharmacodynamic" behavioral and physiological effects of an OP compound begin with the binding of the compound to the active site of the acetylcholinesterase (ChE) molecule, inhibiting inactivation of the neurotransmitter ACh when released from presynaptic neurons and producing overstimulation by the neurotransmitter. Although there is no universal agreement concerning "critical levels" of brain cholinesterase (ChE), most investigators have emphasized that symptoms of acute intoxication and changes in behavior appear only when brain ChE activity is reduced by at least 50-60 percent (Bignami and Michalek, 1978; Bignami et al., 1975; Russell, 1977). In the early phase of acute intoxication, behavioral disturbances are accompanied by reduced brain ChE and elevation of brain ACh levels. The disappearance of the symptoms of intoxication with return of ACh to normal levels occurs considerably earlier than the normalization of ChE activity. Moreover, repeated administration of anticholinesterases (antiChEs) to adult rodents induces the development of tolerance to their toxicity; i.e., behavioral disturbances disappear despite persisting low levels of brain ChE. In recent years a decrease in the density of muscarinic and nicotinic receptor sites has been recognized as one of the main adaptive mechanisms to overstimulation by acetylcholine in adult animals (Costa et al., 1982; Russell, 1982; Russell and Overstreet, 1987).

It is clear from these brief comments that the neurobehavioral effects of even one class of potentially neurotoxic substances involve complex interactions among the chemical processes it initiates upon entry into the body and the outcome it produces in physiological and behavioral functions. For purposes of the present discussion, examples are chosen from research using one typical OP, diisopropyl fluorophosphate (DFP), which has been used extensively as a model compound (Michalek et al., 1978, 1981, 1988; Overstreet and Russell, 1984; Russell and Overstreet, 1987). Among various components involved in the mechanisms of synthesis and degradation of ACh (Russell and Overstreet,

1987), this chapter deals only with the following three markers, all located pre- or postsynaptically:

1. ChE, the primary target of antiChE agents, the enzyme involved in the inactivation by hydrolysis of ACh;
2. choline acetyltransferase (ChAT), whose enzymatic activity is responsible for the synthesis of ACh from its immediate precursors, choline and acetylcoenzyme A; and
3. muscarinic ACh receptors (mAChRs), essential for brain cholinergic neurotransmission and linked to second messenger systems that mediate a subsequent "cascade" of events leading to physiological and behavioral effects.

Changes in these components are discussed first with regard to the phenomena of intoxication and tolerance during critical developmental stages of the rat, i.e., the pre- and early postnatal periods and senescence.

EFFECTS OF PRENATAL EXPOSURE TO DFP: FROM BIRTH TO WEANING

In the initial phase of prenatal subchronic intoxication, i.e., from the sixth to the tenth day of pregnancy, DFP has been shown to cause, in the pregnant female, a syndrome of cholinergic stimulation (tremors, sweating, salivation, lacrimation, and diarrhea) lasting for many hours after each injection. Results of a typical experiment are summarized in Table 1. Maternal weight gain is significantly reduced. The toxic syndrome appears considerably more pronounced than that previously observed in adult males treated similarly (Michalek et al., 1982). Moreover, a great variability in the response of individual dams in terms of severity and duration of the symptoms is evident. Subsequently, the symptoms attenuate markedly in some dams, but remained quite evident in others. Although the treatment does not cause mortality of dams, the pups of DFP-treated litters may be stillborn or die within a few hours after birth. These cases of reproductive wastage are clearly associated with the marked depression of weight and possibly with delayed parturition (by about 24 hours).

After prenatal exposure of mothers to DFP, the body weight in newborns is about 6 percent lower than that of controls and there is a slight retardation of body growth up to day 10. The postnatal pattern of gain in brain weight is not modified by DFP treatment. Data on brain ChE and mAChRs are presented in Table 2. The levels of brain ChE at birth in the DFP group do not differ from those of the controls, and both groups showed similar increases of enzymatic activity until

TABLE 1 Effects of Subchronic Intoxication with DFP in Pregnant Rats on Gestation, Birth Statistics, and Litter Survival

	Control	DFP
Total number of dams	20	20
Length of gestation (days)	21.2 ± 0.2	21.8 ± 0.2
Weight gain of dams (g)		
6th–10th day	14.4 ± 2.8	3.6 ± 2.4[a]
10th–20th day	71.9 ± 4.2	71.8 ± 5.5
Number of pups per litter	11.4 ± 0.8	10.1 ± 0.6[b]
Lost at birth	0	4
Lost within 48 h	1	8
Litters surviving up to weaning	19	8

NOTE: Treatment of Wistar rats (220–240 g) on alternate days: DFP (in arachis oil) first dose of 1.1 mg/kg (subcutaneous) on day 6 of pregnancy, subsequent doses of 0.7 mg/kg until day 20 (corresponding to 25% of LD_{50}).

[a]Significantly different from control $p < 0.001$ as determined by t-test.
[b]Not including four litters with pups stillborn or dead within a few hours after delivery, which were often cannibalized.

SOURCE: Michalek et al. (1985).

TABLE 2 Effects of Subchronic Intoxication with DFP in Pregnant Rats on Brain Total Cholinesterases (ChE) and [^{3}H]Quinuclidinyl Benzilate (QNB) Receptor Binding Sites During Postnatal Development

Age (days)	Brain ChE (nmol AcThCh hydrolyzed/min/mg protein)		[^{3}H]QNB binding (fmol/mg protein)		
	Control	DFP	Control	DFP	% of Control
Newborn	22.9 ± 1.4	21.8 ± 2.0	102 ± 7	70 ± 4[a]	68
5	34.0 ± 1.3	31.0 ± 1.6	142 ± 7	127 ± 9	89
10	39.6 ± 3.0	32.9 ± 1.7	258 ± 7	193 ± 13[a]	74
15	38.1 ± 2.6	38.7 ± 13	335 ± 15	263 ± 19	78
20	40.0 ± 2.0	38.5 ± 2.5	443 ± 36	442 ± 10	100

NOTE: For treatment see Table 1. Mean ± SEM of 8 animals for each age (except newborn $n = 16$) belonging to different litters. [^{3}H]QNB at 1.5 nM concentration; mean ± S.E.M, $n = 10$ animals for each age (except newborns $n = 20$). AcThCh = acetylthiocholine.

[a]Significantly different from control values ($p < 0.01$) as determined by t-test.

SOURCE: Michalek et al. (1985).

weaning. On the other hand, experiments on quinuclidinyl benzilate (QNB) receptor binding show a significantly lower level of mAChRs at birth and at 10 days in DFP pups compared to controls.

Results reported in Table 3 show that exposure to DFP at the end of pregnancy produces a consistent depression of ChE activity in maternal brain during a period of at least 48 hours. The enzyme activity in fetal brain is less inhibited initially and approaches full recovery within the period. These data on fast recovery of fetal brain ChE are in agreement with results reported in the literature for other OP compounds. Subacute exposure of rats to parathion during the third trimester of pregnancy did not modify brain ChE in newborns (Talens and Wooley, 1973). Daily administration of dichlorvos to pregnant rats during the same period lowered ChE levels in newborns, but no substantial delay in postnatal development was subsequently observed (Zalewska et al., 1977). Prenatal exposure of mice to dicrotophos did not alter the postnatal development of brain ChE and ChAT (Bus and Gibson, 1974).

What processes may be involved in these differences between effects of OP on ChE activity in fetal and maternal brain? It is well known that *pharmacokinetic* factors influence the processes by which an antiChE reaches its sites of action. For example, such compounds bind to molecules other than acetyl-ChE (i.e., butyrylcholinesterase and aliesterase) that produce no apparent functional effects on behavioral or physiological variables. These enzymes, found in plasma and erythrocytes, have been described as "scavengers" or "sinks" that can reduce the concentration of an antiChE entering the CNS (Russell et al., 1986). For example, higher levels of plasma ChE in females have been shown to result in lesser brain sensitivity to DFP, as compared to males (Overstreet et al., 1979). In fetal brains after in utero exposure to DFP, cholinesterases present in maternal plasma, erythrocytes, and placenta also play an important role as "scavengers." Other data obtained in our laboratory indicate that total cholinesterases in maternal plasma and amniotic fluid 90 minutes after DFP were inhibited by 95 percent, and those in fetal plasma by 75 percent, i.e., considerably more than maternal and fetal brain ChE (i.e., 80 and 50 percent, respectively). The fast recovery of ChE in the fetus probably depends on the considerably higher protein synthesis rate in fetal compared to adult brains (Gupta and Dettbarn, 1986, 1987; Gupta et al., 1984; Lajtha and Dunlop, 1981). These facts suggest that following exposure to OPs, recovery to normal levels of ChE activity occurs more rapidly in the fetus than in the adult because (1) initial reduction in ChE activity is not as great in the former and (2) de novo synthesis of replacement ChE is more rapid.

TABLE 3 Effects of Subchronic Intoxication of Pregnant Rats by DFP on Maternal and Fetal Brain aChE and Maximal Number of [^{3}H]-QNB Binding Sites (B_{max}) at the End of Pregnancy

	Interval Between Last Treatment and Sacrifice (h)	Brain ChE (nmol AcThCh hydrolyzed/min/mg protein)			mAChRs (B_{max}) (fmol/mgprotein)	
		Control	DFP	% Control Activity	Control	DFP
Dams	1.5	40.9 ± 2.4	7.6 ± 0.1^a	18		
	24	41.7 ± 0.2	8.7 ± 1.4^a	20	1,152 ± 41	780 ± 64^a
	48	42.1 ± 1.6	9.2 ± 1.8^a	22		
Fetuses	1.5	13.9 ± 0.5	7.0 ± 0.5^a	50		
	24	18.0 ± 0.1	11.8 ± 0.3^a	65	187 ± 21	153 ± 11^b
	48	18.1 ± 0.4^c	16.5 ± 0.5^a	91		

NOTE: For treatment see Table 1. Binding constants (B_{max}) of [^{3}H]QNB binding calculated from Scatchard analysis. Mean ± SEM of five animals. Mean ± SEM of 10 pools of two fetuses each. AcThCh = acetylthiocholine.

^{a}Significant difference at $p < 0.01$.
^{b}Significant difference at $p = 0.05$.
^{c}Newborns (in three cases delivery occurred 24 h earlier than in DFP group).

SOURCE: Michalek et al. (1985), modified.

Because *pharmacodynamic* processes leading to behavioral effects begin with the binding of the neurotransmitter ACh to its receptor sites, effects of the above changes in ChE activity on mAChRs are of special interest in the context of the present discussion. Analyses of receptor binding (Table 3) have shown decreases in brain from both DFP-treated dams and 21-day fetuses. The overall pattern for the latter is a significant decrease in levels of mAChR at birth persisting through postnatal day 10, with recovery at subsequent developmental stages. Results of other experiments in which animals were examined before delivery of the young have also shown decreases in numbers of mAChR binding sites in both fetal and maternal brain. Because decreases in numbers of receptor sites appear to be an adaptive mechanism to overstimulation by endogenous ACh, it can be postulated that high levels of ACh in fetal brain must have occurred in spite of the relatively rapid recovery of ChE after treatment. This conclusion is supported by a report (Kewitz et al., 1977) that a single administration of a sublethal dose of DFP to a 20-day pregnant dam elevated free ACh in the fetal brain that lasted considerably longer than in the maternal brain. The temporary delay in the postnatal development of mAChR may indicate that in this time period, the tolerance induced by prenatal OP treatment is gradually being reversed. The time taken by muscarinic receptor sites for recovery (i.e., about three weeks) is similar to that found by Costa and coworkers (1981) in adult rats treated with another OP (disulfoton). A major feature of information now available about prenatal exposure to OPs is the finding of a fetal reduction of mAChRs and a postnatal delay in their development well after the complete recovery of brain ChE inhibition.

EFFECTS OF EARLY POSTNATAL EXPOSURE TO DFP

Effects of exposure to DFP during early postnatal development are summarized in Figure 1. Repeated treatment causes only a weak and short-lasting behavioral syndrome characteristic of cholinergic stimulation, without reduction in body or brain weight gain or modification of protein content in brain tissue. Neurochemically some significant effects of DFP are clearly observable. Brain ChE activity in control animals shows a systematic increase. Levels of ChE in those treated with DFP are consistently lower than controls, being reduced at 14 days by about 45 percent and at 28 days by about 70 percent. Recovery occurs following the end of treatment, but levels are still some 30 percent below control levels after 12 days of withdrawal. These findings

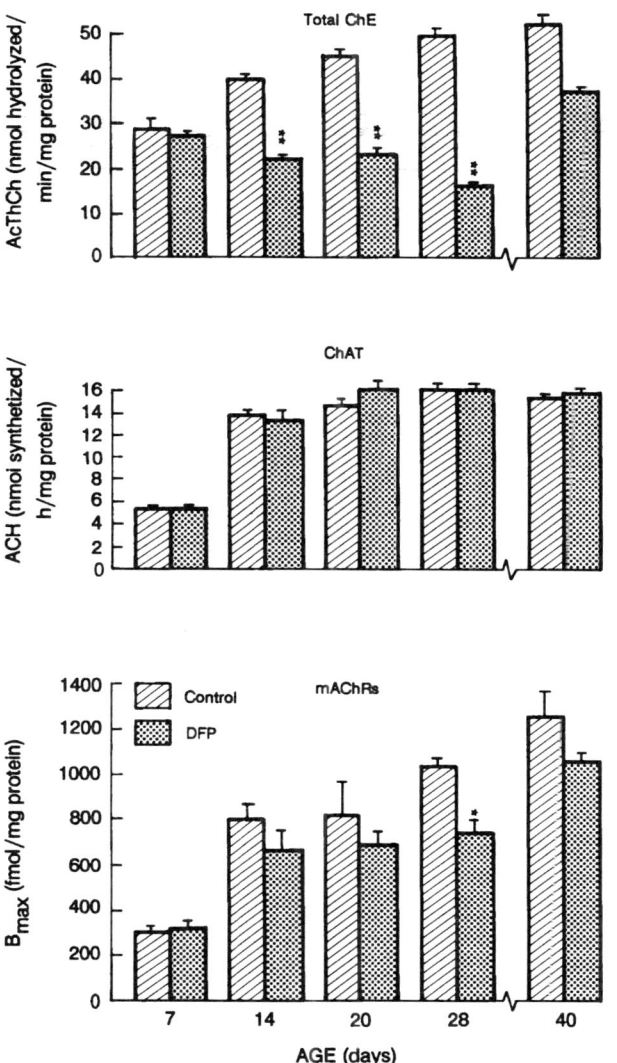

FIGURE 1 Effects of subchronic intoxication by DFP in 7- to 27-day Wistar rats on development of brain ChE, ChAT, and mAChRs of pups belonging to four litters and nursed by their mothers up to weaning. Treatment on alternate days: DFP (in arachis oil) 0.45 mg/kg (subcutaneous) from day 7 up to day 19, and subsequently 0.70 mg/kg up to day 27 (25 percent of LD_{50}). Animals killed 24 hours after the last treatment. Each column represents mean ± SEM from eight animals (two from each litter). Asterisks indicate a significant difference from control (* $p < 0.05$, ** $p < 0.01$) as determined by t-test. AcThCh = acetylthiocholine.

SOURCE: Michalek et al. (1985), adapted.

are generally consistent with reports of investigators using adult animals as subjects, who have reported median recovery times within the range of 10-12 days (Austin and James, 1970; Chippendale et al., 1974; Ehlert et al., 1980). That full recovery may not be complete even after four weeks of withdrawal has also been demonstrated in adult animals (Russell et al., 1989).

The major change in activity of the synthetic enzyme ChAT occurs as a significant increase postnatally, approaching an asymptotic level within two weeks for both DFP and control animals. This pattern has been reported in adult rats made tolerant to DFP and other OPs (Russell et al., 1975; Stavinoha et al., 1969; Wecker et al., 1977).

Changes in mAChR binding in brain tissue from control animals follow a general course similar to that of ChE; i.e., binding increases systematically as postnatal age increases (Figure 1). Receptor binding also increases with chronological age in DFP-treated animals. However, the increases are consistently less than in the controls after one week of treatment, the difference between treatment groups being maximal at 28 days; 12 days after withdrawal from DFP the difference is no longer statistically significant. This general pattern has also been observed in another rodent model (Levy, 1981).

The information discussed above points to some important analogies in mechanisms underlying the effects of DFP on functioning of the cholinergic system. The analogies hold despite the considerable differences in age-related effects of OPs on brain ChE activity.

EFFECTS OF EXPOSURE TO DFP: SENESCENT RATS

Genetic (Strain-Specific) Differences in Effects

Except for data reported by Pintor et al. (1988), the development of tolerance to an OP compound late in the life span has not been investigated. This would seem to be a matter of particular interest because, in spite of some controversial data, most investigators report declines in the density of mAChRs in various brain regions of senescent rodents. Most research results have indicated a decrease of cholinergic markers in the rat striatum, but data concerning age-related alterations in the cerebral cortex and hippocampus are controversial (Bartus et al., 1982, 1985; Michalek et al., 1988). One of the major factors responsible for such discrepancies could be the different genetic strains of rats used in these studies. In fact, behavioral and neurochemical studies of mice, utilizing multiple strain comparisons, have shown that the patterns of age differences are influenced by genotype (Michalek et al., 1988). These findings are important for understanding aging as a product of gene-environment interaction and for identifying strains that offer the greatest potential for studying the interaction. Most investigations of biochemical changes in neurotransmitter systems of aging rats have been performed on animals of only one strain—Wistar, Fischer 344, or Sprague-Dawley being the strains most frequently used. However, knowledge about strain differences is important in defining "How old is old?" (Coleman, 1989).

Results of a recent series of experiments serve as an example of the kinds of information generated by studies of interactions between genetic factors and the aging of neurochemical events in the brain (Michalek et al., 1988; Pintor et al., 1988). The experiments involved comparisons of age-related differences in AChE, ChAT, and mAChRs in tissues from three brain areas of Wistar and Fischer 344 male rats at ages 3 and 24 months. It should be noted that the 50 percent survival rates of the two strains are very similar, i.e., 28-30 months. Results of the experiments are presented in two forms: graphically as histograms and statistically as two-way analyses of variance ANOVAs. The former provide information about each neurochemical variable measured. The latter test the statistical significances of two main factors, i.e., age and strain, and of interactions between them.

Inspection of the upper part of Figure 2 shows an overall similarity between the two strains in levels of ChE activity and in decreases in activity with aging. This conclusion receives general support from results of ANOVA presented in Table 4A. Strain differences are significant only in the hippocampus, Fischer animals having lower levels of activity. Age-related declines in ChE activity are significant in all brain regions, varying from 25 to 40 percent in both strains. Nonsignificant interactions indicate that the aging factors are not strain dependent.

The histograms in the lower part of Figure 2 suggest that levels of ChAT activity in the Fischer strain are consistently higher than those in the Wistar animals at both ages, an observation supported by ANOVAs (Table 4A). Inspection of this figure also suggests that, with one exception, ChAT activity decreases with aging. The ANOVAs reported in Table 4A establish that this trend is statistically significant only in the striatum (approximately 30 percent). The only significant strain × age interaction occurs in the cortical ChAT activity, where decreases (approximately 15 percent) are noted with aging in the Fischer but not in the Wistar strain.

The data on mAChR binding are given in Table 5. In all instances, binding (B_{max}) is higher in Fischer than in age-matched Wistar animals, indicating a larger population of receptor sites. The ANOVAs (Table 4) show that both strain and age differences are highly significant.

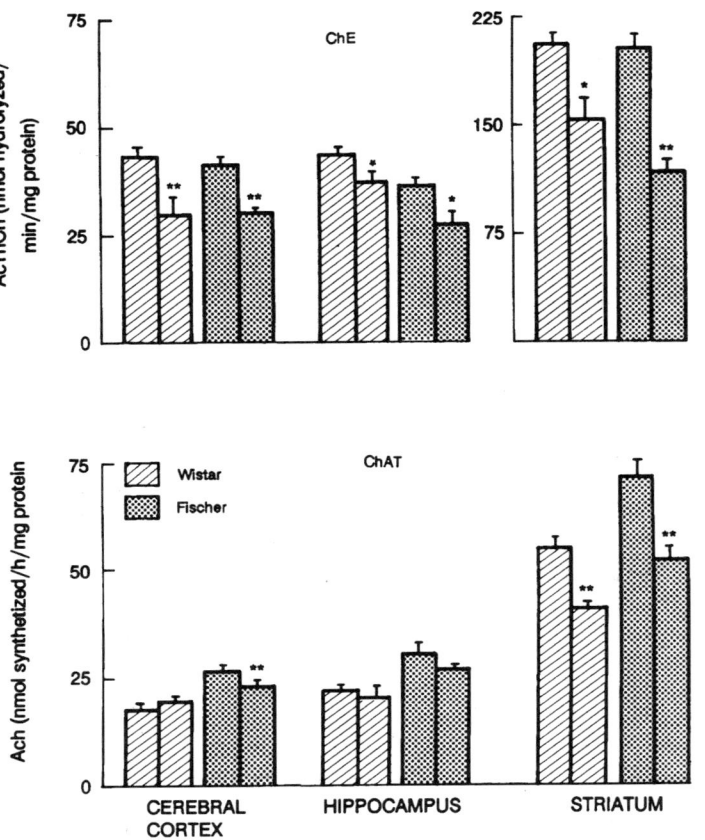

FIGURE 2 Comparisons of age-related changes in cortical, hippocampal, and striatal ChE and ChAT between Wistar and Fischer 344 rats. Numbers in columns indicate age in months; note the different scale used for ChE data in the striatum. Each column represents mean ± SEM from six animals. For factorial analysis of variance (2 strains × 2 ages ANOVA) see Table 4 and text. Asterisks indicate significant differences from previous age (* $p < 0.05$, ** $p < 0.01$), as assessed by post hoc analysis, using Student's t-test with Bonferroni correction. AcThCh = acetylthiocholine.

SOURCE: Michalek et al. (1989).

Densities of mAChRs in the hippocampus and striatum of both strains decreased by 25-39 percent with age. Only in the cerebral cortex is there a significant strain × age interaction. Data in Table 5 indicate that the significance may be attributed to a decline of approximately 30 percent in aging Fischer, but not Wistar, animals. Table 5 also shows that binding in the Fischer animals was consistently associ-

TABLE 4 Factorial Analysis of Variance of Data on ChE, mAChRs (B_{max}), and ChAT of Young and Senescent Male Rats

A. Wistar and Fischer 3- and 24-Month Fischer Rats

	ChE				mAChRs (B_{max})				ChAT			
	F (strain)	F (age)	F (treat-ment)	F (age × treatment interaction)	F (strain)	F (age)	F (treat-ment)	F (strain × age interaction)	F (strain)	F (age)	F (treat-ment)	F (strain × age intraction)
Cerebral cortex	N.S.	46.7		N.S.	143.2	47.7		20.5	49.1	N.S.	N.S.	13.3
Hippocampus	22.0	20.7		N.S.	109.2	33.3		N.S.	33.1	N.S.	N.S.	N.S.
Striatum	N.S.	34.6		N.S.	51.9	58.5		N.S.	26.1	39.0	N.S.	N.S.

B. DFP Treatment of 3- and 24-Month Fischer Rats

	ChE				mAChRs (B_{max})				ChAT			
	F (age)	F (treat-ment)		F (age × treatment interaction)	F (age)	F (treat-ment)		F (age × treatment interaction)	F (age)	F (treat-ment)		F (age × treatment interaction)
Cerebral cortex	18.2	314		N.S.	62.6	171.3		N.S.	54.9	N.S.		N.S.
Hippocampus	11.8	258		N.S.	54.6	65.9		N.S.	8.8	N.S.		N.S.
Straitum	46.1	409		N.S.	14.9	40.6		N.S.	25.0	N.S.		N.S.

NOTE: Degrees of freedom were 1.20 in all calculations. For $F > 14.82$, $p < 0.001$; for $14.8 > F > 8.10$, $p < 0.01$. N.S. = not significant.

SOURCE: Michalek et al. (1989) and Pintor et al. (1988), modified.

TABLE 5 Age-Related Differences in Binding Parameters of Muscarinic Receptor Sites in Brain Regions of Wistar and Fischer Rats

		Age (months)			
		Wistar		Fischer	
		3	24	3	24
Cerebral cortex	B_{max}	1,207 ± 6	1,106 ± 6	1,970 ± 75	1,409 ± 36^a
	K_D	163 ± 14	152 ± 15	263 ± 9	193 ± 7
Hippocampus	B_{max}	1,111 ± 76	682 ± 55^a	1,723 ± 20	1,326 ± 88^a
	K_D	140 ± 24	110 ± 24	253 ± 10	215 ± 2
Striatum	B_{max}	1,292 ± 59	856 ± 32^a	1,874 ± 116	1,211 ± 72^a
	K_D	182 ± 17	110 ± 19	110 ± 16	231 ± 7

NOTE: B_{max} is expressed as femtomoles per milligram of protein; K_D as picomolar. Values are means ± SEM of six experiments.

^{a}Significant difference $p < 0.01$, as assessed by post hoc analysis using Student's t-test with Bonferroni correction.

SOURCE: Michalek et al. (1989).

ated with a lower affinity (higher K_D) of the [^{3}H]QNB ligand for mAChR sites; i.e., the tendency for the ligand to bind to the receptor was less than in the Wistar strain. The ANOVAs supported this conclusion, strain differences being significant at $p < 0.001$ and age differences at $p < 0.01$ (Michalek et al., 1988). There was no significant strain × age interaction.

Such results clearly indicate that the outcomes of studies in neurobehavioral toxicology are likely to be affected significantly by genetic or aging variables, both of which have effects on neurochemical processes that are involved in behavior. The results also suggest hypotheses about the mechanisms by which such effects are mediated. For example, the findings described above are consistent with a loss during aging of pre- and postsynaptic cholinergic neurons in the Fischer 344 strain. Some of the same age-related changes have been reported by other investigators working with the same strain (Lippa et al., 1981; Pedigo and Polk, 1985; Pedigo et al., 1984; Sherman et al., 1981). There also is good agreement between the results described here and those reported by others using Wistar rats as models (Ingram et al., 1981; London et al., 1985; McGeer et al., 1971; Roman et al., 1984).

It is of considerable interest that marked differences in the concentrations of mAChRs have been described by Overstreet et al. (1984) for two of the above regions (hippocampus and striatum) in two selectively bred lines, Flinders sensitive line (FSL) and Flinders resistant line (FRL) rats. Such studies again suggest the importance of genetic factors in major effects on cholinergic function for some strains or lines (Fischer 344 or FSL rats) compared to others (Wistar and FRL rats): an increased cholinergic function may be an important factor contributing to increased sensitivity to DFP (Russell and Overstreet, 1987) and thus influence the rate of aging in terms of deficit of cortical ChAT and mAChRs. Although there are inherent limitations to extrapolations from rodents to humans, some reports indicate deficits in cortical, hippocampal, and striatal ChAT and mAChRs in elderly humans (Bartus et al., 1982; Collerton, 1986; Côté and Kremzner, 1983).

Subchronic Intoxication During Senescence

One approach to testing the hypothesis that hyperfunctioning of the cholinergic system results in greater sensitivity to DFP and to an accelerated rate of aging is to subject senescent Fischer 344 rats to repeated administration of DFP (Pintor et al., 1988). In the initial phase of subchronic intoxication (i.e., treatment 1 to 4), DFP caused a typical syndrome of cholinergic stimulation (tremors, sweating, salivation, lacrimation, and diarrhea) lasting for many hours after each injection. In its severity and duration, the toxic syndrome appeared more pronounced in senescent than in young animals. In particular, tremors in the former lasted until the next DFP injection 48 hours later, whereas they disappeared within 2-3 hours in the latter. At the end of the treatment period the symptoms were attenuated in both young and senescent animals. The mortality rate was significantly higher ($p < 0.02$) among the senescent (60 percent) than among the young rats (14 percent). The ANOVA for repeated measures showed that differences in body weight were significant both for age [$F(2,30) = 28.12, p < 0.001$] and for treatment [$F(1,20) = 43.48, p < 0.001$]. Although both age groups lost weight during the treatment period, the decrease was greater for the senescent (-94 g) than for the young (-30 g) animals. Body weight is a general measure of capability to maintain caloric intake and water balance.

Effects of the subchronic DFP treatment on enzyme activities in three brain areas are presented in Figure 3: ANOVA (2 ages × 2 treatments) confirms the striking difference in both age- and treatment-related effects on ChE (Table 4B). There are no significant interaction factors, indicating that the groups were similarly affected. The ANOVA of the results for ChAT showed a quite different state of affairs: age is the only significant variable.

Effects of the DFP treatment on mAChR binding are summarized

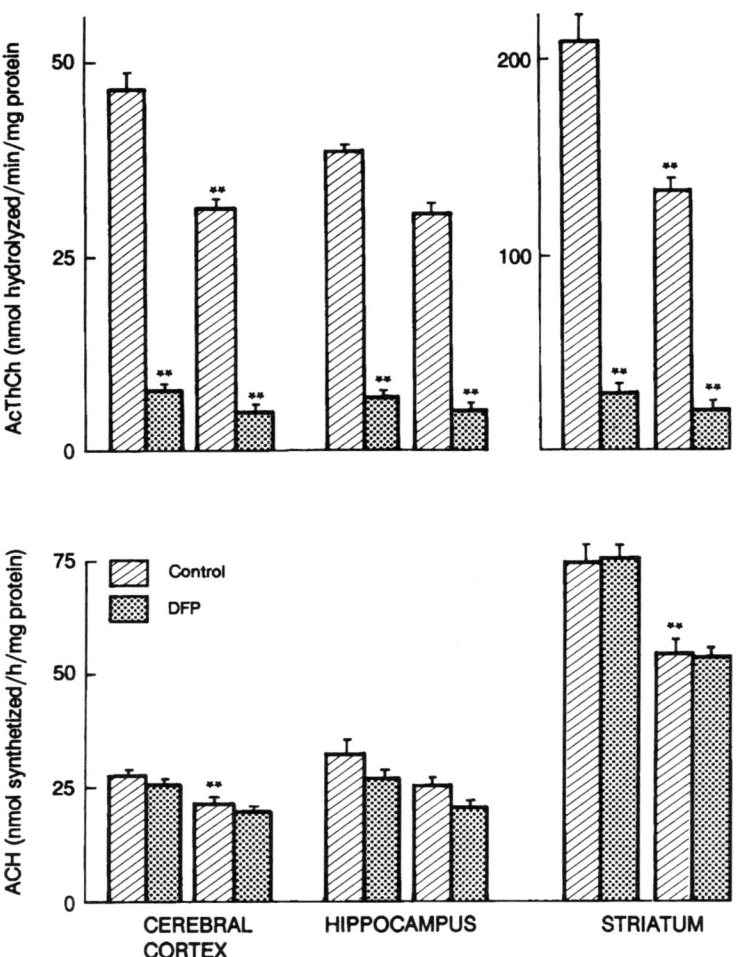

FIGURE 3 Effects of subchronic intoxication by DFP on brain ChE and ChAT of young and senescent Fischer 344 rats (for treatment see Table 6). Numbers above the columns indicate age in months. Numbers in parentheses show percentage inhibition. Each column represents mean ± SEM from six animals (except for 24-month DFP-treated rats, $n = 4$). For factorial analysis of variance (2 ages × 2 treatments ANOVA), see Table 4 and text. Asterisks indicate significant differences for age (* $p < 0.01$, ** $p < 0.001$) and treatment ($p < 0.001$). AcThCh = acetylthiocholine.

SOURCE: Pintor et al. (1988).

TABLE 6 Effects of Subchronic Intoxication by DFP on mAChRs in Brain Regions of Young and Senescent Fischer Rats

Brain Region	Treatment	Age (months)	B_{max} (fmol/mg protein)	K_D (pM)
Cerebral cortex	Control	3	2,028 ± 67	257 ± 19
	DFP	3	1,231 ± 61	165 ± 9
	Control	24	1,483 ± 20	193 ± 10
	DFP	24	938 ± 59	120 ± 5
Hippocampus	Control	3	1,696 ± 31	267 ± 17
	DFP	3	1,238 ± 57	185 ± 3
	Control	24	1,398 ± 45	220 ± 10
	DFP	24	932 ± 65	150 ± 8
Striatum	Control	3	1,903 ± 125	278 ± 25
	DFP	3	1,394 ± 41	260 ± 16
	Control	24	1,334 ± 16	255 ± 35
	DFP	24	811 ± 78	200 ± 3

NOTE: Treatment on alternate days: DFP (in arachis oil) first dose 1.6 mg/kg (subcutaneous); subsequent doses 1.1 mg/kg for two weeks. Means ± SEM from six animals (except for 24-month DFP-treated rats, $n = 4$). Factorial analysis of variance (2 ages × 2 treatments ANOVA, Table 4) showed significant differences in B_{max} for age and treatment in all areas ($p < 0.001$). B_{max} and K_D expressed as means ± SEM from six animals (except for 24-month DFP-treated rats, $n = 4$).

SOURCE: Pintor et al. (1988).

in Table 6. Two-way ANOVAs (Table 4B) confirm what is apparent in the table: i.e., DFP produced down-regulation in all three brain areas assayed. It is also clearly apparent that mAChR levels in the brains of the senescent animals are significantly lower than in the brains of young rats. Despite this difference, percentage decreases in mAChR receptor binding induced by the DFP treatment were very similar for both age groups, as reflected in the lack of significant interaction for any of the three brain areas.

Such comparative effects of subchronic DFP intoxication in young and senescent animals have several implications of potential interest to neurobehavioral toxicologists. For example, the much higher mortality in senescent animals would appear to be inconsistent with the fact that the percent down-regulation of mAChRs during DFP treatment was not significantly different from that in young animals. As discussed earlier, plasticity of mAChRs appears to be an important compensatory mechanism for decreased ChE activity and elevated ACh levels, e.g., in the development of behavioral tolerance to OPs. How is it possible to account for the differences in mortality when receptor plasticity as measured by mAChR binding did not differ? Several hypotheses can be offered. One that is particularly obvious is the possibility that age-related differences in peripheral events (e.g., cardiovascular, respiratory) may be the underlying mechanism(s). The hypothesis that age-related changes in permeability of the blood-brain barrier may be involved is not supported by the experimental data showing a lack of significant age × treatment interactions in brain ChE or ChAT activities. One theoretical model has recently been proposed that could account for the present mortality data and have broader application to behavioral and physiological functions (Russell, 1988; Russell et al., 1986). The central theme of the concept is that behavioral and physiological processes are differentially receptor dependent; i.e., they require different densities of receptor occupancy to function normally. It follows that there are "basal thresholds" in receptor populations below which abnormalities in behavioral and physiological functions, including mortality, occur. Reexamination of Table 6 shows that populations of receptors (as defined by [^{3}H]QNB binding) characteristic of the normal young animal had decreased by 40-55 percent in the DFP-treated senescent rats. Such decrements could significantly affect functions involving cholinergic innervation. Put in more general terms, understanding of the effects of pathologically or xenobiotically induced insult to the nervous system on behavior can benefit significantly by knowledge about the mechanisms of action involved. Animal models are indispensable in generating that knowledge.

Time-Course Recovery of mAChRs Following Down-Regulation in Brain of Senescent Rats

It is well established that following termination of repeated treatment with an antiChE agent in newborn and young animals, down-regulation of mAChRs gives way to almost complete recovery, requiring about two weeks in early postnatal life (Michalek et al., 1985) and three weeks in adulthood (Costa et al., 1981). Given that protein synthesis in the brain declines with age (Dwyer et al., 1980), it may be predicted that muscarinic receptor recovery is slower in senescent DFP-tolerating rats.

Preliminary experiments to test this hypothesis are now in progress in our laboratory, with Sprague-Dawley male 3- and 24-month-old rats serving as subjects. To reduce the mortality rate of aged rats during treatment, lower dosages of DFP are being utilized (first dose 1.1 mg/kg, followed on alternate days by two doses of 0.7 mg/kg and four doses of 0.35 mg/kg). The ChE and ChAT activity and mAChR binding in the cerebral cortex, hippocampus, and striatum are being assayed at weekly intervals up to five weeks after termination of DFP treatment. Most previously reported findings are being confirmed. Differences in mortality rate, although smaller, are still present, i.e., 15 percent for young rats and 40 percent for senescent rats. No differences are detected in brain ChE inhibition or in ChAT activity in any area. The down-regulation of mAChR density (without changes in affinity) in surviving senescent rats at the end of treatment is still present and, in terms of percentage of age-matched control values, is similar in the two age groups. However, the influence of age on the rate of recovery is evident: both brain ChE activity and mAChR density reach pretreatment values in young rats within two weeks, compared to almost five weeks in senescent rats. These results indicate that the synthesis of both ChE and mAChR molecules is impaired in brain tissues as a consequence of aging.

CONCLUSION

It is the basic contention of the present discussion that a complete science and technology of neurobehavioral toxicology cannot be written without knowledge of neurochemical events intervening between exposure to a neurotoxin and the consequent effects on behavior. Although behavioral assays in and of themselves have significant contributions to make to risk assessments, neurobehavioral toxicology has a much broader agenda. Toxins bind to molecules on biologically active tissues within the body in order to produce their effects. Understanding how a toxic compound is transported to its site of action and the nature of the cascade of events it initiates is more than a matter of academic interest. Such knowledge can provide bases for anticipating and identifying molecular structures that are potentially noxious. It can give some insight into means for protecting against toxic risks and into therapeutic procedures by which untoward exposures may be managed.

The specific examples discussed here were chosen to illustrate how neurochemical changes throughout the life span can influence the effects of toxic exposures. Clearly, both risk assessment and clinical procedures, as well as basic knowledge, must take such interactions into consideration. Over 40 years ago, Professor C.L. Hull (1943) commented

> ... any theory of behavior is at present, and must be for some time to come, a molar theory. This is because neuroanatomy and physiology have not yet developed to a point such that they yield principles which may be employed as postulates in a system of behavior theory. ...

These comments have relevance to neurobehavioral toxicology today. It is important for neurobehavioral toxicologists never to forget that the living organism is not empty. It is filled with a multitude of events that are involved whenever toxic exposures induce behavioral malfunctions. In assessing the roles of animal models in neurobehavioral toxicology, it is apparent that they have been, and will continue to be, essential to our understanding of the nature of these events.

REFERENCES

Austin, L., and K. A. C. James. 1970. Rates of regeneration of acetylcholinesterase in rat brain subcellular fractions following DFP inhibition. Journal of Neurochemistry 17:1705–1707.

Bartus, R. T., R. L. Dean, B. Beer, and A. S. Lippa. 1982. The cholinergic hypothesis of geriatric memory dysfunction. Science 217:408–417.

Bartus, R. T., R. L. Dean, M. J. Pontecorvo, and C. Flicker. 1985. The cholinergic hypothesis: A historical overview, current perspective and future directions. Annals of the New York Academy of Science 444:322–358.

Bignami, G., and H. Michalek. 1978. Cholinergic mechanisms and aversively motivated behaviors. Pp. 173–255 in Psychopharmacology of Aversively Motivated Behavior, H. Anisman and G. Bignami, eds. New York: Plenum.

Bignami, G., N. Rosic, H. Michalek, M. Milosevic, and G. L. Gatti. 1975. Behavioral toxicity of anticholinesterase agents: Methodological, neurochemical, and neuropsychological aspects. Pp. 155–215 in Behavioral Toxicology, B. Weiss and V.G. Laties, eds. New York: Plenum.

Bus, J. S., and J. E. Gibson. 1974. Bidrin: Perinatal toxicity and effect on the development of brain acetylcholinesterase and choline acetyltransferase in mice. Food and Cosmetics Toxicology 12:313–322.

Butcher, L. L., and N. J. Woolf. 1986. Central cholinergic systems: Synopsis of anatomy and overview of physiology and pathology. Pp. 73–86 in The Biological Substrates of Alzheimer's Disease, A.B. Scheibel, A.F. Wechsler, and M.A.B. Brazier, eds. New York: Academic Press.

Chippendale, T. C., C. W. Cotman, M. D. Kozar, and G. S. Lynch. 1974. Analysis of acetylcholinesterase synthesis and transport in the rat hippocampus: Recovery of acetylcholinesterase activity in the septum and hippocampus after administration of diisopropylfluorophosphate. Brain Research 81:485–496.

Coleman, P. D. 1989. How old is "old"? Neurobiology of Aging 10:115.

Collerton, D. 1986. Cholinergic function and intellectual decline in Alzheimer's disease. Neuroscience 19:1–28

Costa, L. G., B. W. Schwab, H. Hand, and S. D. Murphy. 1981. Reduced ^{3}H quinuclidinyl benzilate binding to muscarinic receptors in disulfoton-tolerant mice. Toxicology Applied Pharmacology 60:441–450.

Costa, L. G., B. W. Schwab, and S. D. Murphy. 1982. Tolerance to anticholinesterase compounds in mammals. Toxicology 25:79–97.

Côté, L. J., and L. T. Kremzner. 1983. Biochemical changes in normal aging in human brain. Pp. 19–30 in The Dementias, R. Mayeux and W. G. Rosen, eds. New York: Raven Press.

Dwyer, B. E., J. L. Fando, and C. G. Wasterlain. 1980. Rat brain protein synthesis declines during postdevelopmental aging. Journal of Neurochemistry 35:746–749.

Ehlert, F. J., N. Kokka, and A. S. Fairhurst. 1980. Altered (3H)quinuclidinyl benzilate binding in the striatum of rats following chronic cholinesterase inhibition with diisopropylfluorophosphate. Molecular Pharmacology 17:24–30.

Gupta, R. C., J. E. Thornburg, D. B. Stedman, and F. Welsch. 1984. Effect of subchronic administration of methyl parathion on in vivo protein synthesis in pregnant rats and their conceptuses. Toxicology Applied Pharmacology 72:457–468.

Gupta, R. C., R. H. Rech, K. L. Lovell, F. Welsch, and J. E. Thornburg. 1985. Toxicology Applied Pharmacology 77:405–413.

Gupta, R. C., and W. D. Dettbarn. 1986. Role of uptake of (^{14}C) valine into protein in the development of tolerance to diisopropylphosphorofluoridate (DPF) toxicity. Toxicology Applied Pharmacology 84:551–560.

Gupta, R. C., and W. D. Dettbarn. 1987. Interaction of cycloheximide and diisopropylphosphorofluoridate (DPF) during subchronic administration in rat. Toxicology Applied Pharmacology 90:52–59.

Hull, C. L. 1943. The problem of intervening variables in molar behavior theory. Psychological Review. 50:273–291.

Ingram, D. K., E. D. London, M. A. Reynolds, S. B. Waller, and C. L. Goodrick. 1981. Differential effects of age on motor performance in two mouse strains. Neurobiology of Aging 2:221–227.

Kewitz, H., O. Pleul, and E. Mann. 1977. Pre- and postnatal development and drug induced alterations of free and bound acetylcholine in rat brain. Naunyn Schmiedeberg's Archives of Pharmacology 298:149–155.

Koelle, G. B. 1975. Anticholinesterase agents. Pp. 445–466 in The Pharmacological Basis of Therapeutics, L. S. Goodman and A. Gilman, eds. New York: Macmillan.

Lajtha, A., and D. Dunlop. 1981. Turnover of protein in the nervous system. Life Sciences 29:755–767.

Levy, A. 1981. The effect of cholinesterase inhibition on the ontogenesis of central muscarinic receptors. Life Sciences 29:1065–1070.

Lippa, A. S., D. J. Critchett, F. Ehlert, H. I. Yamamura, S. J. Enna, and R. T. Bartus. 1981. Age-related alterations in neurotransmitter receptors: An electrophysiological and biochemical analysis. Neurobiology of Aging 2:3–8.

London, E. D., S. B. Waller, A. T. Ellis, and D. K. Ingram. 1985. Effects of intermittent feeding on neurochemical markers in aging rat brain. Neurobiology of Aging 6:199–204.

McGeer, E. G., H. C. Fibiger, P. L. McGeer, and V. Wickson. 1971. Aging and brain enzymes. Experimental Gerontology 6:391–396.

Michalek, H., A. Meneguz, G. M. Bisso, G. Carro-Ciampi, G. L. Gatti, and G. Bignami. 1978. Neurochemical changes associated with the behavioral toxicity of organophosphate compounds. Pp. 187–201 in Advances in Pharmacology and Therapeutics, Vol. 9: Toxicology, Y. Cohen, ed. Oxford: Pergamon Press.

Michalek, H., A. Meneguz, and G. M. Bisso. 1981. Molecular forms of rat brain acetylcholinesterase in DFP intoxication and subsequent recovery. Neurobehavioral Toxicology and Teratology 3:303–312.

Michalek, H., A. Meneguz, and G. M. Bisso. 1982. Mechanisms of recovery of brain acetylcholinesterase in rats during chronic intoxication by isoflurophate. Archives of Toxicology 5S:116–119.

Michalek, H., A. Pintor, S. Fortuna, and G. M. Bisso. 1985. Effects of diisopropyl fluorophosphate on brain cholinergic systems of rats at early developmental stages. Fundamentals Applications of Toxicology 5:S204–S214.

Michalek, H. S. Fortuna, and A. Pintor. 1989. Age-related differences in brain choline acetyltransferase, cholinesterases and muscarinic receptor sites in two strains of rat. Neurobiology of Aging 10:143–148.

Myers, R. D. 1974. Handbook of Drug and Chemical Stimulation of the Brain: Behavioral, Pharmacological and Physiological Aspects. New York: Van Nostrand Reinhold.

Overstreet, D. H., R. W. Russell, S. C. Helps, P. Runge, and A. M. Prescott. 1979. Sex differences following pharmacological manipulation of the cholinergic system by DFP and pilocarpine. Psychopharmacology 61:49–58.

Overstreet, D. H., and R. W. Russell. 1984. Selective breeding for differences in cholinergic function: Sex differences in the genetic regulation of sensitivity to the anticholinesterase, DFP. Behavioral and Neural Biology 40:227–238.

Pedigo, N. W., Jr., and D. M. Polk. 1985. Reduced muscarinic receptor plasticity in frontal cortex of aged rats after chronic administration of cholinergic drugs. Life Sciences 37:1443–1449.

Pedigo, N. W., Jr., L. D. Minor, and T. N. Krumrei. 1984. Cholinergic drug effects and brain muscarinic receptor binding in aged rats. Neurobiology of Aging 5:227–233.

Pintor, A., S. Fortuna, M. T. Volpe, and H. Michalek. 1988. Muscarinic receptor plasticity in the brain of senescent rats: Down-regulation after repeated administration of diisopropyl fluorophosphate. Life Sciences 42:2113–2121.

Reinert, J.C. 1984. Pesticides in the indoor environment. Pp. 233–238 in Indoor Air, Vol. 1: Recent Advances in the Health Sciences and Technology, B. Berglund, T. Lindvall, and J. Sundell, eds., Stockholm: Swedish Council of Building Research.

Roman, F., O. Della Zuana, M. Lonchampt, G. Saint Romas, and J. Duhault. 1984. Modifications biochimiques chez la rat Wistar age des 24 mois. Comptes Rendus des Seances de la Societe de Biologie et Ses Filiales 178:372–381.

Russell, R.W. 1977. Cholinergic substrates of behavior. Pp. 709–731 in Cholinergic Mechanisms and Psychopharmacology, Advances in Behavioral Biology, Vol. 24, D. J. Jenden, ed. New York: Plenum.

Russell, R.W. 1982. Cholinergic system in behavior: The search for mechanisms of action. Annual Review of Pharmacology and Toxicology 22:435–463.

Russell, R. W. 1988. A basic role of neuromediator receptors in theoretical models of behavior. Psikhologicheskii Zhurnal 9:147–157.

Russell, R. W., and D. H. Overstreet. 1987. Mechanisms underlying sensitivity to organophosphorus anticholinesterase compounds. Progress in Neurobiology 28:97–129.

Russell, R. W., D. H. Overstreet, C. W. Cotman, V. G. Carson, L. Churchill, F. W. Dalglish, and B. J. Vasquez. 1975. Experimental tests of hypotheses about neurochemical mechanisms underlying behavioral tolerance to the anticholinesterase diisopropyl fluorophosphate. Journal of Pharmacology and Experimental Therapeutics 192:73–85.

Russell, R. W., C. A. Smith, R. A. Booth, D. J. Jenden, and J. J. Waite 1986. Behavioral and physiological effects associated with changes in muscarinic receptors following administration of an irreversible cholinergic agonist (BM 123). Psychopharmacology 90:308–315.

Russell, R. W., R. A. Booth, C. A. Smith, D. J. Jenden, M. Roch, K. M. Rice, and S. D. Lauretz. 1989. Roles of neurotransmitter receptors in behavior: Recovery of function following decreases in muscarinic receptor density induced by cholinesterase inhibition. Behavioral Neuroscience (in press).

Sherman, K. A., J. E. Kuster, R. L. Dean, R. T. Bartus, and E. Friedman. 1981. Presynaptic cholinergic mechanisms in brain of aged rats with memory impairments. Neurobiology of Aging 2:99–104.

Stavinoha, W. B., L. C. Ryan, and P. W. Smith. 1969. Biochemical effects of an organophosophorus cholinesterase inhibitor on the rat brain. Annals of the New York Academy of Sciences 160:378–382.

Talens, G., and D. Woolley. 1973. Effects of parathion administration during gestation in the rat on development of the young. Proceedings of the Western Pharmacology Society 16:141–145.

Wecker, L., P. L. Mobley, and W. D. Dettbarn. 1977. Central cholinergic mechanisms underlying adaptation to reduced cholinesterase activity. Biochemical Pharmacology 26:633–637.

Zalewska, Z., I. Rakowska, G. Matraszek, and D. Sitkiewicz. 1977. Effect of dichlorvos on some enzyme activities of the rat brain during postnatal development. Neuropatologia Polska 15:255–262.

Animal Models of Dementia:
Their Relevance to Neurobehavioral Toxicology Testing

David H. Overstreet and Elaine L. Bailey

There is a wealth of evidence from clinical and epidemiological research to document the fact that a wide variety of exogenous chemicals are capable of producing dementia. Dementia involves declines in learning memory and other cognitive processes, (e.g., problem solving), all of which are necessary for normal adaptations to changing physical and psychosocial environments. The development of animal models of dementia is making it possible to supplement knowledge gained from clinical and epidemiological approaches with information obtained by experimental manipulations of the variables involved. The material that follows, briefly reviews the research designs, procedures, and specific paradigms used in such experimental studies.

ANIMAL MODELS OF DEMENTIA

With the increasing recognition of the significance to individuals, and to society generally, of the primary degenerative dementias, interest in the development of animal models of dementia has been growing rapidly. Although some skepticism has been expressed about the possibility of constructing such models, a generally optimistic view has prevailed (Heise, 1984; Overstreet and Russell, 1984). For example, several investigators have used neurotoxins such as ibotenic acid, an excitotoxic amino acid, or AF64A, a putatively specific cholinergic neurotoxin, as tools for creating morphological lesions in the central nervous system (CNS), thereby producing cholinergic deficits analogous to those characteristic of Alzheimer's disease (Bailey et al., 1986; Hepler et al., 1985). There have been approaches using neurochemical methods (see chapters by Michalek and Pintor, and Russell). These studies have routinely shown that treatments which interfere with normal cholinergic neurotransmission lead to disruption in measures of learning and memory, thereby providing support for the "cholinergic hypothesis of memory" (Bartus et al., 1982; Coyle et al., 1983). There is also a growing appreciation that the behavioral measures used in these studies are applicable to the detection of the neurobehavioral effects of suspected environmental toxicants (e.g., Walsh and Chrobak, 1987).

Learning and memory are theoretical constructs that cannot be measured directly. They are inferred from observations of behavior under certain specified conditions. Learning is manifested by systematic changes in behavior as a consequence of repeated exposures to the same stimulus environment; memory, as the preservation of learned behavior over time (Heise, 1984). Any manipulation of an animal's performance may confound the interpretation of the possible effects of that manipulation on learning. As a consequence, an investigator studying animal models of dementia should examine a range of tasks from which learning and memory measures may be extracted, as well as observing, for comparison, other behavioral parameters that are definitely not related to learning and memory. For example, if, during the course of an experiment, food or water is used as a reinforcer, the investigator should measure the effect of the experimental manipulation on the food or water directly. Without elaborating further, it is clear that in order to specify toxic effects on cognitive processes, it is desirable to use batteries of measures rather than to depend upon single assays.

Requirements that animal models of dementia must meet and the characteristics of research designs in which they are put to work have been discussed in detail elsewhere (Heise, 1984; Hepler et al., 1985; Kennett et al., 1987; Olton, 1983; Overstreet and Russell, 1984; Russell and Overstreet, 1984; Tilson and Mitchell, 1984; Willner, 1984). They are mentioned here only very briefly as a general setting within which to project the more detailed discussion to follow. Specifications for the development of animal models and for safeguards in the use of them have been established by international organizations (e.g., World Health Organization, 1975), by national scientific bodies (e.g., Xintaris et al., 1974), and by individual investigators (e.g., Weiss and Laties, 1975). These specifications include systematic manipulation of independent variables, while eliminating effects of other potentially confounding factors; precise measurement of dependent variables by using reliable measuring instruments; attention to the validity of animal models for generalizations to other species, including human; and strict adherence to today's scientific ethics in the care and treatment of animal subjects. With these general points in mind, the use of environmental or pharmacological "challenges" as methodologies in neurobehavioral toxicology is considered.

ENVIRONMENTAL CHALLENGES

MacPhail et al. (1983) have given several examples of the use of an environmental challenge to uncover a debilitating effect of an environmental toxin. They define environmental challenges as "variables that are either known or suspected to affect a baseline of behavior." In effect, the various tasks described above for measuring learning and memory are environmental challenges because they place some demand on the organism. It is precisely for this reason that they may be likely to reveal some effect of the environmental toxin, whereas standard neurobehavioral toxicology testing would not. It is also possible to infer that a limited number of brain regions might be affected if the treatment results in a disruption of memory.

Among other environmental challenges that can be used in an attempt to uncover an effect of a suspected environmental toxin are manipulations of schedule-controlled behavior (MacPhail et al., 1983). As indicated above, spatial or temporal alternation tasks can be used to infer the working memory of an animal. Some years ago, we reported that rats treated chronically with the anticholinesterase diisopropyl fluorophosphate (DFP) did not develop tolerance to its effects on alternation behavior (Overstreet et al., 1974). This finding correlates with the observation of memory disturbances in humans exposed to organophosphate pesticides (Russell and Overstreet, 1987).

A final group of environmental challenges that might be considered for studying neurobehavioral toxicology are paradigms involving stress. As far as we know, no investigator has used this approach as yet, although it has been widely used on animal models of depression (e.g., Willner, 1984) and to study the effects of some drugs (e.g., Weiss et al., 1961). Among the possibilities are the forced swim test (Porsolt, 1982), the inescapable shock ("learned helplessness") paradigm (Maier, 1984), and restraint (e. g., Kennett et al., 1987). After the animals are exposed to these various forms of stress (sometimes, during exposure), measures of their ability to move are taken. It is reasonable to hypothesize that animals exposed to environmental toxicants would be more susceptible to these stressful conditions and would exhibit greater reductions in activity than control animals. This approach has been used to differentiate between two lines of rats that have been selectively bred for varying responses to the anticholinesterase DFP (Overstreet, 1986; Overstreet et al.,1988b).

PHARMACOLOGICAL CHALLENGES

The use of pharmacological challenges to uncover changes in an organism chronically exposed to chemical agents is a recent development in neurobehavioral toxicology testing (see Zenick, 1983); as far as we know, this design has been used only rarely in studies of animal models of dementia. However, the principles underlying these challenges were well known to some investigators much earlier. In our early work on the development of tolerance to DFP, for example, we showed that rats could "become tolerant without acute behavioral changes" (Chippendale et al., 1972). These rats, which received daily low doses of DFP, had reductions in brain acetylcholinesterase activity comparable to other rats treated with higher dosages and showed comparable increased sensitivity to the muscarinic antagonist scopolamine (Chippendale et al., 1972).

In a subsequent study using a daily, low-dose paradigm of DFP treatment and a challenge design, it was found that rats developed subsensitivity to muscarinic agonists within five days of starting treatment. The subsensitivity was complete within nine days, at about the time brain acetylcholinesterase activity was at its lowest (Overstreet, 1974). In still another study using the challenge design, we determined that the muscarinic subsensitivity which follows anticholinesterase treatment can be observed after a single, acute treatment; it first appears at about 48 hours and lasts for about two weeks (Overstreet et al., 1977). In fact, it was these challenge studies which led to the notion that decreases in muscarinic receptor concentrations might be a primary mechanism underlying tolerance development to anticholinesterases (Russell and Overstreet, 1987; Schiller, 1979). Zenick (1983) also called attention to this advantage of the challenge design: by using appropriate challenge agents, some hint of the adapting changes taking place in the central nervous system can be obtained.

The challenge design has also been very useful in understanding the changes that have occurred in our two selectively bred rat lines—the Flinders Sensitive Line (FSL) and the Flinders Resistant Line (FRL). These rats were selectively bred to differ in their responses to DFP (Overstreet et al., 1979). Subsequently, it was found that the FSL rats were more sensitive to muscarinic agonists (Overstreet and Russell, 1982), which correlated with increased concentrations of muscarinic receptors in the hippocampus and striatum (Overstreet et al., 1984).

More recently, it has been found that the FSL and FRL rats differ in their sensitivity to agents acting upon other neurotransmitter receptors (Russell and Overstreet, 1987; see Overstreet et al., 1988b, for reviews). Thus, selective breeding for differences in sensitivity to anticholinesterases has led to changes in sensitivity to a range of other drugs. These findings suggest that investigators should challenge their treated animals with a range of compounds; otherwise, they may make conclusions that are not accurate (i.e., a neurobehavioral toxicant may induce adaptive changes in a number of neurochemical systems).

As far as we know, the challenge approach has not been used with much purpose by investigators studying animal models of dementia, even though it is reasonable to expect adaptive changes in the lesioned animals (Finger, 1978). We would like to describe some of our recent work in which the challenge design has been very useful in exploring the time-dependent changes that occur in rats after hippocampal administration of the neurotoxin AF64A. The challenge approach was used to help answer the question whether AF64A has both pre- and postsynaptic effects at cholinergic synapses because the suggestion has been made that it works mainly presynaptically (Hanin, 1984).

If AF64A destroys cholinergic axons in the hippocampus, one would expect a supersensitivity to develop as an adaptation to the lost cholinergic input. We approached this question by challenging the rats with scopolamine, a muscarinic antagonist, and oxotremorine, a muscarinic agonist, and measuring locomotor activity by direct observation of line crossing in a open field chamber. We were surprised by the initial results, carried out three months after surgery, which showed the AF64A-treated rats to be subsensitive to oxotremorine and supersensitive to scopolamine. These data are consistent with receptor decreases, not increases. Intrigued by these results, we sacrificed the rats and carried out receptor binding assays on the hippocampal homogenates. There was a significant 30 percent reduction in the number of muscarinic receptors in the AF64A-treated rats (Schiller et al., 1990).

In a subsequent experiment we decided to challenge the rats much sooner after the hippocampal administration of AF64A. At three weeks after treatment, the rats were subsensitive to scopolamine, the antagonist, and supersensitive to oxotremorine, the agonist, which suggests that a supersensitivity had developed. These animals were then left for several months and rechallenged. At this time they were supersensitive to scopolamine and subsensitive to oxotremorine, confirming our earlier results. Thus, there are time-dependent changes in cholinergic mechanisms in response to hippocampal injections of AF64A. The early changes are consistent with the expected supersensitivity;

however, a subsensitivity later occurs which is associated with a loss of muscarinic receptors (Schiller et al., 1990).

From this it should be clear that a challenge design may be useful in establishing that changes in sensitivity have occurred as a consequence of some treatment or manipulation. Therefore, information about adaptive changes in animals exposed to neurobehavioral toxicants and the mechanisms underlying these adaptive changes can also be gathered by using similar designs. Zenick (1983) gives a number of examples of how the pharmacological challenge design has been used to uncover an effect of a neurobehavioral toxicant. The use of this design should increase in frequency as more investigators become aware of its utility.

In closing this section, we wish to offer a few words of caution about the pharmacological challenge design. Once one has selected a compound to use, there are still problems about the choice of parameters. Locomotor activity is a useful parameter that is sensitive to a range of compounds, whereas operant responding requires more effort to establish but is more sensitive to drug effects. Another problem is the possibility of choosing only one or a limited range of compounds, which might give the investigators a false picture of the mechanisms underlying the adaptive changes. Whenever possible, a wide range of compounds should be selected. A consequence of multiple compounds is multiple testing; therefore, operant responding is favored over locomotor activity as the dependent variable because it is less subject to shifts in baseline.

MEASURING BEHAVIORAL EFFECTS: SPECIFIC PARADIGMS

It has been said that there is a finite number of measurable behaviors, but that the number is very large. Examination of the research literature on neurobehavioral toxicology indicates that certain paradigms have been favored in studies involving animal models, favored at least in part because of their analogies to human behaviors. The categories in which these paradigms are included is discussed below in some detail.

Inhibitory (Passive) Avoidance

The typical experimental environment in which passive avoidance is generated is a two-compartment box. The animal is placed in the lighted compartment on the first day and given a foot shock upon entering the dark compartment. Memory is inferred from the length of time the rat remains in the lighted compartment when put there 24 hours later; the longer the stay, the better is the memory. This task can be a particularly useful one because parameters such as strength of shock, its duration, and the time between testing and retention can be varied to search for differences between groups. Large numbers of animals can also be tested in a short space of time. However, because this task uses aversive stimulation, a number of potentially confounding variables must be checked before a firm conclusion can be reached. For example, a drug (e. g., vasopressin) that has intrinsic aversive properties may appear to enhance the memory of this task. Similarly, any manipulation that makes the animal more "fearful" or alters its sensitivity to shock will influence its performance on the passive avoidance task. These potential effects must therefore be tested by independent means.

Active Avoidance Tasks

There are a number of variations to test environments used in active avoidance testing. Runways and two-compartment boxes with either one-way or two-way avoidance tasks have been used. It is also possible to alter the conditioned stimulus, which is usually either a tone or a light. The basic procedure is to place the animal in the apparatus and, after a brief interval (30 s), give the conditioned stimulus. This stimulus is followed in 5–10 s with the shock, if the animal has not moved beforehand. Thus, measures of both avoidance and escape are recorded. If a treatment has a general debilitating effect on the animals, then both avoidances and escapes should be affected. If the treatment influences learning only, then only avoidances will be affected.

The active avoidance tasks require more effort because most animals require 50 or more trials to reach some criterion of learning. Although the escape measure provides an index of the motor effects of a treatment, there are other problems of interpretation. For example, drugs that stimulate motor activity, such as scopolamine and amphetamine, are known to facilitate active avoidance responding (Barrett et al., 1974). At the same time, scopolamine disrupts passive avoidance performance, and some investigators have used the scopolamine-treated animal as a model for dementia (Flood and Cherkin, 1986). Another problem with these tasks is that aversive stimuli are used. In conclusion, although active avoidance tasks can be used to measure learning in animals, many other tests must be conducted before other confounding variables can be ruled out.

Operant Conditioning

Operant conditioning tasks require the animal to press a bar to deliver a reward (food or water). A wide range of schedules can be used, varying from the simple continuous reinforcement schedule to the more complex delayed reinforcement of low rates of responding (DRL). It is not our intention here to summarize all of the possible schedules that may be used. Rather, a couple of them will be discussed to give an indication of their usefulness as well as their limitations. In all cases, however, it must be remembered that one is working with a deprived animal. Any manipulation that influences food or water intake will influence performance on the task. Similarly, any manipulation that dramatically alters motor capabilities could also influence the task.

The continuous reinforcement schedule or various fixed ratio schedules (e.g., FR5—five presses per reward) can be very useful in testing the acute effects of various agents, but they are not particularly useful in studying learning and memory measures per se. If a treatment that disrupted passive avoidance did not have any influence on either acquiring or performing an operant task, then a more specific argument could be made about its effects. We have found the FR5 schedule of operant responding to be more sensitive to the effects of cholinergic agonists than open field activity (Overstreet, unpublished observations, 1988).

The DRL schedule of reinforcement involves rewarding the animal for responding at low rates. If the animal responds before a set time (e.g., 20 s), a timer is reset and it must wait another 20 s to obtain a reward. This schedule has been particularly useful for looking at disinhibition, which can be produced by lesioning the hippocampus or injecting cholinergic antagonists such as scopolamine. The reader will note that the two treatments mentioned above are also well known to disrupt memory. Whether the DRL task is a useful measure of memory function in animals is debatable (Heise, 1984). One problem is that classical stimulants such as amphetamine, which can enhance memory under a range of conditions (McGaugh, 1973), may produce a disinhibition of DRL responding similar to scopolamine. Another limitation of the DRL task is the long time required for the animal to reach an acceptable criterion before manipulations can be attempted.

The last operant tasks to be discussed are the alternation paradigms. The one we have used to study the effects of cholinergic agents and anticholinesterase tolerance is the single alternation task (Overstreet et al., 1974). The rats are initially trained to bar-press whenever a light comes on; then, only on every other trial. Any presses during the dark period or during the - trials are counted as errors of commission (i.e., disinhibition), whereas a failure to press during the + trials is an error of omission. Thus, the task can simultaneously obtain measures of general motor function as well as disinhibition or "memory." We found that anticholinesterases not only reduced general motor function, but also produced disinhibition. Although tolerance development was complete for the former effects, it was incomplete for the latter (Overstreet et al., 1974). Spatial alternation tasks, rather than temporal, can also be examined. However, despite their usefulness, they require considerable effort.

Maze Learning

Mazes have been used to test rodents for a very long period of time, and most readers would be aware of the Tryon maze-bright and maze-dull rats obtained by selective breeding. There was a controversy in the 1930s about the ability of mazes to measure "intelligence" or learning, and the controversy continues today. Of the many mazes available to test rodents, this discussion is confined to just two: the T maze and the radial arm maze.

The T maze can be used under a variety of conditions. It has often been used in normal rats to examine spontaneous alternation behavior (e.g., Overstreet et al., 1988a; Scheff and Cotman, 1977). Heise (1984) has reviewed the effects of drugs on this task and has concluded that the task more likely measures short-term memory than habituation. Both cholinergic agonists and antagonists can modify rates of alternation, depending on the conditions of the experiment (Squire, 1969). The T maze can also be used under conditions of food or water deprivation; a common paradigm is rewarded alternation, where the rewards on successive trials are in opposite arms of the T maze (e.g., Karpiak, 1983). Others have used even more sophisticated approaches in order to separate working (short-term) from reference (long-term) memory (Hepler et al., 1985; Olton, 1983).

The radial arm maze has been used extensively by Olton and colleagues to study the effects of hippocampal lesions initially, and lesions to other cholinergic systems later, on spatial learning and memory (e.g., Olton, 1983; Olton et al., 1979). Other investigators have now used it to study quite a number of manipulations of the cholinergic system. Typically, all eight arms of the radial arm maze are baited and a trial continues until the rat consumes seven of the eight rewards. During the trial the rat may return to an alley it has already visited, thus making an error (working memory). Normal rats tend to reach error-free performance within a few days, but rats with lesions in the nucleus basalis, for example, take much longer. Thus, the task has been particularly useful in detecting the memory-disruptive effects of treatments that affect the cholinergic system.

Recently, we have become dissatisfied with the standard radial arm maze paradigm, as have others (Olton, 1983) because it does not provide a measure of long-term or reference memory. One approach to this problem has been the construction of a sixteen-arm maze, with only nine of the arms baited. Another approach has been to remove the rat from the maze after several rewards and return it after a delay of varying intervals. Our approach has been to bait only three of the eight arms of a standard maze. Such a procedure has allowed us to measure both reference memory (entry into never-baited arms) and working memory (reentry into an arm that was baited) during performance. We have found that administration of AF64A into the hippocampus produced a significant effect on working memory, but not on reference memory (Schiller et al., 1990).

The baiting of only three arms of the eight-arm maze permits other procedural manipulations. For example, once both groups have reached asymptotic levels of performance, the location of the rewards can be changed and the ability of the animals to relearn the task can be measured. Such a manipulation also differentiated control rats from AF64A-treated rats. In addition, however, the procedure allowed us to observe the effects of physostigmine, a cholinesterase inhibitor often used experimentally in the treatment of Alzheimer's disease. Both the control and the AF64A-treated rats exhibited improved performance in the maze during daily treatment with 0.15 mg/kg of physostigmine, with the performance of the physostigmine-treated AF64A group resembling the saline-treated control group (Schiller et al., 1990).

In conclusion, both the T maze and the radial arm maze have been extremely useful in detecting disturbances of higher brain function produced by various treatments. Their main limitations are that food or water deprivation is often necessary and that they are time-consuming, requiring daily running in the mazes for up to several weeks. Most investigators using these tasks would agree that the outcome more than makes up for these limitations. They could be useful additions to a behavioral battery designed to detect the neurobehavioral toxicology of a particular agent (see Walsh and Chrobak, 1987).

Other Measures

Some of the paradigms that might be used by neurobehavioral toxicologists to examine the potential effects of a toxin on cognitive processes have been summarized above. This section includes some other measures that investigators have found useful in assessing behavior under toxic conditions which produce dementia.

A commonly employed measure is locomotor activity. Its uses and limitations have been thoroughly reviewed by Reiter and McPhail (1979). We have found that hippocampal AF64A treatment induces hyperactivity, as well as disrupts memory (Bailey et al., 1986). A measure of the effects of a treatment on locomotor activity would be particularly useful as a simple nonmanipulative task, if an investigator used a limited number of tasks to assess memory (e.g., Dunnett et al., 1982). As indicated above, it is common for drugs that stimulate activity to disrupt passive avoidance behavior.

Among other tasks that might be used by both neurobehavioral toxicologists and those studying dementia in animals are measures of reactivity such as startle reactions, measures of food or water intake, and measures of sensory sensitivity. Tilson and Mitchell (1984) have recently reviewed these and commented on their advantages and limitations. The point we wish to make is that, too often, cognitive psychologists overlook these less complex tasks when they study effects of changes in brain structures or functions on learning and memory.

CONCLUSION

Clinically the essential feature of dementia is ". . . a loss of intellectual abilities of sufficient severity to interfere with social or occupational functioning" (American Psychiatric Association, 1980). Such losses are multifaceted, being reflected in a variety of behavioral abnormalities. In the preceding discussion, paradigms have been described which provide measures of analogous behavioral abnormalities in animals exposed to environmental toxicants. It has been suggested that studies involving such animal models can generate information of importance in supplementing knowledge in behavioral toxicology based upon clinical and epidemiological research. Animal models provide means of varying exposures to toxic substances and of controlling potentially confounding variables to extents not usually available in research involving human subjects. They also provide unique approaches in the search for mechanisms and sites of action of such substances.

ACKNOWLEDGMENTS

This work was supported in part by a grant from the Flinders University Research Budget to D. Overstreet. Elaine Bailey was supported by a Flinders University Postgraduate Research Scholarship.

REFERENCES

American Psychiatric Association. 1980. Diagnostic and Statistical Manual of Mental Disorders, DSM. Washington, D.C.
Bailey, E. B., D. H. Overstreet, and A. D. Crocker. 1986. Effects of intrahippocampal injections of the cholinergic neurotoxin AF64A on open-field activity and avoidance learning in the rat. Behavioral Neural Biology 45:263–274.
Barrett, R. J., N. J. Leith, and O. S. Ray. 1974. Analysis of facilitation of avoidance acquisition produced by d-amphetamine and scopolamine. Behavioral Biology 11:189–203.
Bartus, R. T., R. L. Dean, B. Beer, and A. S. Lippa. 1982. The cholinergic hypothesis of geriatric memory dysfunction. Science 217:408–417.
Chippendale, T., G. Zawolkow, R. Russell, and D. Overstreet. 1972. Tolerance to low acetylcholinesterase levels: Modification of behavior without acute behavioral change. Psychopharmacologia 26:127–139.
Coyle, J. T., D. L. Price, and M. R. DeLong. 1983. Alzheimer's disease: A disorder of cortical cholinergic innervation. Science 219:1184–1190.
Dunnett, S. B., W. C. Low, S. D. Iversen, U. Stenevi, and A. Bjorklund. 1982. Septal transplants restore maze learning in rats with fornix-fimbria lesions. Brain Research 252:335–348.
Finger, S. 1978. Recovery from Brain Damage. New York: Plenum.
Flood, J. F., and A. Cherkin. 1986. Scopolamine effects on memory retention in mice: A model of dementia? Behavioral Neural Biology 45:169–184.
Hanin, I. 1984. AF64A: A novel cholinotoxin. Potential tool in cholinergic receptor research. Trends in Pharmacological Science 5:94–97.
Heise, G. A. 1984. Behavioral methods for measuring effects of drugs on learning and memory in animals. Medical Research Review 4:535–558.
Hepler, D. J., D. S. Olton, G. L. Wenk, and J. T. Coyle. 1985. Lesions in nucleus basalis magnocellularis and medial septal area of rats produce qualitatively similar memory impairments. Journal of Neuroscience 5:866–873.
Karpiak, S. E. 1983. Ganglioside treatment improves recovery of alternation behavior after unilateral entorhinal cortex lesion. Experimental Neurology 81:330–339.
Kennett, G. A., C. T. Dourish, and G. Curzon. 1987. Antidepressant-like action of 5-HT1A agonists and conventional antidepressants in an animal model of depression. European Journal of Pharmacology 134:265–274.
MacPhail, R. C., K. M. Crofton, and L. W. Reiter. 1983. Use of environmental challenges in behavioral toxicology. Federation Proceedings 42:3196–3200.
Maier, S. F. 1984. Learned helplessness and animal models of depression. Progress in Neuropsychopharmacology and Biological Psychiatry 8:435–446.
Olton, D. S. 1983. The use of animal models to evaluate the effects of neurotoxins on cognitive processes. Neurobehavioral Toxicology and Teratology 5:635–640.
Olton, D. S., J. T. Becker, and G. E. Handelman. 1979. Hippocampus, space, and memory. Behavioral Brain Science 2:313–365.
Overstreet, D. H. 1974. Reduced behavioral effects of pilocarpine during chronic treatment with DFP. Behavioral Biology 11:49–58.
Overstreet, D. H. 1986. Selective breeding for increased cholinergic function: Development of a new animal model of depression. Biological Psychiatry 21:49–58.
Overstreet, D. H., and R. W. Russell. 1982. Selective breeding for sensitivity to DFP. Effects of cholinergic agonists and antagonists. Psychopharmacology 78:150–154.
Overstreet, D. H., and R. W. Russell. 1984. Animal models of memory disorders. Pp. 257–278 in Animal Models of Psychopathology, N. W. Bond, ed. Sydney: Academic Press.
Overstreet, D. H., R. W. Russell, B. J. Vasquez, and F. W. Dalglish. 1974. Involvement of muscarinic and nicotinic receptors in behavioral tolerance to DFP. Pharmacology Biochemistry and Behavior 2:45–54.

Overstreet, D. H., S. C. Helps, A. M. Prescott, and G. D. Schiller. 1977. Development and disappearance of subsensitivity to pilocarpine following a single administration of the irreversible anticholinesterase, DFP. Psychopharmacology 52:263–269.

Overstreet, D. H., R. W. Russell, S. C. Helps, and M. Messenger. 1979. Selective breeding for sensitivity to the anticholinesterase, DFP. Psychopharmacology 65:15–20.

Overstreet, D. H., R. W. Russell, A. D. Crocker, and G. D. Schiller. 1984. Selective breeding for differences in cholinergic function: Pre- and post-synaptic mechanisms involved in sensitivity to the anticholinesterase, DFP. Brain Research 294:227–232.

Overstreet, D. H., R. A. Booth, and D. J. Jenden. 1988a. Effects of an irreversible muscarinic agonist (BM123) on avoidance and spontaneous alternation performance. Pharmacology Biochemistry and Behavior 31:337–344.

Overstreet, D. H., R. W. Russell, A. D. Crocker, J. C. Gillin, and D. S. Janowsky. 1988b. Genetic and pharmacological models of cholinergic supersensitivity and affective disorders. Experientia 44:465–472.

Porsolt, R. D. 1982. Behavioral despair. Pp. 121–139 in Antidepressants: Neurochemical, Behavioral and Clinical Perpectives, S. J. Enna, ed. New York: Raven Press.

Reiter, L., and R. MacPhail. 1979. Motor activity: A survey of methods with potential use in toxicity testing. Neurobehavioral Toxicology and Teratology. 1:53–66 (suppl.).

Russell, R. W. and D. H. Overstreet. 1984. Animal models of neurobehavioral toxicology. Pp. 23–57 in Animal Models of Psychopathology, N. W. Bond, ed. Sydney: Academic Press.

Russell, R. W. and D. H. Overstreet. 1987. Mechanisms underlying sensitivity to anticholinesterase agents administered acutely or chronically. Progress in Neurobiology 28:97–129.

Scheff, S. W., and C. W. Cotman. 1977. Recovery of spontaneous alternation following lesions of the entorhinal cortex in adult rats: Possible correlation to axon sprouting. Behavioral Biology 21:286–293.

Schiller, G. D. 1979. Reduced binding of ^{3}H-quinuclidinyl benzilate associated with chronically low acetylcholinesterase activity. Life Sciences 24:1150–1154.

Schiller, G. D., D. H. Overstreet, A. D. Crocker, and E. B. Bailey. 1990. Time-dependent changes in cholinergic sensitivity following intrahippocampal AF64A administration: Implications for models of Alzheimer's Disease. Alzheimer's Disease and Associated Disorders (submitted).

Squire, L. F. 1969. Effects of pre-trial and post-trial administration of cholinergic and anticholinergic drugs on spontaneous alternation. Journal of Comparative and Physiological Psychology 69:69–75.

Tilson, H. A., and C. A. Mitchell. 1984. Neurobehavioral techniques to assess the effects of chemicals on the nervous system. Annual Review of Pharmacology and Toxicology 24:425–450.

Walsh, T. J., and J. J. Chrobak. 1987. The use of the radial arm maze in neurotoxicology. Physiology and Behavior 40:799–803.

Weiss, F. and V. G. Laties, eds. 1975. Behavioral Toxicology. New York: Plenum.

Weiss, B., V. G. Laties, and F. L. Blanton. 1961. Amphetamine toxicity in rats and mice subjected to stress. Journal of Pharmacology and Experimental Therapy 132:366–371.

Willner, P. 1984. The validity of animal models of depression. Psychopharmacology 83:1–16.

World Health Organization. 1975. Early Detection of Health Impairment in Occupational Exposure to Health Hazards. Geneva: WHO Technical Report Series No. 571.

Xintaris, C., B. L. Johnson, and J. de Grout, eds. 1974. Behavioral Toxicology. Washington, D.C.: U.S. Department of Health, Education and Welfare.

Zenick, H. 1983. Use of pharmacological challenges to disclose neurobehavioral deficits. Federation Proceedings 42:3191–3195.

Bridging Experimental Animal and Human Behavioral Toxicology Studies

Deborah A. Cory-Slechta

THE SCOPE AND AGENDA OF BEHAVIORAL TOXICOLOGY

Behavioral toxicology can be generally conceptualized as that scientific discipline which strives to understand the mechanisms by which toxicants affect behavior. In this respect, it is similar to its counterpart, behavioral pharmacology, the goal of which is to delineate the mechanisms by which drugs modulate behavior. Since its inception, the scope of behavioral toxicology has expanded considerably, driven in part by the need to ascertain the role of existing environmental contaminants in producing functional impairment, as well as the need to develop procedures to preclude future introduction of neurotoxic chemicals. In this way, it differs from behavioral pharmacology which, instead, seeks to develop compounds or agents with specific behavioral actions for therapeutic purposes.

As the above implies, behavioral toxicology actually has dual, overlapping agendas. One derives from the growing recognition of the need to screen for performance impairment prior to the introduction of new chemicals into the environment, as well as to provide information relating to risk assessment based on neurotoxic endpoints. The second agenda involves the more traditional role of behavioral toxicology as the scientific discipline defined above, whose goal is to understand both the behavioral and the biological mechanisms by which toxicants impact behavioral function. It is primarily the latter agenda to which the comments herein are addressed.

Despite recent advances in neurobiology and the obvious utility of in vitro approaches in the elucidation of biological substrates of behavior, it is difficult to conceive of any substitute for assessing the ultimate impact of a neurotoxicant on behavior, or of assessing behavioral mechanisms of toxicant action, other than in the whole organism. The links between molecular neurobiology or between neuropathological alterations and behavioral impairments are still obscure. Although the relationships between certain neurotransmitters and behavior have become increasingly evident, such as the role of dopamine in parkinsonism, other aspects of those relationships remain puzzling, e.g., the extensive dopaminergic depletions noted before any overt behavioral impairments appear. Thus, there are no substitute or alternative procedures for evaluating the functional impact of a toxicant. Put another way, behavioral toxicity cannot be reliably predicted from molecular events.

STATE OF DEVELOPMENT

Although the experimental capabilities for more precisely delineating behavioral and biological mechanisms of toxicant-induced performance impairment are generally at hand, the discipline remains largely at a characterization or descriptive stage of development. Much of its scientific literature attempts little more than to ascertain whether a particular toxicant alters a particular class of behavior, or to assess performance impairments produced by a toxicant across a range of behavioral endpoints; in some cases, only the barest approximation to a hypothesis may be invoked. Furthermore, little attempt may be made to rationalize the particular behavioral approach chosen, which may, instead, be based predominantly on available apparatus or technology in that laboratory. Nonetheless, owing more to the sheer magnitude of work with certain compounds, rather than to any systematic progression of studies within a laboratory, in certain areas these studies have begun to provide the prerequisite foundation from which more mechanistic approaches can now proceed.

Studies of performance impairments induced by lead exposure provide one example. Lead may be considered a prototypical behavioral toxicant and undoubtedly has been the most extensively studied of such compounds, both at the experimental animal and at the human level. The permanent mental retardation, which in some cases was the residual effect of acute high-dose lead exposure in children, resulted in a subsequent focus of these studies on issues of learning deficits at lower lead exposure levels. Human studies of environmental lead exposure in children have almost invariably focused on age-appropriate IQ and other psychometric tests as their behavioral endpoint. The most recent of such studies have documented decrements in IQ and similar psychometric measures at blood lead concentrations as low as 10 µg/dL (e.g., Bellinger et al., 1987; Fulton et al., 1987). However, even with the separation of verbal and motor subscales, IQ tests represent extremtently global measures of performance that encompass a variety of different behavioral functions, as well as the involvement of multiple sensory systems, many of which may be marginally affected or others of which may be more dramatically affected. Such global measures always present the possibility that particular subtle deficits may be obscured by the sheer multiplicity of concurrently measured behaviors or may be clouded by a reserve capacity of the organism. The specific nature of the IQ decrements in humans thus remains unresolved.

Experimental animal studies can more readily address aspects of lead-induced changes in learning. Table 1 summarizes those studies that have assessed lead-induced changes in learning by using acquisition of a visual discrimination as a behavioral endpoint. The various studies are subdivided on the basis of both the type of visual cue utilized and the developmental period of lead exposure. Plus signs show those experiments reporting an impairment of visual discrimination learning as a result of lead exposure, whereas minus signs accompany those that found no change. As indicated by the preponderance of plus signs in each column, two types of visual discrimination paradigms emerge as those more sensitive to lead exposures: discriminations based on differences in brightness and on size of visual cues. Shape-form discrimination shows little obvious impact of lead. A within-laboratory comparison provides further support for this across-study conclusion. Winneke et al. (1977) reported that lead-treated rats required more trials to acquire a size discrimination than did control rats, but were not impaired in the acquisition of a form discrimination. Although the effects noted with color-based discrimination are suggestive, they are, at present, based on a restricted data set.

The differential lead effects based upon visual cue emphasize the critical importance of the environmental context in modulating the behavioral effects of a toxicant such as lead. No generalized deficit in visual discrimination learning can be ascribed to lead; instead, such deficits depend upon environmental cues.

Studies of visual discrimination learning following lead exposure can direct future efforts aimed at understanding the behavioral and biological mechanisms that might explain such differential effects. With respect to behavioral mechanisms, the possiblity of a generalized performance decrement—for example, an increase in response bias or an alteration in motivation level—obviously fails to accommodate the differential effects of visual cue. One explanation resides in the possibility of differential degrees of control exerted by the stimuli

TABLE 1 Lead-Induced Changes in Visual Discrimination Learning

Developmental Period of Lead Exposure	Brightness	Shape-Form	Size	Color
Prenatal	+Brady et al. (1975)[a] Rat +Zenick et al. (1978) Rat +Driscoll and Stegner (1976) Rat	+Zenick et al. (1978) Rat -Carson et al. (1974)[b] Sheep -Winneke et al. (1977) Rat	+Carson et al. (1974) Sheep +Winneke et al. (1977) Rat +Winneke et al. (1982) Rat	
Postnatal	-Brown (1975) Rat +Bushnell et al. (1977) Monkey -Hastings et al. (1977) Rat +Hastings et al. (1979) Rat -Hastings et al. (1979) Rat	-Carson (1976) Sheep -Overmann (1977) Rat +Rice and Willes (1979) Monkey -Rice (1985) Monkey	-Carson (1976) Sheep +Bushnell and Bowman (1979) Monkey	+Bushnell and Bowman (1979) Monkey +Rice (1985) Monkey
Adult	-Lanthorn and Isaacson (1978) Rat			

[a]Effect of lead reported.
[b]Absence of lead effect reported.

over behaviors, which consequently exhibit differential behavioral sensitivity to disruption. Numerous behavioral pharmacology studies have shown that central nervous system (CNS) drugs may exhibit a much greater magnitude of effect on behavior that is under weak stimulus control (Laties, 1975), i.e., performances which exhibit relatively low overall accuracy levels or require an extensive number of trials to attain criterion performance. In such cases, the stimuli may be less salient, thus failing to generate strong control over performance and rendering it more easily disrupted by other factors.

The fact that lead effects on visual discrimination learning depend upon the type of visual cue, as shown in Table 1, might also reflect differential sensory effects of lead on the visual system. In regard to such a possibility, Fox et al. (1982) have reported persistent decreases in both visual acuity and spatial resolution in 90-day old rats that had been exposed from birth to weaning to 200 ppm of lead acetate via the dam. Although these, as well as other visual system deficits resulting from lead exposure have been reported (Bushnell et al., 1977; Fox and Chu, 1988; Fox and Farber, 1988), their direct impact in mediating behavioral toxicity remains to be systematically investigated.

Experimental animal studies also reveal that lead impairs learning based on the acquisition of spatial discrimination and can be categorized on the basis of both the route and the developmental period of exposure. Table 2 illustrates several additional issues of importance

TABLE 2 Lead-Induced Changes in Spatial Discrimination Learning

Developmental Period of Lead Exposure	Exposure Route	
	Oral	Intraperitoneal
Pre- or postnatal	+Snowdon (1973)[a]	-Brown et al. (1971)[b]
	+Bushnell and Bowman (1979a)	+Klein et al. (1977)
	+Bushnell and Bowman (1979b)	-Rosen et al. (1985)
	-Overmann (1977)	
	+Levin and Bowman (1983)	
	+Laughlin et al. (1983)	
Postweaning	+Geist and Mattes (1979)	-Brown et al. (1971)
Adult	+Avery and Cross (1974)	-Snowdon (1973)
	+Lanthorn and Isaacson (1978)	-Ogilvie (1978)
	+Ogilvie (1977)	-Bullock et al. (1966)
	±Dietz et al. (1979)	-Penzien et al. (1982)

[a]Effect of lead reported.
[b]Absence of lead effect reported.

in understanding the behavioral toxicity of lead, in particular, which may apply equally to other toxicants: for example, the critical importance of the kinetics of lead to its behavioral toxicity, the detrimental effects on spatial learning, and the consistency of the effect across a variety of different behavioral procedures. Lead-induced impairment of spatial discrimination learning has been observed following postnatal, postweaning, and adult exposures, in contrast to brightness discrimination (Table 1) which appears most vulnerable in response to prenatal exposures. Thus, different behavioral performances may exhibit quite different critical periods of exposure to a toxicant, or the critical exposure period for behavioral effects produced by a toxicant may be determined at least partly by the sensitivity of the behavioral procedure.

Comparative changes in schedule-controlled behavior induced by lead exposure reveal additional aspects of its behavioral toxicity, aspects that in turn may impact on, or even underlie, other lead-induced performance effects. The most extensively studied of such reinforcement schedules has been the fixed-interval (FI) schedule, in which the reward for responding is temporally based, with the contingency stipulating that the first response occurring after a specified interval of time elapses produces reinforcement. Figure 1 presents the dose-effect function that summarizes the various studies of lead-induced changes in FI schedule-controlled behavior. It plots a parameter of FI performance (as a percentage of the corresponding control data) in relation to treatment dose.

In constructing this summary figure, the lead exposure dosage or concentration has been recalculated in milligrams per kilogram. Because not all experimenters used the same dependent variables, response rate was used where presented, but in other cases, the outcome was based on total number of reponses, median interresponse time (IRT), or group mean percentage of control reinforcements, all of which can be impacted upon by response rate changes. The data were plotted from the session or sessions in which peak effects occurred, with the exception of data from our studies (Cory-Slechta and Thompson, 1979; Cory-Slechta et al., 1983, 1985), which were restricted to results from the first 30 experimental sessions so as to be comparable to the number of sessions used in most other studies. Prenatal and oral preweaning exposure studies could not be included in this summary figure because it was not possible to ascertain the dose to which the developing animals were exposed. Figure 1 reveals an inverse U-shaped function relating dose of lead to performance on the FI schedule of reinforcement. That is, exposure to lower concentrations or doses of lead produces response rate or output increases on the FI schedule; as the dose or exposure to lead increases, however, the rates of responding

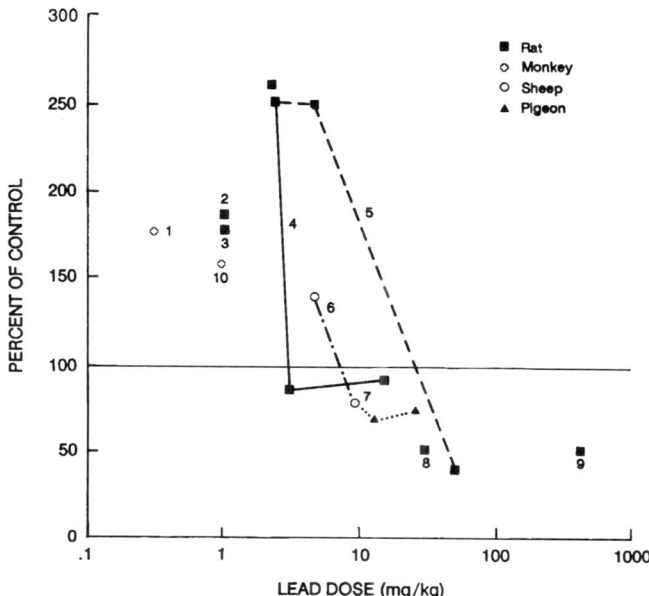

FIGURE 1 Summary of studies investigating changes in fixed-interval performance (plotted as percent of the control group) as a function of lead dosage, taken from the session or sessions in which peak effects occurred. Data from studies involving prenatal or lactational exposures could not be included because it was not possible to ascertain the dose to which the developing organisms were exposed. Different experimental species are indicated by different symbols; numbers refer to different studies: (1) Rice et al. (1979); (2) Cory-Slechta et al. (1985); (3) Cory-Slechta (1989); (4) Cory-Slechta et al. (1983); (5) Cory-Slechta and Thompson (1979); (6) Van Gelder et al. (1973); (7) Barthalmus et al. (1977); (8) Angell and Weiss (1982); (9) Zenick et al. (1979); (10) Rice (1988).

SOURCE: Cory-Slechta (1984).

are depressed below control values. Plotting changes in FI response rate against the reported blood lead values in each study produces a similar function (Cory-Slechta, 1984).

The potential generality of the dose-effect function is evidenced by the similarity of the lead-induced changes in rates of responding that have been described in other temporally based reinforcement schedules. For example, Nation et al. (1983) reported a dose-effect function for lead-induced changes in variable-interval (VI) performance comparable to that shown in Figure 1 for the FI schedule, with a daily dose of 1.0 mg/kg of lead increasing VI response rates, whereas both 5.0 and 10.0 mg/kg suppressed response rates. Rice and Gilbert (1985) reported no effect of exposure to a low level of lead (associated with steady-state blood lead of 11–13 µg/dL) on response rate or on the mean IRT value of monkeys responding on another temporally based schedule of reinforcement, a differential reinforcement of low rate (DRL 10 s or DRL 30 s) schedule. However, on both the DRL 10- and the DRL 30-s schedules, they did report an increase in the number of nonreinforced responses and a decline in the number of reinforcers received by lead-treated monkeys, an effect that would seemingly necessitate an increased overall rate of responding. This pattern of effects would be evident in the distribution of IRTs but might not have impacted the measured index, mean IRT, the value of which could be substantially influenced by a small number of very long IRTs.

The response rate-increasing properties of low-level lead exposure derive from a decrease in the time between successive responses (i.e., interresponse times). In particular, the frequency of short interresponse times (less than 0.5 s) during the fixed interval is increased by lead exposure, such that successive responses occur more rapidly in lead-exposed organisms than in controls; no consistent changes in postreinforcement pause time are noted. Figure 2 shows the proportion of short IRTs of control rats (left panels) and of rats treated with 25 ppm of lead acetate (right panels) over the course of 40 experimental sessions on an FI 60-s schedule of food reinforcement in two separate replications (top panel, Cory-Slechta et al., 1985; bottom panel, Cory-Slechta, 1989). As can be seen, the range of short IRTs exhibited by control and lead-exposed rats was actually quite comparable, but lead exposure yielded a shift of the distribution toward the upper extremes of the range (i.e., higher proportions) in both replications. Thus, control and lead-treated animals begin to respond at the same time during the fixed interval, but lead-exposed organisms then respond at much higher rates than control animals, engendering more responding per unit time.

In contrast, schedules of reinforcement based on number of responses (ratio based), rather than on temporal parameters, exhibit a different pattern of lead effects, another indication that its behavioral toxicity is dependent upon the environmental or behavioral context. Although high-level exposures to lead are reliably associated with decreases in response rate on ratio schedules, evidence for rate-enhancing effects at lower exposure levels is not compelling (Angell and Weiss, 1982; Barthalmus et al., 1977; Cory-Slechta, 1986; Padich and Zenick, 1977; Rice, 1988). Although Angell and Weiss (1982) reported shorter median IRTs on an FR schedule in rats exposed to lead prenatally only, the IRTs were actually not significantly different from those of nonexposed controls.

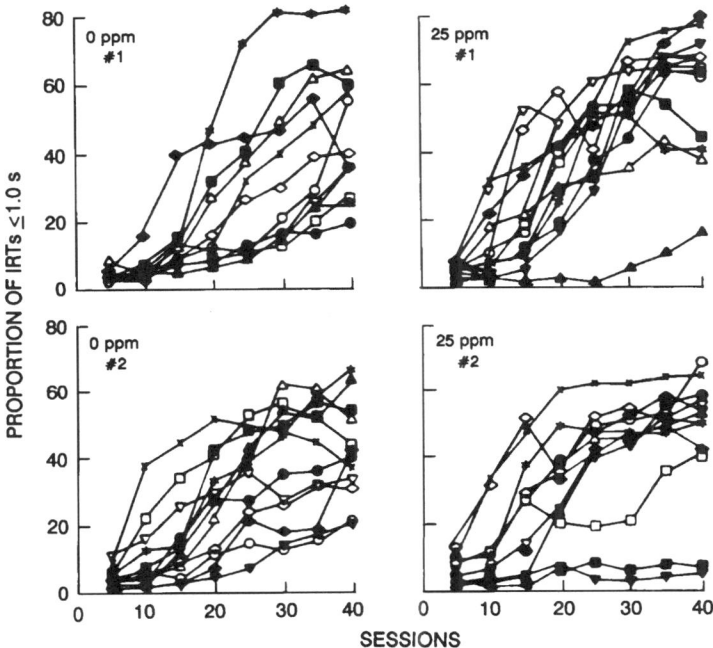

FIGURE 2 Proportion of short interresponse times (less than or equal to 0.5 s) of individual control rats (left panels) and rats exposed to 25 ppm of lead acetate (right panels) on a fixed-interval 60-s schedule of food reinforcement. Data are shown over the course of the first 40 experimental sessions. Top panels from Cory-Slechta et al. (1985); bottom panels, Cory-Slechta (1989).

As with the discrimination learning literature, comparative studies of lead-induced changes in schedules of reinforcement can guide mechanistically based experiments. Although an elevation in the rate of responding results from low-level lead exposure, the effect appears to be restricted to temporally based schedules of reinforcement; no such effect is consistently noted when reinforcement delivery occurs under response-based contingencies. Microanalysis of FI performance reveals that if pause time can be construed as an index of timing behavior, it remains intact. However, once responding begins, response rates of lead-exposed rats greatly exceed those required by the reinforcement schedule, suggesting as one possibility, a decreased responsiveness of lead-exposed organisms to the feedback generated by their own behavior on the schedule. Higher lead exposure levels produce a generalized nonspecific suppression of responding, which has been noted on the fixed-interval, fixed-ratio, and variable-interval schedules, raising the possibility of underlying motivational factors (e.g., a decline in reinforcer efficacy).

Many scientists, including some behavioral scientists have difficulty in understanding the rationale for studying schedule-controlled behavior, both because of its less obvious correspondence to human behavior than, for example, conventional learning paradigms, and because of the ostensible difficulty in interpreting response rate changes. It should be emphasized, however, that in the human environment, as well as in the experimental laboratory, rewards or reinforcers for behavioral performances, including learning, occur under various reinforcement schedules. The human environment obviously entails far greater complexity, with many reinforcement schedules and various reinforcers available concurrently. Nevertheless, human behavior occurs, and is consequated, under schedules of reinforcement. The study of simple schedules of reinforcement in the laboratory represents a simplified approach to such processes in order to provide a more molecular analysis of the variables controlling such performances. Furthermore, the lever-press response used in experimental studies is but an arbitrarily chosen response. What if, instead, the particular response increased by lead exposure involved time out of seat or time off task in a classroom setting? Increased frequencies of such responses would have obvious detrimental consequences for children's scholastic performance.

This raises a further issue, namely, that changes in response rate engendered by a toxicant such as lead might interact with, or even underlie, other behavioral deficits. Consider a situation, for example, in which lead-induced increases in rates of responding engender premature responding to stimulus cues in a learning task. The resulting decreased level of accuracy might then be interpreted as a learning impairment. Furthermore, the acquisition of appropriate response patterns on schedules of reinforcement per se may represent learning deficits, as has already been described.

As this discussion with respect to an intensively studied neurotoxicant such as lead indicates, much of our work remains at a descriptive stage. Nevertheless, the utilization by experimental animal studies of more sensitive and specific behavioral endpoints than have been incorporated into many of the human studies in this area, has led over the past five years or so to a striking correspondence between results in the two areas (shown in Figure 3) in the reported levels of lead in blood at which behavioral deficits are reported. The more recent studies in humans indicate performance impairments in children at levels as low as 10 µg/dL (Bellinger et al., 1987; Fulton et al., 1987); studies in both rats and nonhuman primates document effects at

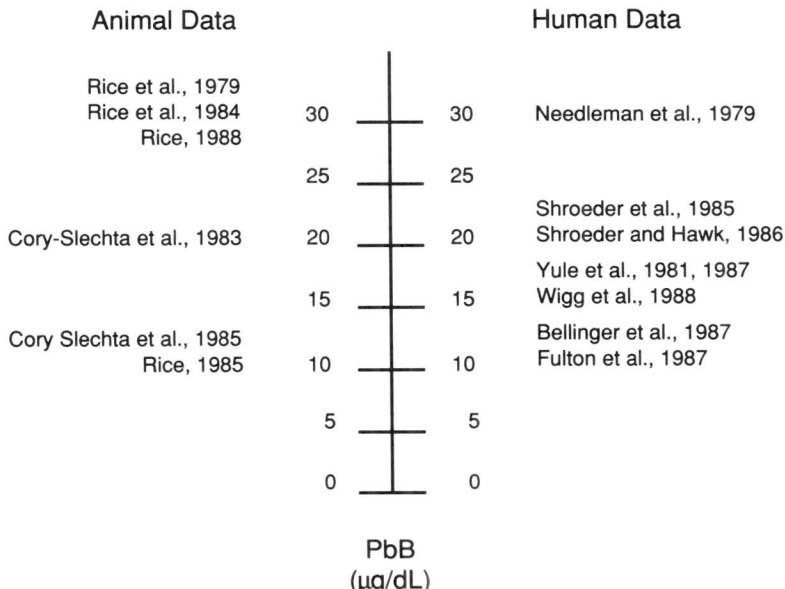

FIGURE 3 Blood-lead levels at which behavioral effects are reported

comparable levels (Cory-Slechta et al., 1985; Rice, 1985). Figure 3 further illustrates the decline in blood lead concentrations associated with performance effects in both humans and experimental animals over the past several years, as earlier studies paved the way for methodological improvements in subsequent efforts.

BARRIERS TO ADVANCEMENT

What has constrained advancement and kept much of the research in behavioral toxicology primarily at the level of characterization? Several factors probably play a role. One is that emerging scientific disciplines such as behavioral toxicology require reliable, systematic characterization studies as a sound base for further efforts. Proceeding to mechanistic-based studies in the absence of such information, or on the basis of unreliable information, would be premature and even counterproductive. Besides the relative youth of this area, another factor that may impose difficulties for a more mechanistically based science is the nonspecificity of most neurotoxicants. Many of the chemicals of interest have a diversity of biological effects, including CNS effects. This poses the distinct possibility that, even in the simplest case, different behavioral effects of a particular toxicant may arise from different behavioral or biological substrates, i.e., from effects on different neurotransmitter systems or pathological lesions in different brain regions. Thus, to fully delineate the gamut of behavioral and biological mechanisms of toxicity for any given neurotoxicant could require an extensive experimental commitment.

A further impediment to more rapid progress in understanding the neurobiological substrates of performance impairment has been the relative lack of systematic experimentation aimed at directly defining the relationships between functional consequences and other neurotoxic effects resulting from exposure. Although many studies utililze a multidisciplinary approach, concurrently measuring various indices of behavioral outcome and changes in neurotransmitter levels in response to a neurotoxicant, for example, few studies undertake the types of definitive experiments required to determine the precise nature of such relationships, which then remain correlational in nature.

Probably one of the primary factors constraining both the scope and the advancement of behavioral toxicology may be the preponderance of "apparatus-driven" research. In many cases, research questions are framed around the available behavioral apparatus within a laboratory, be it a radial arm maze or an open field, rather than upon a hypothesis formulated on the basis of the current scientific literature. This can often be noted in the introduction to published studies, which present only the barest approximation to a hypothesis and an obscure rationale, namely, this toxicant may affect that performance. This is likely also to be one of the factors contributing to the frequent shifts in the toxicant of current scientific interest, known colloquially as the poison of the month: if you cannot change the behavioral apparatus, and thus the behavioral question, you are left with changing the toxicant.

The latter situation arises, no doubt at least in part, from limitations of equipment availability imposed by funding restrictions over the past several years. However, probably more importantly, it reflects minimal or inadequate familiarity with the breadth of behavioral sciences in general and with contemporary state-of-the-art behavioral procedures. In addition, it may reflect a prevalent notion among many nonbehavioral scientists that anyone can conduct behavioral testing, a misconception apparently based on the ostensible simplicity of endpoints such as motor activity with which those outside the behavioral field tend to be most familiar.

ACCELERATING THE PACE

One notion that has been advanced to expedite progress in behavioral toxicology is the development of new behavioral tests. However, the rationale for this argument is not compelling and may even be viewed as counterproductive. For instance, consider the multitude and variety of procedures currently in use to examine learning. The number of new procedures that could be devised just to assess this particular function is almost limitless. However, it must be remembered that every newly developed procedure requires extensive behavioral investigation to ascertain the variables controlling the performance, as well as pharmacological determinations of the comparative sensitivity of the procedure to other learning tasks. Might not a more productive and expeditious approach emphasize systematic and comparative studies of a toxicant's effects across existing, better-understood learning procedures, in terms of both behavioral and pharmacological variables?

A second approach to accelerating the pace would be to incorporate some of the more complex, state-of-the-art behavioral paradigms into behavioral toxicology. Consider again the case of learning, a much emphasized component of neurotoxicology and of neuroscience in general. Many of the more conventional techniques suffer from the limitation that once the organism has mastered the problem to be learned, only performance is being measured. For example, animals running a maze may learn to turn toward the appropriate side or color cue quickly. Similarly an organism may learn to lever-press only in the presence of a red light, and not in the presence of green, within only a few experimental sessions. This presents a particular problem in evaluating a toxicant whose effects have a delayed onset or which accumulates only slowly. It also makes the assessment of reversibility of toxicant-induced learning changes difficult.

A more complex behavioral procedure known as repeated acquisition, or acquisition of response sequences, originated partly because of such a need (Boren, 1963). In this particular task, the organism is required to learn a new sequence of responses of fixed length during each experimental session. A schematic of the apparatus configuration for this paradigm is illustrated in Figure 4. Initially, this paradigm engenders quite high error rates, but as the organism gains experience with the task, the error rate stabilizes from session to session and the organism learns each new sequence at a fairly constant rate. The procedure has several distinct advantages over other learning techniques. First, it allows the measurement of learning on a repeated basis, thus providing a stable baseline rate of learning across sessions from which perturbations can be assessed. In addition, the delineation of various classes and patterns of errors following chemical exposure can provide useful information about the type of learning deficit, facets which are relevant both to the issue of behavioral mechanisms of toxicity and to screening and risk assessment.

Furthermore, the procedure is often run in conjunction with a per-

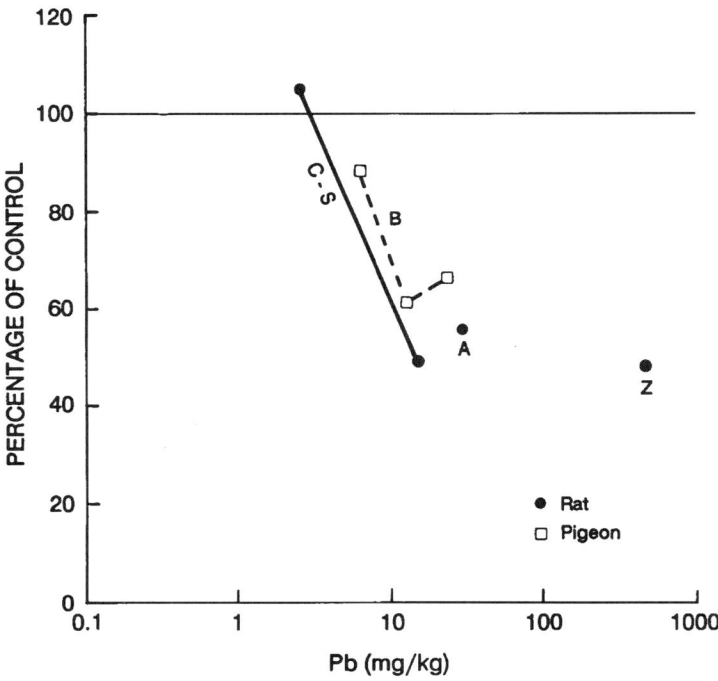

FIGURE 4 Schematic diagram of the intelligence panel used for the repeated acquisition procedure. In the procedure, each three-light unit serves as a discriminative stimulus signaling the next correct response in the sequence of responses leading to reinforcement. The lights have no fixed relationship to the lever; the association between lights and levers changes each session with the new correct sequence of responses.

SOURCE: From Pollard et al. (1981).

formance component in which the response sequence remains constant from experimental session to session. During the daily experimental session, the performance component alternates periodically (e.g., every 10 minutes) with the learning component (as defined above). Thus, both learning and performance are being concurrently assessed during an experimental session. This is especially advantageous because the performance component allows the assessment of nonspecific chemical effects, as would be exemplified by changes that occur in both the learning and the performance components, whereas behavioral alterations observed exclusively in the learning component may more specifically reflect changes in learning. Representative performance of a nonhuman primate responding under such a schedule is shown in Figure 5, in which a separation of the effects of d-amphetamine on the learning and the performance components is evident. Finally, the procedure has recently been further amended (Thompson et al., 1986) to include a memory component. This is accomplished by retesting the acquisition of the learning component sequence at various time intervals following the original learning.

In spite of the emphasis on toxicant and chemical-induced alterations in learning, only two studies to date have utilized the repeated acquisition baseline to evaluate toxicant-induced learning deficits. Paule and McMillan (1986) employed a variant of this procedure to track the time course of trimethyltin (TMT) effects on learning. By separating the various error components of repeated acquisition performance in their analysis, a differential time course of TMT on various classes of errors was noted. It showed that early responses in the sequences

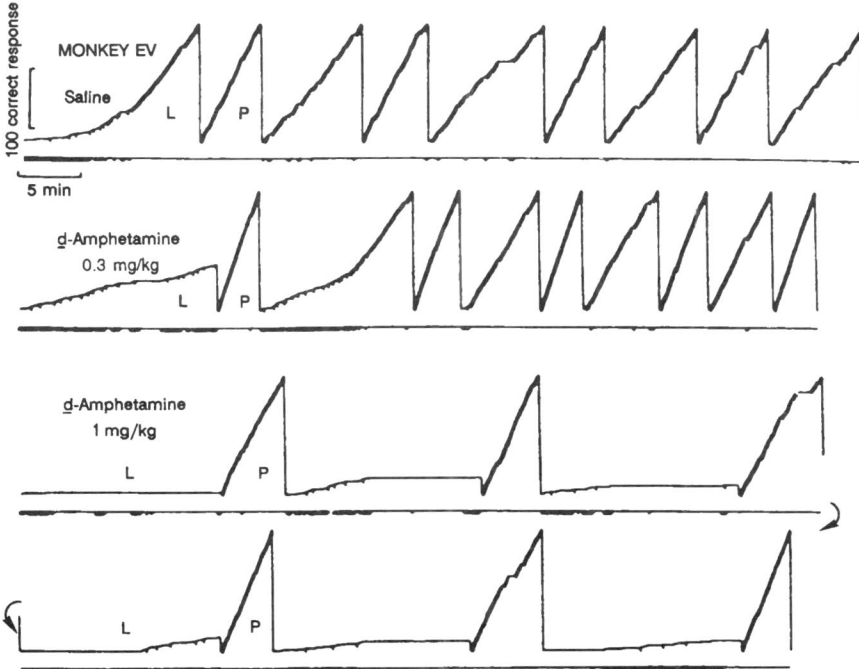

FIGURE 5 Cumulative records illustrating the effects of the administration of d-amphetamine on the performance of a monkey working on a multiple schedule which alternated repeated acquisiton or learning (L) and performance (P) components. Amphetamine affected behavior primarily during the learning component, increasing the number of errors in a dose-related fashion, while leaving the performance component relatively intact.

SOURCE: From Thompson and Moerschbacher (1979).

were disrupted to a greater extent by TMT than were later stages of the sequences, suggesting an effect on learning, whereas the recall necessary for the longer sequences remained relatively intact. In an earlier study, Anger and Setzer (1979) reported that intramuscular carbaryl administration increased both session time and error rates of monkeys working on a four-response sequence repeated acquisition baseline.

The acquisition of response sequences serves as but one example of a more sophisticated procedure for evaluating learning deficits. It emphasizes the point that increased awareness of advancements in behavioral sciences can accelerate the progression of behavioral toxicology toward its ultimate goal. One reason for the hesitancy on the part of many investigators to turn to such techniques may be the more extensive training required to produce stable baseline performances than are required when using simpler procedures. Given the potential of these techniques, however, future work designed to accelerate the training process would be of great benefit to behavioral pharmacology and toxicology.

Another strategy useful to facilitate cross-species extrapolation, that is, to bridge experimental animal and human behavioral toxicology studies, is to strongly emphasize the use of behavioral paradigms that are directly applicable to both populations. Operant schedules of reinforcement exemplify one class of such baselines. Innumerable studies have documented the comparability of schedule-controlled performance in a wide variety of species, including humans (e.g., Dews and Wenger, 1977; Holland, 1958; Kelleher and Morse, 1969; Laties and Weiss, 1963; Richelle, 1969; Tews and Fischman, 1982), a point documented by Figure 6. Behavioral pharmacology studies have further shown the similarity across species of drug effects on schedule-controlled behavior. In fact, the study of schedule-controlled behavior was a strategy suggested by Bornschein et al. (1980) to enhance direct extrapolation of lead-induced behavioral impairments across species.

Some of the more advanced complex techniques may be even more applicable. The repeated acquisition paradigm described above has been used to study learning in rodents (Schrott et al., 1980), pigeons (e.g., Thompson, 1980), and nonhuman primates (Moerschbacher and Thompson, 1980), as shown in Figure 5. The technique has also been utilized with human populations, including those with developmental disabilities (Suessbrick, 1983) and, more recently, victims of Alzheimer's disease (Gershensen, Thompson, and Gisselquist, personal communication, 1988). Figure 7 compares the decline in the number of errors on a five-link response sequence in normal elderly humans (top panel) to those with Alzheimer's disease (bottom panel). The greater number

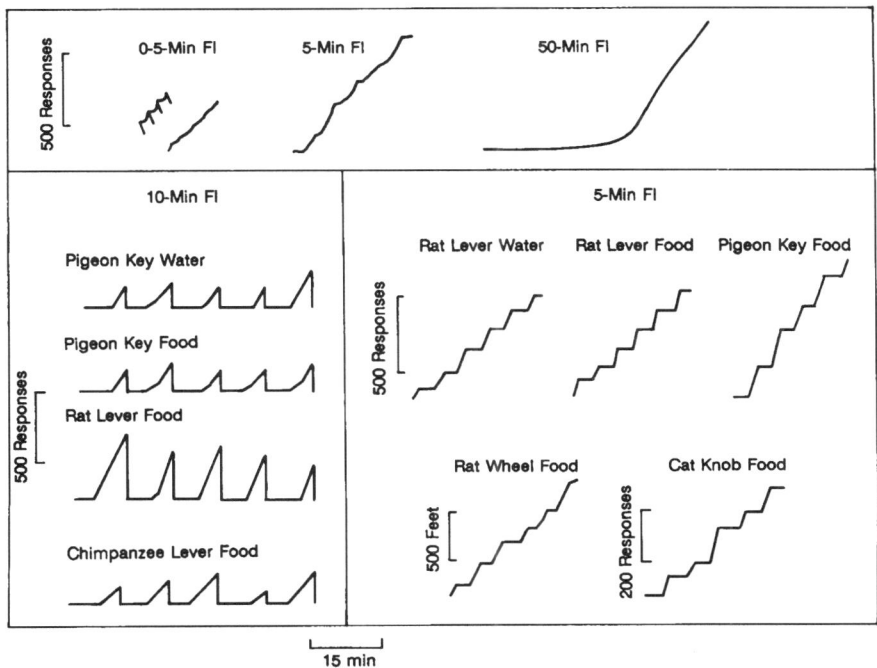

FIGURE 6 Generality of fixed-interval (FI) performance under various conditions. The ordinate shows the cumulative number of responses, whereas time is represented on the abscissa. The typical performance is characterized by little or no responding after reinforcement, followed by a gradually accelerating rate of responding. In all examples, a fixed-interval schedule of food or water presentation was in effect. The lower left frame shows performance under a 10-minute fixed-interval schedule. Each reinforcement delivery resets the recording pen to the baseline. The comparability of performance of the pigeon, rat, and chimpanzee, either pecking a key or pressing a lever, is evident. Likewise, during a 5-minute fixed-interval schedule (lower right frame), the comparability of performance of rats, pigeons, and cats pecking a key, pressing a lever, wheel-running, or pulling a knob is illustrated.

SOURCE: From Kelleher and Morse (1969).

of errors and slower decline in error rates in the patients with Alzheimer's are evident. Thus, although admittedly difficult to train, the response sequence paradigm shows comparable baseline performance across species and evidences sensitivity both to drugs and to neurodegenerative disease. Other learning and memory paradigms with direct cross-species applicability include procedures such as delayed alternation and delayed matching to sample. Obviously, the use of such behavioral tasks in the case of human evaluation may not be easily implemented

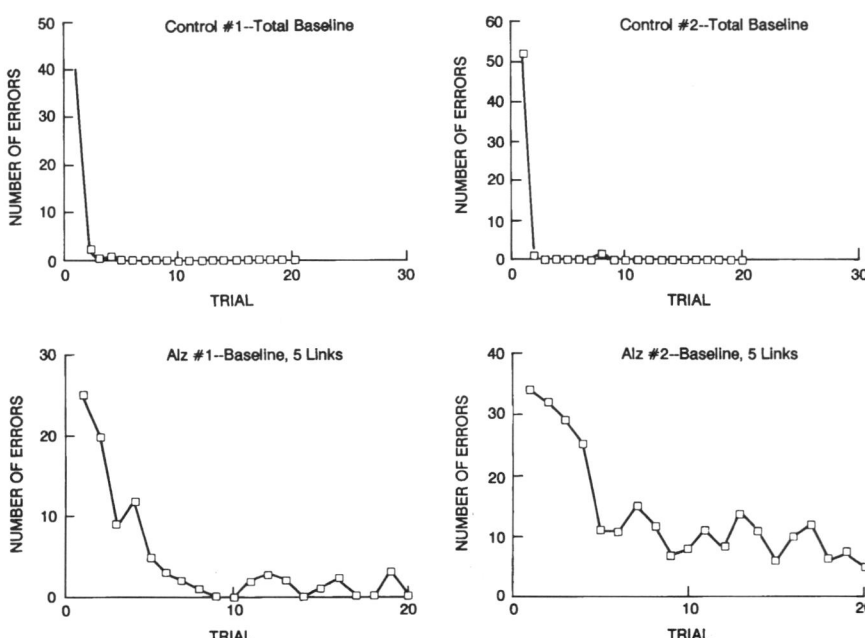

FIGURE 7 Performance of normal elderly humans (top panels) and Alzheimer's patients (Alz; bottom panels) working on a five-link repeated acquisition paradigm. Total number of errors over successive trials of a session are presented. The rapid decline in errors during initial trials of the session and the low subsequent errors over subsequent trials of normals are in stark contrast to the slower initial decline in errors and the higher sustained rate of errors over the remainder of the session in the Alzheimer's patients.

SOURCE: From Gershensen, Thompson, and Gisselquist (personal communication, 1988).

when large sample populations are involved. However, one initial alternative is to study more intensively a smaller proportion of the more highly exposed individuals within the population.

Although multidisciplinary studies pave the way for establishing biological mechanisms, most stop short of experimentally evaluating and defining the nature of the connection between various parameters of neurotoxicity. For example, a toxicant may induce changes in some aspect of behavior, as well as decrease the levels of some neurotransmitter. It is common to hypothesize a causal relationship between the two, but far less frequent to test such a hypothesis directly. More systematic research exploring the relationships between behavioral toxicity and other parameters of neurotoxicity would expedite our understanding of the biological bases of toxicant-induced behavioral effects.

Finally, it should be remembered that both experimental animal and human behavioral toxicology studies have common goals: to understand the mechanisms by which toxicants affect behavior and to implement procedures to screen for neurotoxic properties of chemicals. A better base of communication and interfacing between human and experimental animal research would accelerate progress on both fronts, allowing each to benefit from advances made by the other.

REFERENCES

Angell, N. F. and B. Weiss. 1982. Operant behavior of rats exposed to lead before or after weaning. Toxicol. Appl. Pharmacol. 63:62–71.

Anger, W. K., and J. V. Setzer. 1979. Effects of oral and intramuscular carbaryl administration on repeated chain acquisition in monkeys. J. Toxicol. Environ. Health 5:793–808.

Avery, D. D., and H. A. Cross. 1974. The effects of tetraethyl lead on behavior in the rat. Pharmacol. Biochem. Behav. 2:473–479.

Barthalmus, G. T., J. D. Leander, D. E. McMillan, P. Mushak, and M.R. Krigman. 1977. Chronic effects of lead on schedule-controlled pigeon behavior. Toxicol. Appl. Pharmacol. 42:271–284.

Bellinger, D., A. Leviton, C. Waternaux, H. Needleman, and M. Rabinowitz. 1987. Longitudinal analyses of prenatal and postnatal lead exposure and early cognitive development. N. Engl. J. Med. 316:1037–1043.

Boren, J. J. 1963. Repeated acquisition of new behavioral chains. Am. Psychol. 17:421.

Bornschein, R., D. Pearson and L. Reiter. 1980. Behavioral effect of mederate lead exposure in children and animal models: Part 2, Animal studies. (CRC Critical Revews in Toxicology 7:101–152.

Brady, K., Y. Herrera, and H. Zenick. 1975. Influence of parental lead exposure on subsequent learning ability of offspring. Pharmacol. Biochem. Behav. 3:561–565.

Brown, D. R. 1975. Neonatal lead exposure in the rat: Decreased learning as a function of age and blood lead concentration. Toxicol. Appl. Pharmacol. 32:628–637.

Brown, S., N. Dragann and W. H. Vogel. 1971. Effects of lead acetate on learning and memory in rats. Arch. Environ. Health 22:370–372.

Bullock, J. D., R. L. J. Wey, J. A. Zaia, I. Zarembok, and H. A.Schroeder. 1966. Effect of tetraethyl lead on learning and memory in the rat. Arch. Environ. Health 13:21–22.

Bushnell, P. J., and R. E. Bowman. 1979a. Reversal learning deficits in young monkeys exposed to lead. Pharmacol. Biochem. Behav. 10:733–742.

Bushnell, P. J., and R. E. Bowman. 1979b. Persistence of impaired reversal learning in young monkeys exposed to low levels of dietary lead. J. Toxicol. Environ. Health 5:1015–1023.

Bushnell, P. J. R. E. Bowman, J. R. Allen, and R. J. Marler. 1977. Scotopic vision deficits in young monkeys exposed to lead. Science 196:333–335.

Carson, T. L. 1976. Auditory and Visual Discrimination Learning in Sheep Prenatally and Postnatally Exposed to Lead. Ph.D. thesis, Iowa State University, Ames.

Carson, T. L., G. A. Van Gelder, G. C. Karas, and W. B. Buck. 1974. Slowed learning in lambs prenatally exposed to lead. Arch. Environ. Health 29:154–156.

Cory-Slechta, D. A. 1984. The behavioral toxicity of lead: Problems and perspectives. Pp. 211–255 in Advances in Behavioral Pharmacology, Vol. 4, T. Thompson, P. B. Dews, and J. E. Barrett, eds. New York: Academic Press.

Cory-Slechta, D. A. 1986. Effects of lead exposure beyond weaning on fixed ratio performance. Neurobehav. Toxicol. Teratol. 8:237–244.

Cory-Slechta, D. A. 1989. The lessons of lead for behavioural toxicology. In Lead Exposure and Child Development: An International Assessment. M. A. Smith, L. D. Grant, and A. I. Sors, eds. Dordrecht: Kluwer Academic Publishers.

Cory-Slechta, D. A. and T. Thompson. 1979. Behavioral toxicity of chronic postweaning lead exposure in the rat. Toxicol. Appl. Pharmacol. 47:151–159.

Cory-Slechta, D. A., B. Weiss, and C. Cox. 1983. Delayed behavioral toxicity of lead with increasing exposure concentration. Toxicol. Appl. Pharmacol. 71:342–352.

Cory-Slechta, D. A., B. Weiss, and C. Cox. 1985. Performance and exposure indices of rats exposed to low concentrations of lead. Toxicol. Appl. Pharmacol. 78:291–299.

Dews, P. B., and G. Wenger. 1974. Rate dependency of the behavioral effects of amphetamine. Pp. 167–227 in Advances in Behavioral Pharmacology, Vol. 1, T. Thompson and P. B. Dews, eds. New York: Academic Press.

Dietz, D. D., D. E. McMillan, and P. Mushak. 1979. Effects of chronic lead administration on acquisition and performance of serial position sequences by pigeons. Toxicol. Appl. Pharmacol. 47:377–384.

Driscoll, J. W. and S. E. Stegner. 1976. Behavioral effects of chronic lead ingestion on laboratory rats. Pharmacol. Biochem. Behav. 4:411–417.

Fox, D. A., and L. Chu. 1988. Rods are selectively altered by lead: II. Ultrastructure and quantitative histology. Exp. Eye Res. 46:613–625.

Fox, D. A., and D. B. Farber. 1988. Rods are selectively altered by lead: I. Electrophysiology and biochemistry. Exp. Eye Res. 46:597–611.

Fox, D. A., A. A. Wright, and L. G. Costa. 1982. Visual acuity deficits following neonatal lead exposure: Cholinergic interactions. Neurobehav. Toxicol. Teratol. 4:689–693.

Fulton, M., G. Thompson, R. Hunter, G. Raab, D. Laxen, and W. Hepburn. 1987. Influence of blood lead on the ability and attainment of children in Edinburgh. Lancet 1:1221–1226.

Geist, C. R., and B. R. Mattes. 1979. Behavioral effects of postnatal lead acetate exposure in developing laboratory rats. Physiol. Psychol. 7:399–402.

Hastings, L., G. Cooper, R. L. Bornschein, and I. A. Michaelson. 1977. Behavioral effects of low level neonatal lead exposure. Pharmacol. Biochem. Behav. 7:37–42.

Hastings, L., G. Cooper, R. L. Bornschein, and I. A. Michaelson. 1979. Behavioral deficits in adult rats following neonatal lead exposure. Neurobehav. Toxicol. 1:227–231.

Holland, J. G. 1958. Human vigilance. Science 128:61–67.

Kelleher, R. T., and W. H. Morse. 1969. Determinants of the behavioral effects of drugs. Pp. 383–405 in Importance of Fundamental Principles in Drug Evaluation, D. J. Tedeschi and R. E. Tedeschi, eds. New York: Raven Press.

Klein, A. W., R. T. Louis-Ferdinand, and D. R. Brown. 1977. Measurement of behavioral, chemical and anatomical changes in the developing rat brain from exposures to low levels of lead. Anat. Rec. 187:626.

Lanthorn, T. and R. L. Isaacson. 1978. Effects of chronic lead ingestion in adult rats. Physiol. Psych. 6:93.

Laties, V. G. 1975. The role of discriminative stimuli in modulating drug action. Fed. Proc. 34:1880–1888.

Laties, V. G., and B. Weiss. 1963. Effects of a concurrent task on fixed-interval responding in humans. J. Exp. Anal. Behav. 6:431–436.
Laughlin, N., R. E. Bowman, E. D. Levin, and P. J. Bushnell. 1983. Neurobehavioral consequences of early exposure to lead in rhesus monkeys: Effects on cognitive behaviors. Pp. 497–516 in Reproductive and Developmental Toxicity of Metals, T. W. Clarkson, G. F. Nordberg and P. R. Sager, eds. New York: Plenum.
Levin, D. E., and R. E. Bowman. 1983. The effect of pre- or post-natal lead exposure on Hamilton search task in monkeys. Neurobehav. Toxicol. Teratol. 5:391–394.
Moerschbacher, J. M., and D. M. Thompson. 1980. Effects of d-amphetamine, cocaine, and phencyclidine on the acquisition of response sequences with and without stimulus fading. J. Exp. Anal. Behav. 33:369–381.
Nation, J. R., A. E. Bourgeois, and D. E. Clark. 1983. Behavioral effects of chronic lead exposure in the adult rat. Pharmacol. Biochem. Behav. 18:833–840.
Needleman, H. D., C. E. Gunnoe, A. Leviton, R. Reed, H. Peresie, C. Maher, and P. Barrett. 1979. Deficits in psychologic and classroom performance of children with elevated lead levels. N. Engl. J. Med. 300:689–695.
Ogilvie, D. M. 1977. Sublethal effects of lead acetate on the Y maze performance of albino mice (Mus musculus L.). Can. J. Zool. 55:771–775.
Ogilvie, D. M. 1978. Effect of lead acetate on memory in mice. Bull. Environ. Contam. Toxicol. 19:143–146.
Overmann, S. R. 1977. Behavioral effects of asymptomatic lead exposure during neonatal development in rats. Toxicol. Appl. Pharmacol. 41:459–471.
Padich, R., and H. Zencik. 1977. The effects of developmental and/or direct lead exposure on FR behavior in the rat. Pharmacol. Biochem. Behav. 6:371–375.
Paule, M. G., and D. E. McMillan. 1986. Effects of trimethyltin on incremental repeated acquisition (learning) in the rat. Neurobehav. Toxicol. Teratol. 8:245–254.
Penzien, D. A., D. R. C. Scott, and J. P. Motiff. 1982. The effects of lead toxication on learning in rats. Arch. Environ. Health 37:83–87.
Pollard, G. T., S. T. McBennet, K. V. Rohrbach, and J. L. Howard. 1981. Repeated acqusition of three-response chains for food reinforcement in the rat. Drug Develop. Res. 1:67–75.
Rice, D. C. 1984. Behavioral deficits (delayed matching to sample) in monkeys exposed from birth to low levels of lead. Toxicol. Appl. Pharmacol. 75:337–345.
Rice, D. C. 1985. Chronic low-level lead exposure from birth produces deficits in discrimination reversal in monkeys. Toxicol. Appl. Pharmacol. 77:201–210.
Rice, D. C. 1988. Schedule-controlled behavior in infant and juvenile monkeys exposed to lead from birth. Neurotoxicol. 9:75–88.
Rice, D. C, and S. G. Gilbert. 1985. Low lead exposure from birth produces behavioral toxicity (DRL) in monkeys. Toxicol. Appl. Pharmacol. 80:421–426.
Rice, D. C., and R. F. Willes. 1979. Neonatal low-level lead exposure in monkeys (Macaca fascicularis): Effect on two choice non-spatial form discrimination. J. Environ. Pathol. Toxicol. 2:1195–1203.
Rice, D. C., S. G. Gilbert, and R. F. Willes. 1979. Neonatal low-level lead exposure in monkeys: Locomotor activity, schedule controlled behavior, and the effects of amphetamine. Toxicol. Appl. Pharmacol. 51:503–513.
Richelle, M. 1969. Combined action of diazepam and d-amphetamine on mixed interval performance in cats. Exp. Anal. Behav. 12:989–998.
Rosen, J. B., R. F. Berman, F. C. Beuthin, and R. T. Louis-Ferdinand. 1985. Age of testing as a factor in the behavioral effects of early lead exposure in rats. Pharmacol. Biochem. Behav. 23:49–54.
Schroeder, S. R., and B. Hawk. 1986. Child-caregiver environmental factors related to lead exposure and IQ. In Toxic Substances and Mental Retardation, S. R. Schroeder, ed. AAMD Monograph, Washington, D.C.

Schroeder, S. R., B. Hawk, D. A. Otto, P. Mushak, and R. E. Hicks. 1985. Separating the effects of lead and social factors on IQ. Environ. Res. 38:144–154.

Schrott, J., J. R. Thomas, and R. A. Banaurd. 1980. Repeated acquisition of four-member response sequences in rats. Psych. Reports 47:503–509.

Snowdon, C. J. 1973. Learning deficits in lead-injected rates. Pharmacol. Biochem. Behav. 1:599–603.

Suessbrick, A. 1983. A procedure for studying repeated learning by mentally retarded individuals. Dissertation Abstracts International 43(8-B):2733.

Tews, P. A., and M. W. Fischman. 1982. Effects of d-amphetamine and diazepam on fixed-interval, fixed-ratio responding in humans. J. Pharmacol. Exp. Therap. 221:373–383.

Thompson, D. M. 1980. Selective antagonism of the rate-decreasing effect of d-amphetamine by chlorpromazine in a repeated acquisition task. J. Exp. Anal. Behav. 34:87–92.

Thompson, D. M., and J. M. Moerschbacher. 1979. Drug effects on repeated acquisition. Pp. 229–259 in Advances in Behavioral Pharmacology, Vol. 2, T. Thompson and P. B. Dews, eds. New York: Academic Press.

Thompson, D. M., J. Mastropaolo, P. J. Winsauer, and J. M. Moerschbacher. 1986. Repeated acquisition and delayed performance as a baseline to assess drug effects on retention in monkeys. Pharmacol. Biochem. Behav. 25:201–207.

Van Gelder, G. A., R. Carson, R. M. Smith, and W. B. Buck. 1973. Behavioral toxicologic assessment of the neurologic effect of lead in sheep. Clin. Toxicol. 6:405–418.

Wigg, N. R., G. V. Vimpani, A. J. McMichael, P. A. Baghurst, E. F. Robertson, and R. J. Roberts. 1988. Port Pirie Cohort Study: Childhood blood lead and neuropsychological development at age two years. J. Epid. Comm. Health 42:213–219.

Winneke, G., A. Brockhaus, and R. Baltissen. 1977. Neurobehavioral and systemic effects of long-term blood lead elevation in rats. I. Discrimination learning and open field behavior. Arch. Toxicol. 37:247–263.

Winneke, G., H. Lilienthal, and W. Werner. 1982. Task dependent neurobehavioral effects of lead in rats. Arch. Toxicol. Suppl. 5:84–93.

Yule, W., R. Landsdown, I. B. Millar, and M. A. Urbanowicz. 1981. The relationship between blood lead concentrations, intelligence and attainment in a school population: A pilot study. Dev. Med. Child. Neurol. 23:567–576.

Yule, W., M. A. Urbanowicz, R. Lansdown, and I. B. Millar. 1984. Teachers' ratings of children's behavior in relation to blood lead levels. Br. J. Dev. Psychol. 2:295–305.

Zenick, H., R. Padich, T. Takarek, and P. Aragon. 1978. Influence of prenatal and postnatal lead exposure on discrimination learning in rats. Pharmacol. Biochem. Behav. 8:347–350.

Zenick, H., W. Rodriquez, J. Ward, and B. Elkington. 1979. Deficits in fixed-interval performance following prenatal and postnatal lead exposure. Dev. Psychobiol. 12:509–514.

Methods and Issues in Evaluating the Neurotoxic Effects of Organic Solvents

Beverly M. Kulig

The term "organic solvents" refers to a chemically heterogeneous class of compounds and mixtures which are typically liquid between 0 and 250° C and are used industrially to extract, dissolve, or suspend materials not soluble in water. Solvents are derived from many chemical classes and are present in products such as paints, glues, adhesives, coatings, and degreasing agents. Solvents are also used to manufacture a wide variety of products including polymers, dyes, plastics, textiles, printing inks, agricultural products, and pharmaceuticals (World Health Organization, 1985).

Because of the widespread application of these compounds in industrial products and processes, the volume of organic solvents produced and the number of workers exposed are considerable. In the United States, for example, some 49 million tons is produced on an annual basis, and the National Institute for Occupational Safety and Health (NIOSH) estimates that approximately 9.8 million American workers are potentially exposed (NIOSH, 1987). Understandably, much of the concern regarding the effects of organic solvents focuses on the occupational setting where exposures can be relatively high. As a result, recommended occupational exposure limits for many of these compounds have been proposed (American Conference of Governmental Industrial Hygienists, 1986; NIOSH, 1987). Despite the emphasis on the occupational setting, however, solvent exposure is not limited to the workplace. Organic solvents are ubiquitous in the environment and are detectable in air, drinking water, and foodstuffs, including human milk (Packard, 1985).

Considerable controversy exists regarding the extent to which organic solvents are toxic to the nervous system and at what exposure levels (Cranmer and Goldberg, 1986; Grasso et al., 1984). For several compounds (e.g., *n*-hexane, methyl *n*-butyl ketone, and carbon disulfide), a sufficiently large data base has been established whereby the consequences of long-term overexposure can be evaluated (Spencer, 1985). For the vast majority of organic solvents, however, very few data regarding nervous system effects are available.

Until recently, the majority of animal studies examining the neurotoxicity of organic solvents concentrated either on the involvement of the nervous system in the acute lethality of these compounds or on the morphological changes that may result from chronic overexposure. Of growing concern, however, is the possibility that long-term solvent exposure is accompanied by subclinical changes which reflect a reduction in the adaptive capabilities of the nervous system. Occupational studies, for example, have suggested that long-term exposure may be associated with a number of neurobehavioral changes including psychomotor slowing, attention and memory impairments, and changes in affective behavior (Baker and Fine, 1986). Although occupational studies are an important source of information regarding the possible long-term effects of overexposure, results from such studies are often difficult to interpret. By applying animal neurobehavioral methods to the study of solvents, the possibility of addressing many of the questions regarding solvent neurotoxicity is becoming increasingly feasible. The present chapter considers some of the interpretative issues that have developed regarding the risks to human health associated with solvent exposure and examines the possible use of animal behavioral methods in this area.

ACUTE SOLVENT EFFECTS

The central nervous system (CNS) is a primary target organ for the acute toxic effects of organic solvents. The acute toxicity of inhaled solvent vapors is characterized, both in animals and in humans, by reversible signs of CNS depression. At moderate levels of overexposure, human subjects often complain about nausea, incoordination, and feelings of intoxication. At sufficiently high exposure levels, unconsciousness and death by respiratory arrest can occur (NIOSH, 1987).

A number of factors contribute to the susceptibility of the CNS to the acute effects of organic solvents. First, during the initial stages of absorption of an inhaled dose of solvent vapors, distribution to different body organs is related to regional blood flow, and as a result, entry into the brain proceeds rapidly (Baker and Rickert, 1981). Further, because of their lipophilic nature, organic solvents accumulate in tissues with a high lipid content, thus making brain and nerve potential depots for these compounds and, in some cases, their toxic metabolites (Bus et al., 1981).

One experimental approach to determining behaviorally effective levels is to expose human volunteers to controlled atmospheres of solvent vapors and to examine changes in psychometric measures designed to assess different aspects of motor, sensory, and cognitive performance (Dick and Johnson, 1986). In general, the results of experimental exposure studies have demonstrated that one of the most consistent effects of inhaled organic solvents in humans is a slowing in behavioral performance evidenced by increased response latencies in simple and complex reaction time tasks (Gamberale, 1985). However, the results of such studies also seem to indicate that the level at which behavioral effects occur tends to be on the higher end of the scale for human occupational exposures. As pointed out by Dick and Johnson (1986), there are a number of difficulties in conducting human experimental exposure studies. In addition to the narrow dose-response range that can ethically be examined, the duration of exposure which is acceptable to human volunteer subjects is often less than that encountered in the occupational situation. Thus, human exposure studies rarely employ an 8-hour exposure regimen even at Threshold Limit Values (TLVs) and, in most instances, are restricted to 2-4 hours of actual exposure. This may explain, in part, the modest effects seen in human exposure studies and the limited dose-response data obtained.

Application of Animal Behavioral Methods

Acute Effects of Single Exposures

Animal studies using operant techniques are being employed increasingly to examine the effects of inhaled organic solvent vapors on behavioral performance, and guidelines for the use of schedule-controlled operant techniques for neurotoxicity evaluation have recently been proposed by the U.S. Environmental Protection Agency (USEPA, 1985). Because the aim of many of the studies employing operant techniques has been to demonstrate the usefulness of these methods in neurotoxicity screening, experimental protocols involving relatively high-level, short-duration exposure schedules have typically been employed (e.g., Glowa and Dews, 1983; Moser et al., 1985). Although such studies can provide highly reliable, quantitative information on which to judge the relative potency of different compounds to affect behavior in that animal test system, they also give the impression that very high concentrations of inhaled solvents are necessary to affect learned behavior in rodents. Perhaps as a result, very few animal studies have examined occupationally relevant concentrations and exposure durations for the purposes of risk assessment.

Although it is unclear at present just how sensitive measures of learned behavior are to the acute effects of inhaled solvents, there is some evidence that animal studies using low-level exposure may be a worthwhile approach for estimating concentrations at which behavioral effects can be expected to begin to occur in humans. Kishi and his coworkers (1988), for example, have recently demonstrated that inhalational exposure to toluene at 125 ppm for 4 hours produced significant effects on signaled avoidance both during the initial stages of exposure and for several hours following the end of exposure.

Results from acute exposure studies conducted in our laboratory using positive reinforcement also indicate measurable effects of solvents at occupationally relevant levels. In a recent study, for example, the effects of low aromatic white spirits were examined in rats working on a two-choice visual discrimination task for water reward. Rats were first trained in operant chambers equipped with two levers, two light panels located above each lever, and a pump for delivering water reward in daily sessions consisting of 100 trials. The rat's task was to depress the lever under the illuminated panel in order to obtain a drop of water. Following stabilization of performance, rats were randomly assigned to one of four groups and exposed by inhalation to low aromatic white spirits at 0 (controls), 1,200 mg/m^3 (~200 ppm), 2,400 mg/m^3 (~400 ppm), and 4,800 mg/m^3 (~800 ppm) for 8 hours and tested immediately following the termination of exposure.

As the left panel in Figure 1 demonstrates, exposure to white spirits at these concentrations produced no observable effects on discrimination accuracy. Speed of responding (right panel), however, was affected at all concentrations tested, with exposure to 1,200 mg/m^3, 2,400 mg/m^3, and 4,800 mg/m^3 producing a mean increase in trial response latency of 47, 96, and 156 percent, respectively. The effects on two-choice response speed did not appear to be the result of changes in motivation for water reward: All groups consumed the same number of reinforcements and the latency to obtain reinforcement following a correct trial response was similar for all groups.

Although only several human experimental exposure studies have been conducted with white spirits and the length of exposure in the human studies was considerably less than 8 hours, the data that are

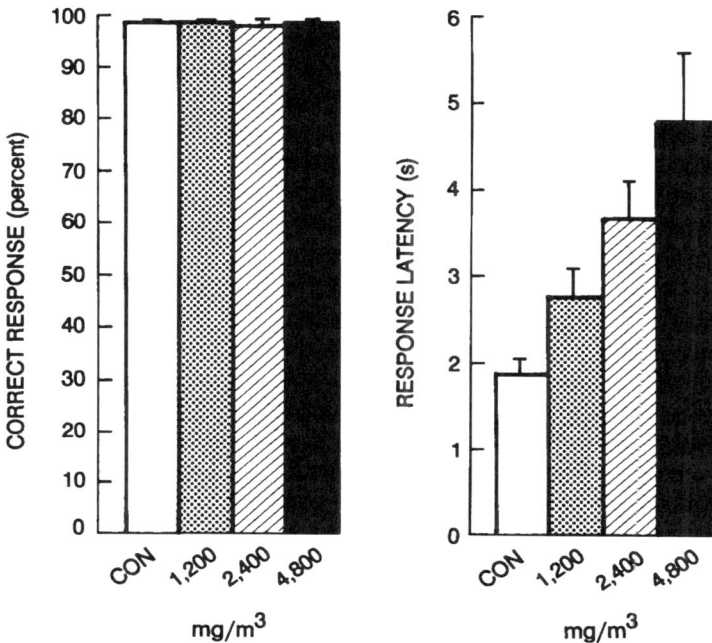

FIGURE 1 Effects of a single 8-hour exposure to white spirits on accuracy (left panel) and response speed (right panel) of rats performing on a two-choice visual discrimination task.

available indicate relatively good agreement between the levels of exposure producing changes in learned performance in both types of studies. For example, in studies conducted with young healthy male volunteers, Gamberale and his coworkers (1975) reported that exposure to white spirits at 4,000 mg/m^3 for 50 minutes produced a small (i.e., 10 ms) but significant increase in response latencies measured in a simple reaction time task. Further, in a study employing a total exposure duration of 7 hours and concentration levels of 34, 100, 200, and 400 ppm, significant effects on response speed measures were found at 100 ppm and higher, with effects on other cognitive and motor tasks appearing at higher concentrations (Cohr et al., 1980).

Thus, although the number of studies aimed at examining the acute behavioral effects of low-level solvent exposure is limited, the data available suggest that the rat may prove to be a more useful model for estimating human observable effect levels than might be expected on the basis of the results obtained in high-level, short-duration exposure studies.

Acute Effects in the Context of Chronic Exposures

Except for instances of accidental poisoning in which lethal solvent concentrations are reached, all signs of CNS depression, even those resulting from a single high-level intoxication, appear to be readily reversible and no evidence exists to indicate that acute solvent exposure is accompanied by neuropathological or persistent neurofunctional sequelae. As a result, acute solvent effects are usually mentioned only in passing in discussions of solvent neurotoxicity (Baker and Fine, 1986; Grasso et al., 1984; Spencer, 1985). Different lines of evidence from the human literature, however, indicate that a need exists for more careful consideration of acute solvent effects particularly in the context of chronic exposure.

First, there is a considerable amount of evidence from the occupational literature demonstrating deficits in behavioral functioning in exposed workers. Discussions of these effects often imply that because neurobehavioral effects were measured in persons exposed on a chronic basis, the effects themselves are chronic in nature. Behavioral testing in occupational studies, however, is often conducted in workers who have just left an acute exposure situation (i.e., the worksite) or in workers tested within a day or two following an exposure period (e.g., the work week). Thus, it is not unlikely that acute effects could contribute to the changes in behavior often reported in the human literature.

Further, there are indications from the human literature that sensitivity to the acute effects of solvents can change in a repeated exposure situation and that such changes may be important in monitoring the potential hazards of these compounds. On one hand, there are both anecdotal and experimental reports indicating that tolerance develops to the effects of organic solvents, i.e., that acute effects are attenuated with repeated exposures (Gotell et al., 1972). On the other hand, patients with suspected solvent-induced toxic encephalopathy often complain of an increased sensitivity to the acute effects of solvents. In such patients, signs of acquired intolerance are usually characterized by dizziness and nausea when they are exposed to even very low concentrations, despite the fact that they had been occupationally exposed for years to higher concentrations without subjective symptoms (Gyntelberg et al., 1986).

From the vast literature on the development of tolerance to ethyl alcohol and other CNS depressants, a number of behavioral paradigms are available for examining tolerance development to industrial solvents as well. Himnan (1984), for example, demonstrated rapid development of tolerance to the effects of repeated short-duration, high-level toluene

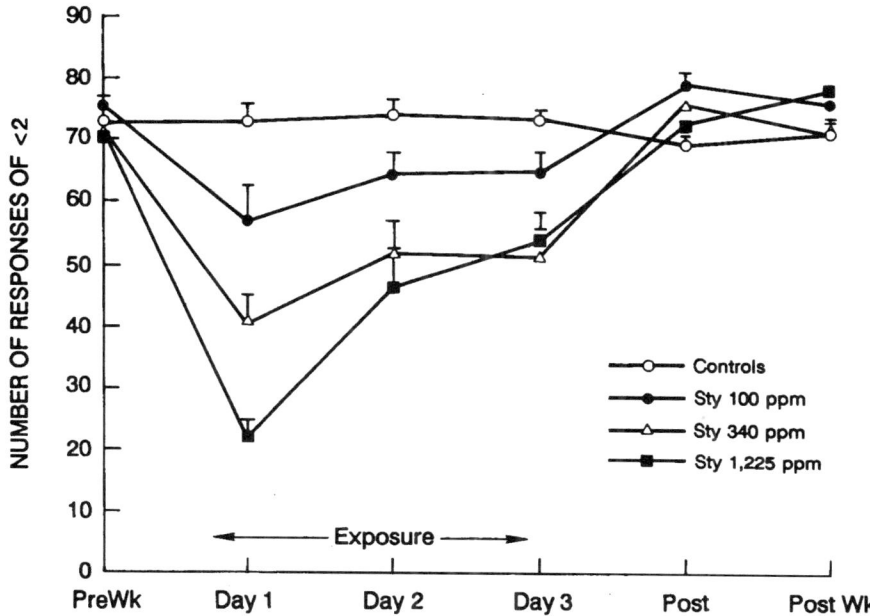

FIGURE 2 Effects of repeated inhalational exposure to styrene on the number of short-latency two-choice responses of rats performing on a visual discrimination task.

exposures on measures of ataxia and rearing. However, the development of tolerance appeared to depend on the behavioral functions examined, with increased activity and head shakes showing a slowly developing reverse tolerance.

In studies investigating the effects of styrene on learned discrimination performance, we have also found a rapidly developing tolerance to the effects of styrene on speed and accuracy measures of discrimination performance during the first week of exposure. As Figure 2 shows, rats exposed to 100, 350, and 1,225 ppm of styrene for 18 hours a day and tested on a visual discrimination task (described above) all showed a reduction on Day 1 in the number of two-choice responses made within 2 s of trial onset. However, as exposure continued, the effects of styrene on short-latency responding, particularly in the highest concentration group, showed a marked attenuation. Further, results from a chronic exposure study (Kulig, 1988) indicated that the rapidly developing tolerance to styrene seen in the first week of exposure persisted throughout chronic exposure.

In contrast, results from studies examining the effects of trichloroethylene (TCE) on discrimination indicated a quite different profile

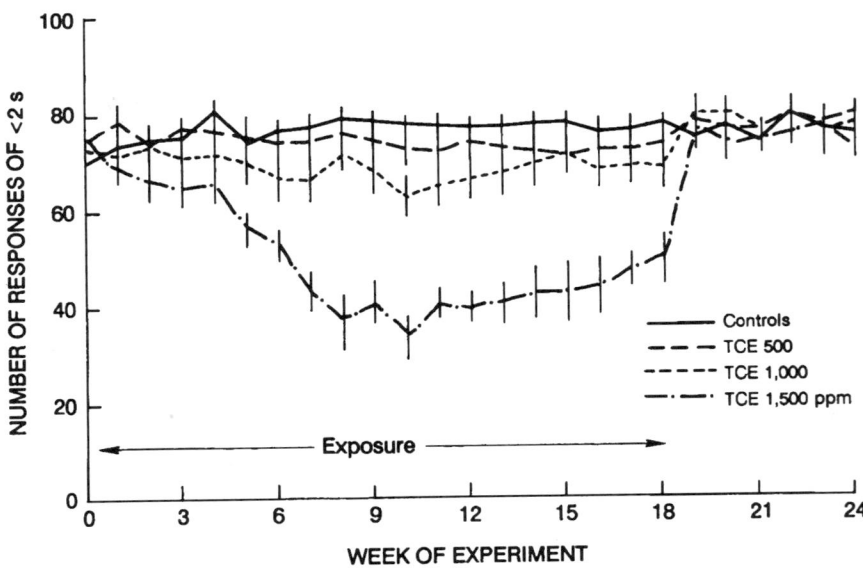

FIGURE 3 Changes in the acute effects of trichloroethylene (TCE) on the number of short-latency two-choice responses of rats during and following 18 weeks of exposure. Redrawn with permission from Kulig, 1987, Pergamon Press PLC.

in the time course of effects during chronic exposure. Groups of rats were exposed to TCE by inhalation at 0, 500, 1,000 and 1,500 ppm for 16 hours a day, 5 days per week, for 18 weeks (Kulig, 1987). Similar to styrene, exposure to TCE led to a within-week development of tolerance to the effects of TCE on response speed of discrimination performance but only during initial stages of exposure. As exposure became chronic, within-week tolerance was lost, with the result that the acute effects of TCE on response speed became more pronounced as exposure continued (Figure 3). Despite the very marked disturbances seen in the behavior of rats chronically exposed to TCE, the effects were apparently acute in nature because no evidence for a carryover of effects into the postexposure period could be demonstrated.

Taken together, the results of our studies as well as those of other investigators indicate that different profiles of tolerance and reverse tolerance develop for different behavioral effects and for different organic solvents. Given the indications from the human literature that similar phenomena also occur in exposed workers, and the problems associated with the interpretation of the nature of the deficits seen in occupational behavioral studies, it appears that animal studies examining acute effects in the context of the chronic exposure situation might provide a worthwhile approach to resolving some of these issues.

Reinforcing Properties of Organic Solvents

Unlike heavy metals and pesticides, organic solvents possess a behavioral property that is not seen with other industrially used chemicals, namely, the potential for abuse. Although it may be argued that possible self-exposure to organic solvents for their euphoric effects has little to do with evaluating the neurotoxicity of these compounds, the reinforcing effects of these compounds and the possible cross-tolerance with recreational drugs such as ethyl alcohol are important considerations in evaluating the potential risk to human health posed by these chemicals both in exposed workers and in the general population. Similar to the effects of ethyl alcohol, acute high-level exposure to solvents can produce feelings of euphoria, and intentional inhalational solvent abuse has become a serious health problem, particularly among children and adolescents (Johnston et al., 1984). The most commonly abused products include glues, paints and paint thinners, gasoline, lighter refills, and cleaning fluids (King, 1982).

Although toluene-containing products appear to be particularly popular, the chemical diversity of the different compounds which are abused suggests that the potential for abuse is not limited to a single compound. In addition to the social, psychological, and economic consequences of any addiction, the long-term abuse of solvent-containing products has been associated with a number of neurotoxic effects, including psychosis, hallucinations, sensory and motor disturbances, and convulsions. Although the acute encephalopathy produced by solvent inhalant abuse appears reversible in most cases, reports indicate that in some patients severe CNS effects may persist indefinitely (Boor and Hurtig, 1977; Grabsky, 1961; King, 1982; Knox and Nelson, 1966; Satran and Dodson, 1963; Weisenberger, 1977). Moreover, although children appear to be the high-risk group in the general population, solvent abuse can occur in the occupational setting as well, and many of the case studies reporting persistent CNS effects have involved adults whose initial experience with the euphoric effects of organic solvents occurred in the occupational setting.

Systematic animal studies of the relative abuse potential of different organic solvents have not yet been conducted; however, animal models for examining the stimulus properties of these chemicals have been described. Nonhuman primates, for example, will self-administer inhaled vapors (Wood, 1979). Further, in studies using rodents, drug discrimination procedures have been used to evaluate the stimulus properties associated with acute intoxication (Overton, 1984). Rees et al. (1987) have shown the similarity in stimulus properties of toluene and barbiturates. If similar results are found with other solvents that are abused by humans, it may be possible to develop a systematic framework for evaluating the abuse potential of industrial and commercial compounds.

The issue of abuse is so different from the usual concerns of toxicology, neuropathology, and occupational medicine that one may question whether this unique behavioral property of solvents belongs in the toxicological picture of risk assessment. When one considers, however, that compounds producing acute euphoric effects may lead to repeated near-lethal (and sometimes lethal) exposure situations that result in irreversible brain damage, an evaluation of which solvents possess the potential for abuse would seem warranted.

CHRONIC SOLVENT EFFECTS ON THE PERIPHERAL NERVOUS SYSTEM

The Role of Animal Studies

The effects of organic solvents on the peripheral nervous system provide some of the strongest evidence for the potential of these compounds to produce irreversible nervous system damage, and animal studies have played an important role in both helping to identify the causative agent in outbreaks of human disease (Allen, 1980) and elucidating the underlying mechanisms of action. In what has now become an almost classic example of neurotoxicological detective work, animal studies, initiated following an outbreak of peripheral neuropathy in a plastics coating plant in Columbus, Ohio in 1973, not only helped identify methyl *n*-butyl ketone (MnBK) as the causative agent, but also demonstrated the role of the solvent methyl ethyl ketone in causing the outbreak. Further, animal experimental studies helped clarify the role of the gamma-diketone pathway in the neurotoxicity of MnBK, identified hexacarbons as a general class of potential neurotoxic agents, and stimulated research into the pathological processes involved in dying back neuropathies (see Spencer and Schaumberg, 1980).

In addition, human data together with animal experimental studies have identified solvents other than those involved in the gamma-diketone pathway as toxic to peripheral nerve. Carbon disulfide, for example, is a metabolic poison which produces a wide array of effects including psychosis and peripheral neuropathy in man and lesions in the brain and peripheral nerve in experimental animals (Wood, 1981). Exposure to trichloroethylene (TCE), a compound used extensively in degreasing operations, was originally thought to be responsible for causing cranial neuropathies. However, when TCE is exposed to light or heat, it breaks down easily into dichloroacetylene. Animal studies helped establish the neurotoxicity of dichloroacetylene, and it is this compound that is now generally recognized as the causative agent in producing TCE-induced trigeminal neuropathy (Spencer, 1985).

More recently, 2-*tert*-butylazo-2-hydroxy-5-methylhexane (BHMH), a solvent used in the manufacture of polyester-based plastics, was removed from the market when central and peripheral nervous system dysfunction developed in workers after several weeks of exposure (Horan et al., 1985). Although premarket testing demonstrated the acute CNS toxicity of this compound in rodents, no chronic neurotoxicity studies were carried out. Following the outbreak of human disease at a manufacturing plant in Texas, animal studies were undertaken which showed that dermal exposure to BHMH produced functional signs of peripheral neuropathy within three weeks of exposure (Spencer et al., 1985). In addition, neuropathological studies indicated that BHMH exposure produced axonal degeneration of the optic tracts, ascending and descending spinal tracts, and peripheral nerve (Spencer et al., 1985). Although BHMH is a six-carbon straight-chain structure, its decomposition does not seem to involve the gamma-diketone pathway (Horan et al., 1985) as with *n*-hexane or methyl *n*-butyl ketone. It is, however, a potent neurotoxic agent both in animals and in humans. What, of course, is particularly unfortunate in the case of BHMH is that the human disease resulting from exposure to this compound need not have occurred if adequate premarket animal testing had been conducted.

Neurobehavioral Methods for Screening Sensory and Motor Effects

Although morphological evidence of nervous system changes has historically been the accepted endpoint in determining neurotoxicity, there is increasing interest in quantitative functional measures that could be used in conjunction with neuropathological evaluations for screening purposes (Buckholtz and Panem, 1986). In the United States, for example, the U.S. Environmental Protection Agency, under the Toxic Substances Control Act, has published guidelines for the use of behavioral methods in neurotoxicity testing (USEPA, 1985), and at an international level, the World Health Organization (WHO) is currently sponsoring various activities in the field of neurotoxicity (WHO, 1986).

Because of the well-documented potential of organic solvents and other industrial compounds to affect sensory and motor function in humans, many of the neurobehavioral techniques thus far developed at the animal level have concentrated on the quantitative assessment of these functional domains. Measurements of grip strength (Meyer et al., 1979), hindlimb splay (Edwards and Parker, 1977), walking patterns (De Medinacelli et al., 1982), coordinated movement (Kulig et al., 1985), and motor activity (Reiter, 1978) have all been successfully applied to evaluating the effects of chemical exposures on different aspects of motor function. Further, the quantitative measures of sensory thresholds thus far proposed also appear to be promising tools in the detection of neurotoxicity. Using a multisensory conditioned avoidance paradigm, for example, Pryor and his colleagues were able to uncover a neurotoxic effect not previously noted with other methods, namely, the ability of toluene, xylene, and styrene to produce irreversible high-frequency hearing loss in weanling rats (Pryor et al., 1984, 1987). Further, the prepulse inhibition of startle would also appear to be a good candidate for examining sensory threshold changes in rodents (Wu et al., 1985; Young and Fechter, 1983). In addition to behavioral evaluations of sensory function, electrophysiological techniques suitable for neurotoxicity evaluation have also been described (Rebert, 1983).

Although no consensus exists as to exactly which tests should be used, there is growing general agreement that comprehensive neurotoxicity assessment will require a battery of neurobehavioral tests aimed at assessing different sensory and motor functions. For the purposes of initial evaluation of a new compound, for example, the use of a standardized functional observational battery, simple tests of motor function, and automated methods of activity have been proposed (MacPhail, 1987; USEPA, 1985). For more comprehensive evaluations of neurotoxic potential of new compounds or for the purposes of risk assessment of compounds presently on the market, a combination of simple testing methods together with more sophisticated techniques would seem to be the most logical approach. Pryor and his colleagues (Pryor et al., 1983), for example, have demonstrated the utility of such an approach to study differences in the neurotoxic profiles of various compounds and to evaluate their relative neurotoxic potential.

In our own laboratory, a battery of tests was developed to examine different types of disturbances in sensory-motor function, including changes in spontaneous activity, grip strength, coordinated hindlimb movement, and peripheral nerve conduction velocity (Kulig, 1989). In this test battery, spontaneous activity is measured in an open field with an automated television camera and capacitance system for detecting both ambulation and rearing (Tanger et al., 1978). To measure fore- and hindlimb grip strength, a technique similar to that described by Meyer et al. (1979) is used. Changes in coordination are evaluated by using an automated television/microprocessor system capable of detecting and describing the placement and movement characteristics of one of the rat's hindpaws, as the rat moves its paw from one rung to the next as it walks along in a rotating motor-driven wheel (Kulig et al., 1985). Finally, the integrity of peripheral nerve function is assessed by measuring the peak latency and amplitude of the compound nerve action potential measured noninvasively from the caudal nerve with techniques similar to those described by Rebert and his colleagues (Rebert et al., 1983).

In order to evaluate whether these tests were sufficiently sensitive and reliable for use in extended exposure studies, the effects of carbon disulfide (CS_2) during 36 weeks of exposure were investigated. In this experiment, rats were exposed either to air or CS_2 to 75, 225, or 700 ppm for 8 hours per day, 5 days per week and examined at predetermined intervals by using the test battery described above.

As Figure 4 demonstrates, CS_2 produced a decreased level of ambulation throughout the course of exposure. Hindlimb grip strength

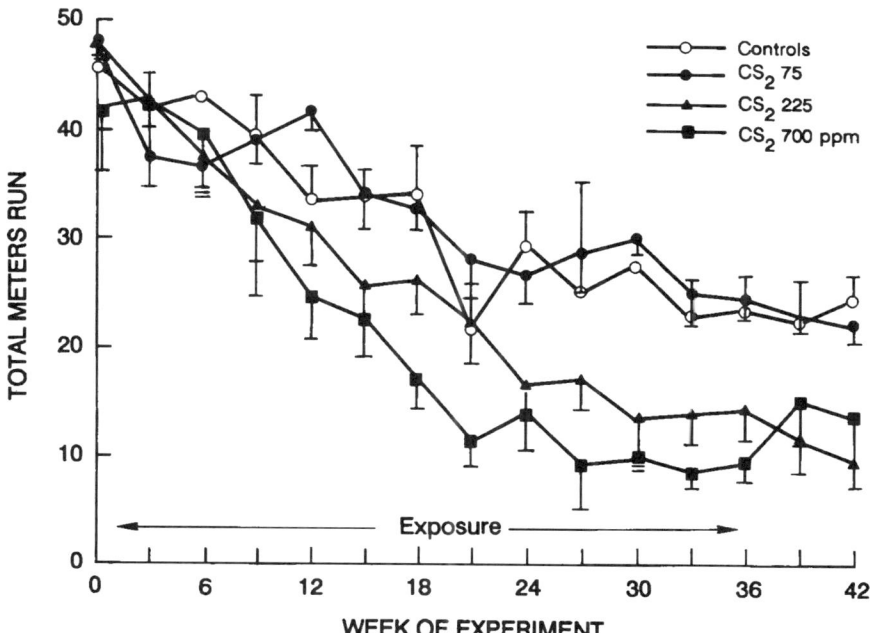

FIGURE 4 Effects of carbon disulfide on open field ambulation during and following 36 weeks of exposure.

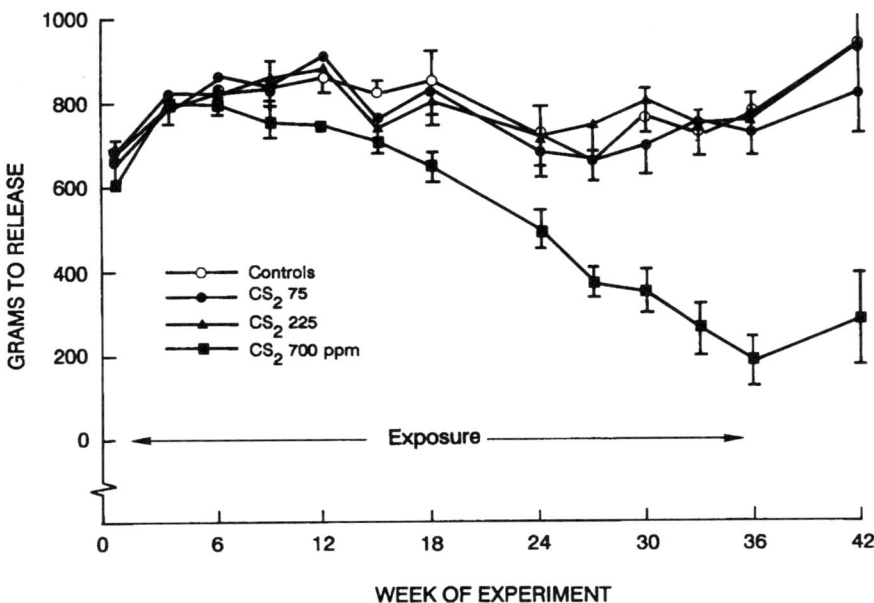

FIGURE 5 Effects of carbon disulfide on hindlimb grip strength during and following 36 weeks of exposure.

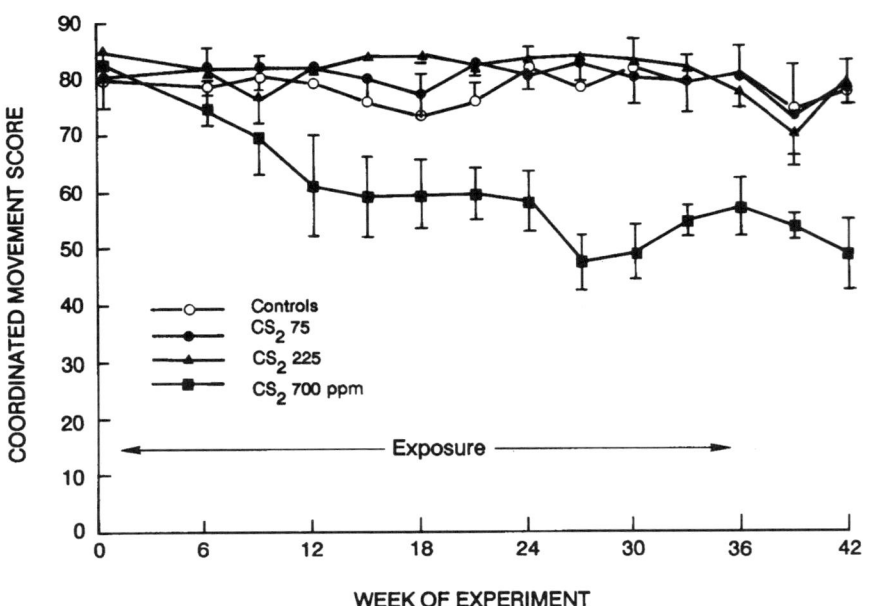

FIGURE 6 Effects of carbon disulfide on coordinated movement during and following 36 weeks of exposure.

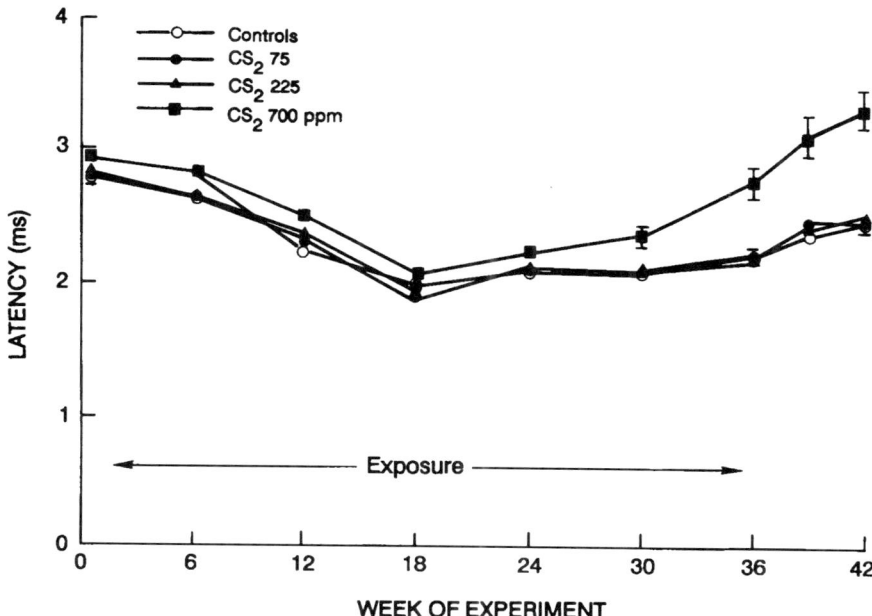

FIGURE 7 Effects of carbon disulfide on the latency of the compound nerve action potental measured from the caudal nerve during and following 36 weeks of exposure.

was also affected beginning in Week 9 of exposure (Figure 5) and was accompanied by deficits in coordinated movement (Figure 6). Furthermore, the behavioral evidence for disturbed peripheral nerve function was supported by electrophysiological changes in peripheral nerve conduction velocity (Figure 7). For all measures, CS_2-induced changes persisted well beyond the end of the exposure period, indicating that these effects were chronic in nature. In order to evaluate possible structural changes in these animals, neuropathological examinations were conducted by J. B. Cavanagh at the University of London at the termination of the study. Results indicated swollen axons and nerve fiber degeneration in the 700-ppm group in the sciatic, tibial, and caudal nerves as well as in the spinocerebellar tracts and the superior colliculus.

Taken together, these data demonstrate the ability of currently available neurobehavioral methods to quantify the progressive development of chemically induced changes in peripheral nerve function and to study the relationship between neurofunctional changes and the morphological changes. In addition to providing information regarding the time course of effects, the repeated testing of chronically exposed animals can also be used to operationally define appropriate time points for conducting morphological and chemical investigations. Further, the ability to quantitate progressive changes in nervous system function during a time in which no observable signs of dysfunction are evident is also an important consideration in the use of these methods for screening new compounds. Decisions regarding exposure schedules and the total duration of exposure, especially in inhalational studies, are often based on practical or rule-of-thumb considerations. However, there is nothing inherent in a 90-day exposure study with a particular exposure schedule to ensure the frank expression of observable signs of neurotoxicity or easy-to-identify light-microscopical changes. In such cases, small, but reliable, quantitative neurofunctional changes may be the only indication that closer examination of the compound using different exposure schedules and a longer duration of exposure is warranted.

Despite the advantages of neurobehavioral testing, an examination of control baseline performance in the CS_2 study demonstrates some of the considerations that must be taken into account in designing neurobehavioral methods for use in chronic exposure experiments. For example, decreased levels of behavior such as that seen in tests of spontaneous activity resulting from repeated testing or perhaps combined with the effects of aging in nonexposed control animals can lead to a floor effect and diminish the usefulness of the test in the latter stages of prolonged studies. Conversely, difficulties with caudal nerve conduction time measurements are more likely to occur during the early stages of exposure. Caudal nerve conduction time improves with age until 150–300 days after birth, and remains relatively stable until an advanced age when it again shows signs of prolongation (Schmelzer and Low, 1987). Because most chronic exposure studies begin when animals are young adults, the first months of exposure correspond to the time of greatest improvement in conduction velocity, making detection of compound-related effects during the early stages of exposure difficult to detect. The need to consider age-related changes is not unique to neurofunctional approaches to neurotoxicity evaluation because morphological and neurochemical changes can also be expected to occur. There is, however, a need in the further development of neurobehavioral methods to evaluate the long-term operating characteristics of any given test in order to better understand its strengths and limitations in the chronic exposure situation.

CHRONIC TOXIC ENCEPHALOPATHY

In addition to their effects on peripheral nerve, organic solvents have also been shown to produce irreversible effects on brain function. Examination, for example, of CS_2-poisoned workers conducted during the first 40 years of this century when occupational exposures were apparently very high, demonstrated the potential of this compound to produce severe manic-depressive psychosis and other signs of CNS dysfunction (see Wood, 1981). In later studies using psychological test instruments for quantifying the degree of psychological changes, Hanninen (1971) demonstrated the ability of these techniques to describe the range and pattern of cognitive and affective changes associated with exposure in workers with symptoms of CS_2 poisoning. In addition, compared to nonexposed control subjects, exposed workers with no subjective symptoms or clinical signs of overexposure also showed changes in psychological test performance. The finding of psychological changes in the absence of overt signs and symptoms together with epidemiological studies indicating that occupational exposure to carbon disulfide in the viscose rayon industry was associated with higher rates of suicide (Mancuso and Locke, 1972), was particularly disturbing because it raised the possibility that organic solvents as a general class of industrial compounds could produce cognitive and affective changes of a significant nature which were virtually undetectable by clinical methods and unrecognizable by the exposed person himself as being associated with chemical exposure.

As a consequence, an increasing number of studies were initiated to examine the psychological functioning of workers in different industries who were exposed to different types of organic solvents. In general, results from these studies repeatedly have shown a higher incidence of subjective complaints related to CNS effects in solvent-exposed workers, changes in objective measures of psychological functioning, and in some cases, a higher prevalence of EEG abnormalities and reduced peripheral conduction velocities (see Baker and Fine, 1986; WHO, 1985). In a series of studies examining persons occupationally exposed to mixed solvents, for example, painters and other occupational groups have been identified as being at risk for developing irreversible changes in brain function based, at least in part, on the results of behavioral evaluations (Arlien-Soborg et al., 1979; Bruhn et al., 1981).

As a result of the growing number of cross-sectional occupational studies demonstrating solvent-related neurofunctional changes and the number of case reports of chronic encephalopathy produced by solvent abuse, two workshops were convened to develop internationally acceptable diagnostic criteria applicable to solvent-induced CNS disease (Cranmer and Goldberg, 1986; WHO, 1985). Because toxic encephalopathy produced by nervous system poisons has been recognized as a clinical entity for many years, diagnostic criteria based on the DSM-III classification of mental disorders (American Psychiatric Association, 1980) served as a basis for differentiating three different levels of psychological impairment produced by chronic neurotoxic overexposure. The mildest level of impairment, termed "organic affective syndrome," was defined as one in which chronic chemical exposure was accompanied by subjective complaints of fatigue, mild memory and concentration difficulties, and affective changes. The term used to describe the second level of impairment was "mild chronic toxic encephalopathy"; it includes both subjective neurotoxic symptoms and sustained changes in personality or mood, as well as deficits in performance on formal neuropsychological testing. Finally, the third level of solvent-induced toxic encephalopathy refers to a severe neuropsychiatric condition characterized by global deterioration of intellectual and emotional functioning such as that described in the turn of the century literature for carbon disulfide poisoning or in present-day case studies of solvent abuse (Cranmer and Goldberg, 1986; NIOSH, 1987; WHO, 1985).

Problems in the Interpretation of Human Studies

The ability of drugs and chemicals to produce toxic encephalopathy is widely recognized, and there is little disagreement regarding the potential of high-level exposure to lead, thallium, alcohol, or drugs to produce severe signs of central nervous system poisoning that may be irreversible or only slowly reversible. The potential of organic solvents to produce persistent changes in brain function, however, has become, for a number of reasons, an issue of considerable controversy.

First, from the discussion above, it is obvious that acute effects can have important consequences for behavioral functioning. However, it is often difficult or impossible to design occupational studies in which the possibility of acute solvent effects contributing to changes in psychological performance has been ruled out. Given the fact that the test instruments sensitive to the effects of acute solvent exposures in the experimental exposure situation are often the same as those that are sensitive in detecting changes in psychological functioning in cross-sectional occupational studies (Gamberale, 1985), a differentiation of acute neurotoxicant effects from mild toxic encephalopathy based on the selection of the test instrument does not seem feasible.

Moreover, occupational environments often contain many different organic solvents, and workers are not necessarily aware of the level and type of their present or previous occupational exposures. Thus, conclusive proof as to the identification of the causative agent(s) producing psychological changes based on the results of these studies is often difficult to obtain.

Another issue in both the conduct and the interpretation of occupational behavioral studies is the selection of appropriate nonexposed control subjects who are suitably matched on the basis of age, sex, educational level, socioeconomic level, and other demographic variables known to affect psychological test performance. When such variables are taken into account in the statistical analyses of group differences of exposed and nonexposed workers, some studies have indicated that originally large differences in psychological test results can become borderline or disappear altogether (Baker et al., 1988; Cherry et al., 1985).

Problems stemming from the use of inappropriate control data are illustrated by a recent report by Gade and his colleagues (Gade et al., 1988). In this study, solvent-exposed workers were examined with clinical psychological tests and diagnosed as having solvent-induced chronic toxic encephalopathy. However, the original evaluation of these patients was apparently made without reference either to population norms for the determination of within-subject profiles of cognitive deficits or to estimates of premorbid levels of functioning, despite the fact that both types of information are necessary for a valid neuropsychological evaluation (Lezak, 1983). When these patients were retested at a later date and compared with nonexposed persons of similar age and education, no differences in psychological test performances could be seen. As a result, the authors were forced to revise their earlier diagnoses of solvent-induced dementia and to conclude that the poor test performance of these patients was related not to solvent exposure, but to the lower level of intelligence and education in their subject sample.

It is apparent from the discussion above that if appropriate controls are used, well-designed neurobehavioral studies can be used to evaluate acute neurotoxicant effects, to monitor the safety of workers exposed to known neurotoxic agents, and to identify possible occupational hazards. However, it is doubtful whether cross-sectional studies at the human level can or should be used as screening tools for the initial identification of compounds that possess CNS neurotoxic properties. In all fairness to the behavioral toxicologists and neuropsychologists working at the human level, they have received little help from their counterparts at the animal level in addressing the issues surrounding the possible adverse effects of long-term solvent exposures on cognitive functioning. In part, this may be due to the difficulty most psychologists working at the animal level have in evaluating and interpreting the sometimes diffuse effects reported in the human literature. However, it is more likely due to the lack of adequate test instruments for examining the effects of chemical exposures on learning, memory, and emotional functioning which can be applied to the chronic exposure situation.

Animal Models of Cognitive Effects

Although the study of learning and memory has occupied the interest of psychologists for many years, some of the paradigms developed to study memory function, such as one-trial passive avoidance learning, are obviously unsuitable for repeated evaluation of cognitive changes in chronic exposure studies. There are, however, techniques described in the literature which, with further study, may provide useful approaches to examining those behavioral processes that would seem to be most likely affected by exposure to centrally acting neurotoxic agents. Heise (1983), for example, has described several discrete-trial operant procedures involving delayed response and delayed comparison paradigms, which can be easily acquired by normal rats and can be used repeatedly to assess changes in memory. Recent studies employing radial arm maze techniques (Peele and Baron, 1988) also indicate that repeated acquisition paradigms may prove useful in assessing memory changes.

Even with the further development of behavioral methods to address more fully the issue of possible cognitive changes accompanying solvent exposure, it still remains to be seen whether these methods can provide sufficiently stable control baselines such as those needed for long-term studies. Moreover, if agent-related effects can be measured on cognitive performance in the absence of clear-cut neuropathological changes in appropriate brain structures, it will be necessary to seek possible underlying mechanisms of action either at the neurochemical level or with morphological techniques more sensitive than those that are used routinely for neurotoxicity assessment.

One approach that may prove fruitful is the combined study of neurobehavioral and neurochemical changes accompanying long-term solvent exposure. Investigators at the Karolinska Institute examining the effects of low-level (80 ppm) toluene exposure, for example, have recently reported reductions in catecholamine turnover rates in rat striatum and increased catecholamine levels in hypothalamus (Fuxe et al., 1982), as well as changes in central receptor binding properties (Fuxe et al., 1987) during subchronic exposure. Studies with styrene have also demonstrated an effect on catecholamine function (Husain et al., 1980), and a common mechanism at the neurochemical level has been proposed (Mutti and Franchini, 1987) based on the ability of dopamine to condense nonenzymatically with solvent metabolites from different chemical groups. Whether such neurochemical effects are acute or chronic in nature and whether they can be directly related to measurable neurofunctional changes have not yet been studied. However, efforts to examine the role of possible changes in transmitter function in producing central neurofunctional effects would appear to offer a promising approach.

CONCLUSION

There seems to be a growing acceptance in toxicology of animal neurofunctional methods for use in screening for neurotoxicity. The development of neurobehavioral methods for assessing motor and sensory function which has occurred in the last 10 years, the growing empirical data base demonstrating both the sensitivity and the applicability of these methods to chronic studies, and the increasing possibility of designing studies to examine neurofunctional changes along with the underlying neurochemical and neuromorphological changes that accompany them, will continue to provide evidence for the importance of neurobehavioral methods in the identification and further understanding of the actions of chemicals on the nervous system.

There are, however, many industrially and commercially used organic solvents already on the market about which little or no information regarding their potential effects on the nervous system is available (McMillan, 1987). Apparently, even the setting of occupational exposure limits to avoid acute, intoxicating effects on the nervous system has eluded an experimental basis. Moreover, psychologists working at the human level have been virtually left on their own to identify neurotoxic agents in the workplace and to sort out, as best they can, the complex issues surrounding chronic human exposures. The subject matter of many of these issues is not the domain of classical toxicology or neuropathology, it is uniquely behavioral in nature. Despite the fact that the potential for chemicals to alter memory, learning, and performance or to produce addiction may not be issues of primary concern in toxicity screening, they are nonetheless important considerations in evaluating the risks to human health associated with long-term chemical exposures. With a better understanding of the issues faced by investigators working at the human level and a greater collaboration with scientists working at the cellular and subcellular levels, behavioral toxicologists may be able to supply the necessary methods and conceptual framework to bridge the rather formidable gap that has evolved in neurotoxicity risk assessment.

REFERENCES

Allen, N. 1980. Identification of methyl *n*-butyl ketone as the causative agent. Pp. 834–845 in Experimental and Clinical Neurotoxicology, P. S. Spencer and H. H. Schaumberg, eds. Baltimore: Williams and Wilkins.

American Conference of Governmental Industrial Hygienists. 1986. Documentation of the Threshold Limit Values and Biological Exposure Indices, fifth edition. Cincinnati, Ohio.

American Psychiatric Association. 1980. Diagnostic and Statistical Manual of Mental Disorders, third edition. Washington, D.C.

Arlien-Soborg, P., P. Bruhn, C. Gyldensted, and B. Melgaard. 1979. Chronic painters syndrome: Chronic toxic encephalopathy. Acta Neurol. Scand. 60:149–156.

Baker, E. L., and L. J. Fine. 1986. Solvent neurotoxicity: The current evidence. J. Occup. Med. 28:126–129.

Baker E. L., R. E. Letz, E. A. Eisen, L. J. Pothier, D. L. Plantamura, M. Larson, and R. Wolford. 1988. Neurobehavioral effects of solvents in construction painters. J. Occup. Med. 30:116–123.

Baker, T. S., and D. E. Ricker. 1981. Dose-dependent uptake, distribution and elimination of inhaled n-hexane in the Fischer 344 rat. Toxicol. Appl. Pharmacol. 61:414–422.

Boor, J. W., and H. I. Hurtig. 1977. Persistent cerebellar ataxia after exposure to toluene. Ann Neurol. 2:440–442.

Bruhn, P., P. Arlien-Soborg, C. Gyldensted, and E. L. Christensen. 1981. Prognosis in chronic toxic encephalopathy: A two-year follow-up study in 26 house painters with occupational encephalopathy. Acta Neurol. Scand. 64:259–272.

Buckholtz, N. S., and S. Panem. 1986. Regulation and evolving science: Neurobehavioral toxicology. Neurobehav. Toxicol. Teratol. 8:89–96.

Bus, J. S., E. L. White, P. J. Gillies, and C. S. Barrow. 1981. Tissue distribution of n-hexane, methyl n-butyl ketone and 2,5-hexanedione in rats after single or repeated inhalation exposure to n-hexane. Drug Metab. Disp. 9:386–387.

Cherry, N., H. Hutchins, T. Pace, and H. A. Waldron. 1985. Neurobehavioral effects of repeated occupational exposure to toluene and paint solvents. Br. J. Indus. Med. 42:291–300.

Cohr, K. H., J. Stokholm, and P. Bruhn. 1980. Neurologic response to white spirit exposure. Pp. 95–102 in Mechanisms of Toxicity and Hazard Evaluation, B. Holmstedt, R. Lauwerys, M. Mercier, and M. Roberfroid, eds. Amsterdam: Elsevier/North Holland.

Cranmer, J. M., and L. Goldberg, eds. 1986. Proceedings of the workshop on neurobehavioral effects of solvents. Neurotoxicol. 7(4).

De Medinacelli, L., W. J. Freed, and R. J. Wyatt. 1982. An index of the functional condition of rat sciatic nerve based on measurements made from walking tracks. Exp. Neurol. 77:634–643.

Dick, R. B., and B. L. Johnson. 1986. Human experimental studies. Pp. 348–387 in Neurobehavioral Toxicology, Z. Annau, ed. Baltimore: Johns Hopkins University Press.

Edwards, P. M., and V. H. Parker. 1977. A simple, sensitive and objective method for early assessment of acrylamide neuropathy in rats. Toxicol. Appl. Pharmacol. 40:589–591.

Fuxe, K., K. Andersson, O. G. Nilsen, R. Toftgard, P. Eneroth, and J.A. Gustafsson. 1982. Toluene and telencephalic dopamine: Selective reduction of amine turnover in discrete DA nerve terminal systems of the anterior caudate nucleus by low concentrations of toluene. Toxicol. Lett. 12:115–123.

Fuxe, K., M. Martire, G. von Euler, L. F. Agnati, T. Hanson, K. Andersson, J. A. Gustafsson, and A. Harfstrand. 1987. Effects of subacute treatment with toluene on cerebrocortical alpha- and beta-adrenergic receptors in the rat. Evidence for an increased number of and a reduced affinity of beta-adrenergic receptors. Acta Physiol. Scand. 130:307–311.

Gade, A., E. L. Mortensen, and P. Bruhn. 1988. "Chronic painters syndrome." A

reanalysis of psychological test data in a group of diagnosed cases, based on comparisons with matched controls. Acta Neurol. Scand. 77:293–306.
Gamberale, F. 1985. Use of behavioral performance tests in the assessment of solvent toxicity. Scand. J. Work Environ. Health 11(Suppl. 1):65–74.
Gamberale, F., G. Annwall, and M. Hultengren. 1975. Exposure to white spirit. II. Psychological functions. Scand. J. Work Environ. Health 1:31–39.
Glowa, J. R., and P. B. Dews. 1983. Behavioral toxicology of organic solvents. II. Comparison of results on toluene by flow-through and closed chamber procedures. J. Am. Coll. Toxicol. 2:319–323.
Gotell, P., O. Axelson, and B. Lindelof. 1972. Field studies of human styrene exposure. Work Environ. Health 9:76–83.
Grabsky, D.A. 1961. Toluene sniffing producing cerebellar degeneration. Am. J. Psychiatr. 118:461.
Grasso, P., M. Sharrat, D. M. Davies, and D. Irvine. 1984. Neurophysiological and psychological disorders and occupational exposure to solvents. Fd. Chem. Toxicol. 22:819–852.
Gyntelberg, F., S. Vesterhauge, P. Fog, H. Isager, and K. Zillstorff. 1986. Acquired intolerance to organic solvents and results of vestibular testing. Am. J. Indus. Med. 9:363–370.
Hanninen, H. 1971. Psychological picture of manifest and latent carbon disulfide poisoning. Br. J. Indus. Med. 28:374–381.
Heise, G. 1983. Toward a behavioral toxicology of learning and memory. Pp. 27–37 in Application of Behavioral Pharmacology in Toxicology, G. Zbinden et al., eds. New York: Raven Press.
Himnan, D.J. 1984. Tolerance and reverse tolerance to toluene inhalation: Effects on open field behavior. Pharmacol. Biochem. Behav. 21:625–631.
Horan, J. M., T. L. Kurt, P. J. Landrigan, J. M. Melius, and M. Singal. 1985. Neurologic dysfunction from exposure to 2-t-butylazo-2-hydroxy-5-methylhexane (BHMH): A new occupational neuropathy. Am. J. Pub. Health. 75:513–517.
Husain, R., S. P. Srivastava, M. Mushtaq, and P. K. Seth. 1980. Effect of styrene on levels of serotonin, noradrenaline, dopamine and activity of acetylcholinesterase and monoamine oxidase in rat brain. Toxicol. Lett. 7:47–50.
Johnston, L. D., P. M. O'Malley, and J. G. Backman. 1984. Highlights from Drugs and American High School Students 1975–1983. National Institute on Drug Abuse, DHEW Pub. No. (ADM) 84–1317. Washington, D.C.: U.S. Government Printing Office.
King, M. D. 1982. Neurological sequelae of toluene abuse. Human Toxicol. 1:281–287.
Kishi, R., J. Harabuchi, T. Ikeda, H. Yokota, and H. Miyake. 1988. Neurobehavioural effects and pharmacokinetics of toluene in rats and their relevance to man. Br. J. Indus. Med. 45:396–408.
Knox, J. M., and J. R. Nelson. 1966. Permanent encephalopathy from toluene inhalation. New Engl. J. Med. 275:1494–1496.
Kulig, B. M. 1987. The effects of chronic trichloroethylene exposure on neurobehavioral functioning in the rat. Neurotoxicol. Teratol. 9:171–178.
Kulig, B. M. 1988. The neurobehavioral effects of chronic styrene exposure in the rat. Neurotoxicol. Teratol. 10:511–517.
Kulig, B. M. 1989. A neurofunctional test battery for evaluating the effects of long-term exposure to chemicals. J. Amer. Cov. Toxicol. 8:71–83.
Kulig, B. M., R. A. P. Vanwersch, and O. L. Wolthuis. 1985. The automated analysis of coordinated movement in rats during acute and prolonged exposure to toxic agents. Toxicol. Appl. Pharmacol. 80:1–10.
Lezak, M. D. 1983. Neuropsychological Assessment, second edition. New York: Oxford University Press.

MacPhail, R. C. 1987. Observational batteries and motor activity. Zbl. Bakt. Hyg. B. 185:21–27.
Mancuso, T. F., and B. Z. Locke. 1972. Carbon disulfide as a cause of suicide. Epidemiological study of viscose rayon workers. J. Occup. Med. 14:595–606.
McMillan, D. E. 1987. Risk assessment for neurobehavioral toxicity. Environ. Health Perspect. 76:155–161.
Meyer, O. A., H. A. Tilson, W. C. Byrd, and M. T. Riley. 1979. A method for the routine assessment of fore- and hindlimb grip strength of rats and mice. Neurobehavioral Tox. 1:233–236.
Moser, V. C., E. M. Coggeshall, and R. L. Balster. 1985. Effects of xylene isomers on operant responding and motor performance in mice. Toxicol. Appl. Pharmacol. 80:293–298.
Mutti, A., and I. Franchini. 1987. Toxicity of metabolites to dopaminergic systems and the behavioural effects of organic solvents. Br. J. Indus. Med. 44:721–723.
National Institute for Occupational Safety and Health. 1987. Organic Solvent Neurotoxicity. Current Intelligence Bulletin 48. DHHS (NIOSH) Publication 87–104.
Overton, D. A. 1984. State-dependent learning and drug discrimination. Pp. 60–127 in Handbook of Psychopharmacology, Vol. 18, L. L. Iversen, S. D. Iversen, and S. H. Snyder, eds. New York: Plenum.
Packard, V. S. 1985. Contaminants in human milk—An update. J. Food Protect. 48:724–729.
Peele, D. B., and S. P. Baron. 1988. Effects of scopolamine on repeated acquisition of radial-arm maze performance by rats. J. Exp. Anal. Behav. 49:275–290.
Pryor, G. T., E. T. Uyeno, H. A. Tilson, and C. L. Mitchell. 1983. Assessment of chemicals using a battery of neurobehavioral tests: A comparative study. Neurobehav. Toxicol. Teratol. 5:91–117.
Pryor, G. T., J. Dickenson, E. Feeney, and C. S. Rebert. 1984. Hearing loss in rats first exposed to toluene as weanlings or as young adults. Neurobehav. Toxicol. Teratol. 6:111–119.
Pryor, G. T., R. A. Howd, and C. S. Rebert. 1987. Hearing loss in rats caused by inhalation of mixed xylenes and styrene. J. Appl. Toxicol. 7:55–61.
Rebert, C.S. 1983. Multisensory evoked potentials in experimental and applied neurotoxicology. Neurobehav. Toxicol. Teratol. 5:659–671.
Rees, D. C., J. S. Knisely, S. Jordan, and R. L. Balster. 1987. Discriminative properties of toluene in the mouse. Toxicol. Appl. Pharmacol. 88:97–104.
Reiter, L. 1978. Use of activity measures in behavioral toxicology. Envir. Health Persp. 26:9–20.
Satran, R., and V. N. Dodson. 1963. Toluene habituation: Report of a case. New Engl. J. Med. 268:719–721.
Schmelzer, J. D., and P. A. Low. 1987. Electrophysiological studies on the effect of age on caudal nerve of the rat. Exper. Neurol. 96:612–620.
Spencer, P. S. 1985. Organic solvent neurotoxicity: Facts and research needs. Scand. J. Work Environ. Health. 11(Suppl. 1):53–60.
Spencer, P. S., and H. H. Schaumberg, eds. 1980. Experimental and Clinical Neurotoxicology. Baltimore: Williams and Wilkins.
Spencer, P. S., C. M. Beaubernard, M. C. Bischoff-Fenton, and T. L. Kurt. 1985. Clinical and experimental neurotoxicity of 2-*t*-butylazo-2-hydroxy-5-methylhexane. Ann. Neurol. 17:28–32.
Tanger, H. J., R. A. P. Vanwersch, and O. L. Wolthuis. 1978. Automated TV-based system for open field studies: Effects of methamphetamine. Pharmacol. Biochem. Behav. 9:555–557.

U.S. Environmental Protection Agency. 1985. Health Effects Testing Guidelines. Fed. Reg. 50:39458–39471.

Weisenberger, B. L. 1977. Toluene habituation. J. Occup. Med. 19:569–570.

Wood, R. W. 1979. Reinforcing properties of inhaled substances. Neurobehav. Toxicol. 1(Suppl 1):67–72.

Wood, R.W. 1981. Neurobehavioral toxicity of carbon disulfide. Neurobehav. Toxicol. Teratol. 3:397–405.

World Health Organization. 1986. Principles and methods for the assessment of neurotoxicity associated with exposure to chemicals. Environmental Health Criteria Document 60. Geneva.

World Health Organization and the Nordic Council of Ministers. 1985. Chronic Effects of Organic Solvents on the Central Nervous System and Diagnostic Criteria. Copenhagen: WHO Environmental Health Series.

Wu, M-F., J. R. Ison, J. R. Wecker, and L. W. Lapham. 1985. Cutaneous and sudatory function in rats following methyl mercury poisoning. Toxicol. Appl. Pharmacol. 79:377–388.

Young, J. S., and L. D. Fechter. 1983. Reflex inhibition procedures for animal audiometry: A technique for assessing ototoxicity. J. Acoust. Soc. Am. 73:1686–1693.

Animal Models:
What Has Worked and What Is Needed

Robert C. MacPhail

Biological continuity between species is the very foundation of modern biomedical science. Animal models are therefore indispensable in evaluating a wide range of chemicals long before significant human exposures may occur. Animal models are also critical in unraveling disease processes, as well as identifying and evaluating prophylactic and therapeutic treatments. Animal models offer the advantage of flexibility, precision, and reproducibility, but at the same time they raise nagging questions regarding their significance and generality. It is therefore quite fitting that this volume should deal with the topic of animal models of neurobehavioral toxicity.

Hanna Michalek has described an extensive series of experiments using rats as an animal model for human exposure to organophosphate (OP) compounds. Rats appear to be very useful in understanding many of the actions of OPs, and also in developing therapeutic approaches to OP intoxication. More specifically her work focuses on the receptor changes accompanying acute and subchronic exposures, and the importance of age and genetic variables. Although acute OP toxicity is generally considered to be an inverse function of age, there are many exceptions. It remains to be determined how important metabolic factors are in determining age-dependent OP toxicity.

Michalek has also shown differences in cholinergic function between Fischer 344 and Wistar rats. With few notable exceptions, genetic considerations have rarely been addressed in neurobehavioral toxicology. Nevertheless, there are enough data to suggest that it is entirely too simplistic to refer to "the" rat, mouse, or monkey in describing one's research. What is needed is a broad-based program of research to determine directly the extent of differences between strains of stocks of commonly used laboratory species, in terms of both the basic neurobehavioral processes and the effects of chemicals.

The evidence suggests that the mechanisms underlying acute OP intoxication, and the development of tolerance, may be very general across species. Therefore, we may be justified in pursuing research with rats, with due regard of course to metabolic, age, and genetic considerations. However, a notable effect of OP exposures in humans, and several other species is delayed neurotoxicity. Organophosphate-induced delayed neurotoxicity (OPIDN) is a permanent neuromuscular disorder involving the peripheral nervous system and spinal cord that can occur after acute exposures to many OPs. The syndrome has been clearly established in a variety of species, including humans. Rats have been widely considered refractory to the development of OPIDN, so it was a great surprise to find that Long Evans rats developed the neuropathy without displaying the clinical signs (e.g., Padilla and Veronesi, 1988). Here too, genetic variables cannot be overlooked, because no evidence of OPIDN in Fischer 344 rats was recently reported (Somkuti et al., 1988). The adequacy of rats as a broad-based model for OP toxicity would be greatly enhanced if functional effects could be revealed in OPIDN.

The work of Russell, Overstreet, and several others has shown that tolerance develops to many of the behavioral and physiological effects of OPs with continued exposure. Recent evidence suggests, however, that learning and memory impairments may be present in rodents made tolerant to OPs (McDonald et al., 1988; Upchurch and Wehner, 1987). These findings may be of tremendous importance and warrant a thorough systematic follow-up. If such findings can be substantiated, they would point to a basic complementarity between measures of learning and memory on the one hand and performance on the other. In addition, similar effects could be looked for in exposed populations of humans. In this way a much better appreciation could be gotten of the risks associated with exposure to OP compounds.

David Overstreet and Elaine Bailey have reviewed some data on animal models of dementia. They rightfully point to the importance of using several tests to evaluate learning and memory, owing to the diversity of phenomena subsumed by these terms, as well as being able to eliminate confounds in interpreting test results. They have also highlighted the importance of using pharmacological and environmental challenges in neurobehavioral toxicology research. What is now needed, in addition to a lot more data, is a systematic evaluation of many of these methods by using standard treatments known to affect learning and memory. Work should also focus on evaluating those chemicals that have been thought to produce learning and memory deficits in humans (e.g., volatile organic solvents, chlordane). A much better appreciation also needs to be gained of the interplay between environmental and pharmacological challenges because this would be of great benefit in both uncovering "silent" toxicity and identifying behavioral mechanisms of toxicant action. Caution must be exercised, however, in evaluating pharmacological challenge data to ensure that an altered drug effect following toxicant exposure is not due to dispositional, rather than functional, variables.

Deborah Cory-Slechta has indicated two major emphases in behavioral toxicology. One has to do with screening and risk estimation on chemicals that may compromise behavioral or neurological integrity. The other has to do with developing fundamental information on the behavioral actions of chemicals and chemical classes. The two emphases are not entirely distinct, although they are characteristically supported by different funding sources. Cory-Slechta also points out that much of the work in behavioral toxicology has been of a "show-and-tell" nature. Given the youthful nature of the field, this is not altogether inappropriate. There are vast numbers of chemicals that have never been adequately evaluated for neurobehavioral toxicity, and many more are coming to market each year. The field will, however, suffer in the long run from a sporadic accumulation of facts and effects. Unifying principles are badly needed to integrate the vast array of data that will be obtained in a vast number of species by using a vast number of testing paradigms.

Cory-Slechta next reviews what is known about lead toxicity. Her results indicate that some lead effects are very general across species and testing laboratories. The finding that low-level lead effects on operant performance depend on the schedule is intriguing and suggests that other drug-behavior interactions may accompany lead exposures. In addition to variations in exposure, data on the consequences of the rate changes in lead-exposed rats—for example, by determining how they adjust to alterations in the prevailing contingencies—will be very helpful in more fully understanding the functional impact of these exposures.

Beverly Kulig has reviewed much of what is known about the adequacy of animal models in better understanding the consequences of solvent exposures. The overriding theme of her chapter is evaluation of sensory, motor, and cognitive functions throughout repeated exposures. Such studies are very demanding in terms of resources and logistics, so the clarity of her results is very encouraging.

Kulig presents results to support the conclusion that different temporal patterns of action may emerge during repeated exposures, which depend on the particular behavior and the particular solvent under investigation. Tolerance developed to some of the effects of styrene, whereas sensitization developed to some of the effects of trichloroethylene (TCE). The TCE results are exemplary of the type of data behavioral toxicologists should be striving to collect. Prominent behavioral disruptions are obtained with repeated exposure that were not apparent initially or that could not be predicted from acute exposure. (Of course, it is possible that duration of exposure could have been substituted to some extent by a greater level of exposure.) Nevertheless, the results clearly indicate the feasibility of finding effects only after repeated exposures. The styrene data, on the other hand, highlight an important problem that has so far been ignored in risk assessment. What are we to make of data showing tolerance to toxicant effects? Does this mean that the organism is no longer at risk from exposure? Do we pay homage to the inherent redundancies in the nervous system that give rise to behavioral and neurological repair mechanisms, and dismiss further concerns over exposure? These questions have yet to be addressed in health and regulatory arenas, but they are by no means trivial.

Kulig also points to another area of research that has escaped the understanding of the risk assessor, namely, the possible reinforcing properties of toxicant exposures. We need only look, however, at the human and animal psychopharmacology literature to appreciate how salient such an effect can become, and what dire health and economic consequences can ensue. The topic may not be restricted to organic solvents. Recent data from our laboratory (Crofton et al., 1989) indicate that a fungicide widely used in the United States, triadimefon, has many behavioral and biochemical actions in common with psychomotor stimulants, most notably methylphenidate. It remains to be determined whether triadimefon can be shown to have reinforcing properties similar to the stimulants, but my considered guess is that it will.

Finally, in a masterful exercise in understatement, Kulig states that scientists working "at the human level . . . have received little help from their counterparts at the animal level in addressing the issues surrounding the possible adverse effects of long-term solvent exposures on cognitive functioning." Although I wholeheartedly endorse this position, I do not believe it is due to a lack of test methods that can be applied to the problem. There are several techniques readily available that evaluate many different aspects of cognitive function which can be applied immediately to this problem. The bigger problem has to do with the resources and logistics required to undertake long-term exposure and assessment studies, and the delay of reinforcement associated with finding effects (if indeed they are to be gotten) only after prolonged exposure.

REFERENCES

Crofton, K. M., V. M. Boncek, and R. C. MacPhail. 1989. Evidence for monoaminergic involvement in triadimefon-induced hyperactivity. Psychopharmacol. 97:326–330.

McDonald, B. E., L. G. Costa, and S. D. Murphy. 1988. Spatial memory impairment and central muscarinic receptor loss following prolonged treatment with organophosphates. Toxicol. Lett. 40(1):42–56.

Padilla, S., and B. Veronesi. 1988. Biochemical and morphological validation of a rodent model of organophosphorus-induced delayed neuropathy. Toxicol. Ind. Health 4(3):361–371.

Somkuti, S. G., H. A. Tilson, H. R. Brown, G. A. Campbell, D. M. Lapadula, and M. B. Abou-Donia. 1988. Lack of delayed neurotoxic effect after tri-*o*-cresyl phosphate treatment in male Fischer 344 rats: Biochemical, neurobehavioral and neuropathological studies. Fundam. Appl. Toxicol. 10(2):199–205.

Upchurch, M., and J. M. Wehner. 1987. Effects of chronic diisopropyl fluorophosphate treatment on spatial learning in mice. Pharmacol. Biochem. Behav. 27(1):143–151.

PART III
Chemical Time Bombs:
Environmental Causes of Neurodegenerative Diseases

On the Identification and Measurement of Chemical Time Bombs:
A Behavior Development Perspective

Norman A. Krasnegor

As a society, we in the United States take it as a given that our children have the right to develop normally. Moreover, our government takes a keen scientific and legislative interest in how to protect our population, both young and old, from the ill effects of chemicals, environmental pollutants, and contaminants of the food supply. The tragedies of thalidomide and diethylstilbestrol (DES) sensitized basic scientists, clinicians, and legislators to the dangers associated with the administration of drugs prenatally and to the health and well-being of the developing neonate (Krasnegor, 1986). Further, evidence is accumulating that not all the damage suffered by the developing fetus exposed to toxicants during gestation results in physical or neurochemical anomalies. Rather, the effects upon the perinate, exposed to chemicals, may well be manifested in psychological or behavioral changes such as irritability, impaired learning ability, hyperactivity, or reduced capacity for information processing (Krasnegor, 1986). It is therefore incumbent upon public health officials, clinicians, and scientists to discover those substances that may be harmful to the fetus and thereby affect an individual's behavioral development from birth. This chapter focuses upon an elucidation of new and proposed approaches for identifying substances (chemicals, drugs, etc.) that may put the developing human at risk for developmental disability.

METHODOLOGICAL CONSIDERATIONS

Developmental behavior toxicology is a field of research devoted to the goals of discovering and elucidating abnormalities in development as a consequence of exposure to drugs or other chemicals (Thompson, 1986). Researchers in this domain of science are faced with formidable methodological problems. In the course of their studies, they are constantly faced with the task of separating developmental variables from toxicological and other environmental ones. The developing organism's rapidly shifting behavioral baseline poses special mensurational difficulties. For example, an organism may exhibit one set of behaviors early in development. These may subsequently disappear from the repertoire only to reappear later in ontogeny. This state of affairs is commonly observed, particularly in the perinatal period of development. Without detailed knowledge of this phenomenon, one may erroneously conclude that a toxicant has produced the change.

Another common problem concerns the issue of developmental delay. Large individual differences in *when* behaviors of interest appear in an organism's repertoire are to be expected. Therefore, precise knowledge of the expected variability is essential to help differentiate between conclusions of a substance's toxic effects on behavior and its natural ontogenesis. Experimental paradigms that are appropriate for assessing the effects of a drug on mature organisms may not suffice for very young ones. Because the effects of a putative toxicant may result in damage to the formation of central nervous system (CNS) structures, which in turn may affect the development of behavioral processes later in life, researchers may be forced to adopt a longitudinal design. This tactic, although deemed appropriate for the problem under study, can significantly increase the cost of research and delay the publication of data.

Another set of questions involves the choice of baselines that should be used to assess whether a substance has the potential for being behaviorally toxic. Should experimental paradigms be employed or should "naturalistic" behaviors be used? Is it better to study learning (classical or operant conditioning), conduct open field studies, or investigate the social and emotional attachment of the neonate to its care giver? Answers to these queries are contingent to some extent on the questions being posed and the extant knowledge concerning the developmental trajectory of the behaviors in question. These tactical judgments are also based in part upon the availability of experimental paradigms that can be employed with perinates and the behavioral mechanisms believed to be affected by the putative toxicant.

A critical issue that confronts researchers in this field of inquiry is what subject should be employed to assess toxicity. Clearly, studies which employ prospective designs with substances that are suspected of having behaviorally toxic activity must employ animal models. A traditional choice for behavioral toxicity studies has been laboratory rats. The rationale for their use is based upon the enormous literature available on aspects of their behavior, physiology, and neurobiology. However, depending on the question, other organisms may be much better suited than rats. For example, caffeine use by pregnant women has been questioned as a potential behaviorally toxic agent. More specifically, its first metabolite, methylxanthine, has been the subject of recent studies designed to determine its potential for affecting behavioral development. The subject chosen for the study was the female rabbit and her offspring. The decision to use this animal was based upon the fact that rabbits metabolize caffeine much as does man (Denenberg, personal communication, 1988; Denenberg et al., 1986). Further, the natural behaviors of the offspring and its interaction with its mother early in development have been well studied in this animal. Thus, rabbits became the logical choice for studying this important question concerning exposure of the fetus to methylxanthine during gestation.

A number of other methodological issues are also unique to the study of behavioral toxicity in the developing organism. Dosing parameters, including amount, route, and when during gestation, are all important considerations in undertaking studies of the developing organism. The planning for dose-effect relationships, although not unique to the study of young organisms, should be included in any comprehensive study of behavioral toxicity. Cross-fostering controls must be employed to obviate the effects that toxic substances may have on maternal behavior, which therefore affect normal mother-offspring interactions. Also of importance is the issue of when, after dosing (the developmental stage), the offspring should be tested.

In summary, the methodological issues associated with the detection of behaviorally toxic substances are both numerous and complex. Meticulous attention must be paid to these methodological details because failure to do so could lead to erroneous conclusions about the behavioral toxicity of a chemical or drug that may be quite beneficial.

APPROACHES TO DETECTING BEHAVIORALLY TOXIC SUBSTANCES

Two approaches are generally employed to determine whether a substance of interest has behavioral toxicity. The first depends upon the availability of epidemiological data that can provide a statistical basis for evaluating the morbidity and mortality associated with a substance. Based upon such knowledge, one might design experimental studies, employing animal models, that can assess dose-response relationships between a substance and putative effects upon behavioral development. The second approach employs animal models to screen substances for behavioral toxicity. This tactic is the one most frequently used because regulatory procedures require testing prior to the release of a drug or chemical for use. Screening is a much more complex strategy because one is not sure what types of behavioral effects to expect or when during development they will be manifest.

A typical approach to identifying substances that may impair normal behavioral development is to employ an animal model (e.g., laboratory rats). Pregnant females are administered the substance of interest. Dosing parameters, when dosing occurs, route of administration, how often, etc., are all variables that are predicated upon the best scientific information concerning when the substance is believed to undermine mechanisms that affect behavioral development. Typically, too, the knowledgeable behavioral toxicologist will cross-foster the offspring postnatally. He will raise the pups, who were exposed prenatally, until they attain the developmental stage of interest and then test them to ascertain whether the substance of interest produces effects upon behavior.

Two categories of substance are of great interest to those who study developmental behavioral toxicology. These are drugs given to pregnant females for preexisting medical conditions or drugs associated with pregnancy and anesthetics associated with delivery. A short review of the literature associated with one class of drugs the barbiturates, given in association with quality medical care to pregnant human females, is provided below.

Reyes et al. (1986) reported that pregnant rats given high doses of phenobarbital (10 times the therapeutic dose) had significant increase in pup mortality and decrease in birth weight of offspring. Voorhees (1985) and Middaugh (1986) both reported biochemical and behavioral anomalies in rat and mouse pups, respectively, after prenatal exposure to low levels of barbiturates. Changes in activity level, learning capacity, and seizure threshold in rodents were reported after early exposure to barbiturates (Chapman and Cutler, 1983; Diaz, 1978; Middaugh et al., 1981; Yanai et al., 1981).

In addition to these findings, researchers also report changes in sexual maturation and behavior after prenatal exposure to barbiturates. The presumed mechanism is alteration in brain loci responsible for sexual differentiation. For example, Clemens et al. (1979) found that the adult sexual behavior of male hamsters was altered compared with controls after prenatal exposure to barbiturates. Females, in the same study, showed no difference as adults after receiving the same dosing regimen. Prenatally administered barbiturates are capable of changing reproductive functions of male and female pups so exposed. Further, female offspring have lower birth weights at puberty, show lower fertility, and have delayed onset of puberty (Gupta et al., 1980). Testosterone concentrations in plasma and brain of neonatal males exposed to barbiturates were lowered by prenatally administered barbiturates (Gupta et al., 1982). This finding suggests that early testosterone deficits could be instrumental in altering masculine development and have a negative impact upon reproductive function in the adult.

A POTENTIAL TIME BOMB?

Barbiturates, as studied in rodents, have been shown to have the potential for being time bombs in that the behavioral deficits observed do not show up until late in development. Physicians have long prescribed barbiturates to their patients for anxiety, sedation, and seizure disorders. In the 1960s and 1970s, barbiturates were frequently prescribed to pregnant women. For example, among the subjects in the Collaborative Perinatal Project consisting of 50,000 pregnancies, some 25 percent of the women were prescribed barbiturates at some time during gestation (Heionen et al., 1977). Based upon Medicaid data from Michigan, Rosa (personal communication, 1988) estimates that approximately 1.5 percent of women receive phenobarbital during the first trimester of their pregnancy. Prenatal exposure of the fetus to barbiturates is not the only time in early development when this class of drugs is given. Neonates are also prescribed the drug as a sedative or anticonvulsant. From an epidemiological perspective then, the number of children exposed, and therefore potentially at risk, is high.

Although the prescription of barbiturates to pregnant women or newborns has been considered safe, recent studies employing animal models suggest that barbiturates may have the potential for neural or behavioral toxicity (Smith, 1977). More recent literature reviews also support this conclusion (Coyle et al., 1980; Fishman and Yanai, 1983; Ornoy and Yanai, 1980; Reinisch and Sanders, 1982; Yanai, 1984).

Although there is a rich literature on animal studies concerning the putative behavioral toxicology of barbiturates, the research findings on humans are meager (van den Berg, personal communication, 1988). A rigorous analysis of behavioral and biomedical data, by using a case control matching design, is currently underway (Reinisch, personal communication, 1988). The investigators are employing a retrospective design, that is, studying a group of young adults (in their early 20s) whose mothers were prescribed barbiturates during pregnancy. The subjects comprise a cohort listed in a Danish birth registry. The study, which is still underway, will be, when completed, the most rigorously designed and comprehensive one of humans exposed prenatally to barbiturates. The data set includes records from school, the army, the criminal justice system, and parents. Psychological and behavioral test scores along with medical records concerning physical development are being amassed. The researchers will then be able to pose questions related to the findings from the animal literature to determine whether behavioral toxicity can be demonstrated in people who were exposed prenatally to barbiturates.

The example provided by barbiturates is illustrative of the dilemma posed when the risk/benefit ratio is examined critically. The animal model data suggest the potential for behavioral toxicity, the behavioral epidemiology data are not yet complete, and the drug class is seen to be beneficial to both the mother and her offspring. Until there are some definitive findings on barbiturates, prescribing practice is unlikely to change.

NEW METHODS FOR MEASURING BEHAVIORAL TOXICITY

Although the usual approach to measuring behavioral toxicity is to dose prenatally and measure changes later in life, an alternative tactic is to measure behavior as early as possible after exposure. Carried to its extreme, this approach implies measurement of fetal behavior.

During the early part of this century, there was considerable scientific interest in prenatal behavioral development. The questions of interest focused upon *when* during life learning can first be demonstrated. More specifically, scientists began to query whether learning could be shown to exist in the fetus. Learning for the purposes of the present discussion is defined as associative or Pavlovian conditioning.

Workers during the 1930s attempted to classically condition the human fetus. Ray (1932), for example, paired a neutral stimulus (vibrotactile stimulus) with an unconditioned stimulus (UCS) that was known to produce movement in the fetus. The UCS, a loud noise, if made suddenly in the presence of a fetus is reliably followed by a startle movement. This latter response can be detected by placing one's hand on the abdomen of a pregnant woman. The UCS was paired with the neutral stimulus (CS) for a number of trials deemed sufficient for the CS alone to elicit the startle response. This study did not confirm the capacity for classical conditioning in the fetus.

A little over 15 years later, Spelt (1948) employed similar procedures and claimed success in demonstrating classical conditioning in the human fetus. He also concluded from the analysis of his data that the fetus has the capacity for extinction and retention of the classically conditioned response. It should be pointed out that more recently, other behavioral scientists have sharply criticized these findings on methodological grounds (Sameroff and Cavanaugh, 1979), thereby leaving open the question of whether classical conditioning of the fetus is possible. Significant progress on this topic had to await methodological innovations that emerged at the start of the current decade (Krasnegor et al., 1987).

The studies of interest employed fetal rats. The breakthrough was predicated upon methodological innovations that allowed researchers to directly observe and manipulate the fetus and thereby rigorously test questions of learning. Blass and Pedersen (1980) and Stickrod (1981) developed procedures for externalizing the uterus of pregnant female rats late in gestation and ways to inject substances into the amniotic fluid of the developing fetus. These new methodologies allowed the investigators to make their observations, the uterine horns to be reinserted, and the fetus to complete its development to term. At that time, fetuses could be delivered vaginally or taken by cesarean section and be cross-fostered to recently delivered mothers.

At a workshop sponsored by the National Institute of Child Health and Human Development (NICHD), these and other techniques for viewing and manipulating the mammalian fetus were summarized (Kolata, 1984). Attending that meeting was William Smotherman who, along with his coworkers, has carried out a number of studies on fetal behavior and fetal learning. In the first of a series of investigations on prenatal learning, Stickrod et al. (1982a) demonstrated that late in development, rat fetuses have the capacity for associative learning. On day 20 of gestation, the uterine horn of a female rat was externalized into a warm saline bath. Apple juice (CS) was injected into the amniotic fluid surrounding the exposed fetuses, and lithium chloride (UCS) was injected into their peritoneum. A single pairing of an aversive stimulus (LiCl) with a novel taste or odor (apple juice) causes adult rats so treated to avoid the taste or odor on subsequent presentations.

The externalized uterus of the dam was reinserted into her peritoneum; she was sutured; and the fetuses, treated as described above, were delivered at term. When the pups were 2 weeks old and were allowed to suckle from an anesthetized dam, they were observed to show preferential nipple attachment in accordance with their prenatal experience. Those pups which had been exposed to the associative conditioning paradigm as fetuses attached less often to nipples that were painted with apple juice compared to control pups (Smotherman and Robinson, 1987). In a follow-up study, Stickrod et al. (1982b) demonstrated that pups which had been conditioned prenatally showed greater delays in crossing a runway, where the air contained the odor of apple juice, to gain access to their mother. In a variant of the procedure, these same authors demonstrated that prenatally conditioned pups preferred to stay at the low-concentration end of a box containing the odor of apple juice. These findings are quite important because they indicate both that conditioning took place before birth and that the learned response was retained postnatally.

Smotherman and his coworkers continued their research on several fronts. They examined the influence of uterine position, a variable feature of the prenatal environment, upon conditioned taste aversion in adult rats (Babine and Smotherman, 1984; Smotherman, 1983). They critically evaluated two existing techniques for the preparation of the female rat for fetal observation and demonstrated that the two procedures, chemomyelotomy and spinal transection, are not equivalent in their effects upon spontaneous fetal activity (Smotherman, 1984; Smotherman et al., 1984). They also developed a new reversible anesthetic procedure for preparing the dam and observing the fetal rat in utero (Smotherman et al., 1986). This method has the advantage of allowing the longitudinal study of behavior of the same subjects before and after birth. Further, Smotherman and Robinson (1986) made critical observations on age-related changes in fetal behavior during the last third of gestation. This work also documents fetal responsiveness to naturally occurring changes within the uterine environment. Similarly, it demonstrates the feasibility of observing fetuses after removal from the uterus and amniotic sac and the quantification of fetal behavior from day 16 to gestation.

By combining the new observation techniques with the knowledge gained on the ontogenesis of movement patterns, Smotherman was able to study the capacity for conditioning to emerge during gestation. In a series of elegant experiments (Smotherman and Robinson, 1987) which included rigorous control procedures, the investigator and his colleagues demonstrated that rat fetuses exposed to a single-trial pairing of a neutral stimulus (mint) and an interperitoneal injection of LiCl on day 17 of gestation, are conditioned by day 19 of gestation. This was shown to be the case because the mint solution alone does not suppress endogenous movement patterns on day 17 or 19 of gestation, but when paired with the LiCl injection on day 17, it markedly suppresses movement by itself in 19-day-old fetuses who had received the conditioning trial.

The work described above clearly and rigorously documents the capacity of the fetus for learning during gestation. It also demonstrates that the organization of fetal behavior patterns has an ontogenetic trajectory. Both of these conclusions indicate that a new window of opportunity exists for the behavioral toxicologist. Associative learning can be established prenatally and followed postnatally. Conditioning, in terms of its acquisition and consolidation, can be studied during gestation. Thus, substances of interest can be studied for their effect both at the time of gestation when they have their putative action upon the developing brain and postnatally during neonatal development. Indeed, Smotherman and his colleagues (Smotherman and Robinson, 1987; Smotherman et al., 1986a, 1986b; also, Baron et al., 1986) have conducted studies of fetal behavior after chronic maternal exposure to ethanol. These studies are exemplars that point the way for demonstrating the power of the approach to identify both substances and mechanisms that may interfere with behavioral development.

ADDITIONAL APPROACHES AND NEW DIRECTIONS

Two questions with which developmentalists and behavioral toxicologists alike must constantly grapple revolve around the issues of prediction and validity. The prediction issue is brought out when studies of development are undertaken with children who are born at risk (e.g., intrauterine growth retardation, low birth weight). What researchers interested in such questions would like to know is whether behavior measured early in life (e.g., the neonatal or infancy phase of development) is predictive of behavioral development in childhood (e.g., at school entry). If reliable and valid measures could be established, children born at risk who would develop normally, from a behavioral perspective, could be accurately separated from those who may evidence behavioral deficits (Bornstein and Krasnegor, 1989). Prediction of a different, albeit equally important, type is sought by behavioral toxicologists. They endeavor to predict whether a substance of interest has behavioral toxicity and whether early exposure will lead to behavioral deficits later in development. Further, they employ animal models which they believe validly relate to the human condition. Developmentalists and behavioral toxicologists also strive to pose questions that can elucidate the putative behavioral or neurobehavioral mechanisms which may subsume the observed deficits.

Is it possible to address these multiple concerns and thereby advance both fields of inquiry alluded to above? Research in the field of eyelid conditioning has much to recommend it conceptually and experimentally as a paradigm for asking questions of relevance to developmental behavioral toxicologists (Gormezano et al., 1983; Harvey and Gormezano, 1986; Solomon and Pendlebury, 1988). There are three main factors that make the nictitating membrane response (NMR) an attractive one for addressing developmental questions in general and behavior toxicology ones in particular.

Multiple Behaviors Can Be Studied

There are at least 10 behavioral factors that can be studied utilizing the NMR/eye blink model system (Harvey and Gormezano, 1986). These are (1) habituation and sensitization, (2) stimulus selection, (3) mediation associations, (4) motivation, (5) memory traces and short- or long-term memory, (6) simple and conditional discriminations, (7) transfer of training, (8) timing, (9) stimulus generalization, and (10) extinction and conditioned inhibition.

Comparative Developmental Studies Can Be Undertaken

Work on the NMR/eye blink has been carried out by using the rabbit as a model system. Research on humans has also been undertaken to study the conditioned eye blink. This has recently involved developmental studies that compared the acquisition of the classically conditioned eye blink across the age span ranging from childhood (8 years of age) to the eighth decade of life (Solomon et al., 1989). Such work holds out the promise that specific comparative studies of the model system and the same response system in humans can be accomplished.

Knowledge Concerning Neurocircuitry and Neurochemistry of the Response Is Accumulating

Research to date has implicated two different brain circuits in this model system. These are found respectively in the cerebellum (McCormick and Thompson, 1984) and the hippocampus (Berger et al., 1986; Moore and Solomon, 1980). The data collected by these investigators suggest that the cerebellum may be the CNS site of simple plasticity for simple delay conditioning. The hippocampus, on the other hand, has been implicated in trace conditioning (Solomon and Gottfried, 1981); discrimination reversal (Berger and Orr, 1983); and as a modulator of simple delay conditioning (Solomon et al., 1983). Moreover, what is known concerning the pharmacology of the conditioned response in rabbits suggests strongly that the cholinergic system is involved in mediating the response (Moore et al., 1976; Solomon et al., 1983).

These three factors provide a convincing argument that the conditioning model described may be a powerful tool for developmental behavior toxicology (see, for example, Yokel, 1983). Additional research is needed to obtain developmental data for this model system. This should be conducted in neonatal humans and young rabbits to fill in the knowledge gap on the ontogeny of the response. The availability of such baseline data will provide developmentalists and behavioral toxicologists with the information needed to evaluate behavioral development and the effects of substances over an impressive age span. It will allow prospective questions to be asked from a developmental perspective. Most importantly, it will allow researchers to connect with a model system (the rabbit NMR/eye blink) which can help differentiate between CNS mechanisms that may be involved in impaired development. This can in turn provide clues to scientists who work with babies born at risk, or with those who were exposed to substances as fetuses, concerning the behavioral and neurobehavioral mechanisms that may have gone awry.

Some progress toward the goal of collecting data in the neonatal human baby has already been made. (Although the work described below is not directly on eye blink conditioning, it is sufficiently related that inclusion is warranted.) Howard Hoffman and his colleagues have been studying the development and characteristics of the startle response for the past two decades. After analyzing this response by using animal models they discovered that the response could be modified. In experiments carried out in adult humans, Hoffman found that when an exteroceptive stimulus precedes one that elicits the glabella response by 100–200 ms, the resultant eye blink is reduced in amplitude (Hoffman and Ison, 1980). If the same stimulus is presented simultaneously with the eliciting stimulus, the eye blink amplitude is enhanced compared to a control condition in which no stimulus is presented (Hoffman and Stitt, 1980; Hoffman et al., 1981).

Hoffman and his coworkers next turned their attention to a comparison of adults and newborns to undertake a developmental analysis of these augmentation-inhibition results. They found that newborns (16–65 hours old) exhibited reflex augmentation to the simultaneous pairing of an exteroceptive stimulus and a gentle, calibrated tap between the eyes (Hoffman et al., 1985). Although the comparative data are of interest, the most compelling information relates to the methodology. Eye blinks can be reliably measured on the first day of life, and systematic data on this response can be collected. This strongly suggests that an approach can be worked out to collect eye blink conditioning data at this time in development and in older infants as well.

In summary, new approaches for measuring simple learning during the perinatal period offer an opportunity to assess the potential for substances to affect behavioral development. Research on these new windows for observation can provide the field of behavioral toxicology with additional tools to effectively evaluate, early in an organism's development, whether—and, potentially, how—it may become impaired later in life. Investigations along these lines should be pursued and vigorously encouraged.

ACKNOWLEDGMENT

Thanks are due to Marsha Sotzsky for preparing this manuscript.

REFERENCES

Babine, A. M., and W. P. Smotherman. 1984. Uterine position and conditioned taste aversion. Behavioral Neuroscience 96:461–466.

Barron S., E. P. Riley, and W. P. Smotherman. 1986. The effect of prenatal alcohol exposure on umbilical cord length in fetal rats. Alcoholism: Clinical and Experimental Research 10:493–495.

Berger, T. W., and W. B. Orr. 1983. Hippocampectomy selectively disrupts discrimination reversal conditioning of the rabbit nictitating membrane response. Behavior Brain Research 8:49–68.

Berger T. W., S. Berry, and R. Thompson. 1986. Role of the hippocampus in classical conditioning of aversive and appetitive behaviors. Pp. 203–240 in The Hippocampus, R. L. Isaacson and K. H. Pribram, eds. New York: Plenum.

Blass E. M., and P. E. Pedersen. 1980. Surgical manipulation of the uterine environment of rat fetuses. Physiology and Behavior 25:993–995.

Bornstein, M. H., and N. A. Krasnegor. 1989. Stability and Continuity in Mental Development: Behavioral and Biological Perspectives. Hillsdale, N.J.: Lawrence Erlbaum Associates.

Chapman, J. B., and M. G. Cutler. 1983. Behavioral effects of phenobarbitone. I. Effects in the offspring of laboratory mice. Psychopharmacology 29:155–160.

Clemens L. G., T. V. Popham, and P. H. Ruppert. 1979. Neonatal treatment of hamsters with barbiturate alters adult sexual behavior. Developmental Psychobiology 12:115–125.

Coyle, I., A. Wagner, and G. Singer. 1980. Behavioral teratogenesis: A critical evaluation. In Advances in the Study of Birth Defects, Neural and Behavioral Teratology, T. V. N. Persaud, ed. Baltimore: University Park Press.

Denenberg, V. H., E. B. Thomas, P. Kramer, and J. R. Raye. 1986. Sleep and wake behavioral state as a developmental assessment procedure. In Advances in Behavioral Pharmacology: Developmental Behavioral Pharmacology, N. A. Krasnegor, D. B. Gray, and T. Thompson, eds. Hillsdale, N. J.: Lawrence Erlbaum Associates.

Diaz, J. 1978. Phenobarbital: Effects of long-term administration on behavior and brain of artificially reared rats. Science 199:90–91.

Fishman, R. H. B., and J. Yanai. 1983. Long lasting effects of early barbiturates on central nervous system and behavior. Neuroscience Biobehavioral Review 7:19–28.

Gormezano, I., E. J. Kehoe, and B. Marshall. 1983. Twenty years of classical conditioning research with the rabbit. Pp. 197–275 in Progress in Psychobiology and Physiological Psychology, Vol. 10, J. M. Sprague and A. N. Epstein, eds. New York: Academic Press.

Gupta C., B. R. Sonawane, and S. F. Yaffe. 1980. Phenobarbitol exposure in utero: Alterations in female reproductive function in rats. Science 208:508–510.

Gupta C., S. F. Yaffe, and B. H. Shapiro. 1982. Prenatal exposure to phenobarbital permanently decreases testosterone and causes reproduction dysfunction. Science 216:640–642.

Harvey, J. A., and I. Gormezano. 1986. The assessment of drug effects on learning and stimulus processing by means of classical conditioning. In Developmental Behavioral Pharmacology, Advances in Behavioral Pharmacology, N. A. Krasnegor, D. B. Gray and T. Thompson, eds. Hillsdale, N.J.: Lawrence Erlbaum Associates.

Heinonen, N. P., D. Slone, and S. Shapiro. 1977. Birth Defects and Drugs in Pregnancy, Chap 24. Littleton, Mass.: John Wright.

Hoffman, H. S., and J. R. Ison. 1980. Reflex modification of startle: I. Some empirical findings and their implications for how the nervous system processes sensory input. Psychological Review 87:175–189.

Hoffman, H. S., and C. L. Stitt. 1980. Inhibition of the glabella reflex by monaural and binaural stimulation. Journal of Experimental Psychology: Human Perception and Performance 6:769–776.

Hoffman H. S., M. E. Cohen, and C. Stitt. 1981. Acoustic augmentation and inhibition of the human eyeblink. Journal of Experimental Psychology: Human Perception and Performance 7:1357–1362.

Hoffman H. S., M. E. Cohen, and L. M. English. 1985. Reflex modification by acoustic signals in newborn infants and in adults. Journal of Experimental Child Psychology 39:562–579.

Kolata, G. 1984. Learning in the womb. Science 225:302–303.

Krasnegor, N. A. 1986. Introduction: Perspectives and new directions. In Advances in Behavioral Pharmacology, Vol. 5, Developmental Behavioral Pharmacology, N. A. Krasnegor, D. B. Gray, and T. Thompson, eds. Hillsdale, N.J.: Lawrence Erlbaum Associates.

Krasnegor, N. A., E. M. Blass, M. A. Hofer, and W. P. Smotherman, eds. 1987. Perinatal Development: A Psychobiological Perspective. Orlando, Fla.: Academic Press.

McCormick, D. A., and R. F. Thompson. 1984. Cerebellum: Essential involvement in the classically conditioned eyelid response. Science 223:296–299.

Middaugh, L. D. 1986. Prenatal maternal barbiturates effects on offspring. In Advances in Behavioral Pharmacology, Vol. 5, Developmental Behavioral Pharmacology, N. A. Krasnegor, D. B. Gray, and T. Thompson, eds. Hillsdale, N.J.: Lawrence Erlbaum Associates.

Middaugh, L. D., L. W. Simpson, and T. N. Thomas. 1981. Prenatal maternal phenobarbital increases reactivity and retards habituation of mature offspring to environmental stimuli. Psychopharmacology 74:349–352.

Moore, J. W., and P. R. Solomon, eds. 1980. The role of the hippocampus in learning and memory. Physiological Psychology (special edition).

Moore J. W., N. A. Goodell, and P. R. Solomon. 1976. Central cholinergic blockade by scopolamine and habituation, classical conditioning, and latent inhibition of the rabbit's nictitating membrane response. Physiological Psychology 7:224–232.

Ornoy, A., and J. Yanai. 1980. Central nervous system teratogenicity: Experimental models for human problems. In Advances in the Study of Birth Defects, Vol. 4,

Neural and Behavioral Teratology, T. V. N. Persaud, ed. Baltimore: University Park Press.

Ray, W. S. 1932. A preliminary study of fetal conditioning. Child Development 3:173–177.

Reinisch, J. M., and S. A. Sanders. 1982. Early barbiturate exposure: The brain, sexually dimorphic behavior and learning. Neuroscience Biobehavioral Review 6:311–319.

Reyes, E., K. Garcia, and J. Wolfe. 1986. Effects of in utero administration of phenobarbital on gamma-glutamyl-transpeptidase. Alcohol 3:153–155.

Sameroff, A. J., and P. J. Cavanaugh. 1979. Learning in infancy: A developmental perspective. In Handbook of Infant Development, J. D. Osofsky, ed. New York: John Wiley & Sons.

Smith, D. W. 1977. Teratogenicity of anticonvulsant medications. American Journal of Disease of Children 131:1337–1339.

Smotherman, W. P. 1983. Mother-infant interaction and the modulation of pituitary-adrenal activity in rat pups after early stimulation. Developmental Psychobiology 16:169–176.

Smotherman, W. P. 1984. Letter to the editor: Learning in the womb. Science 225:1093.

Smotherman, W. P. 1985. Glucocorticoids and other hormonal correlates of conditioned taste aversion. In Experimental Assessments and Clinical Applications of Conditioned Food Aversions, N. S. Braverman and P. Bronstein, eds. Annals of the New York Academy of Sciences, Vol. 443. New York.

Smotherman, W. P., and S. R. Robinson. 1986. Environmental determinants of behavior in the rat fetus. Animal Behaviour 34:1859–1873.

Smotherman, W. P., and S. R. Robinson. 1987. Psychobiology of fetal experience in the rat. In Perinatal Development: A Psychobiological Perspective, N. A. Krasnegor, E. M. Blass, M. A. Hofer and W. P. Smotherman, eds. Orlando, Fla.: Academic Press.

Smotherman, W. P., L. S. Richards, and S. R. Robinson. 1984. Techniques for observing fetal behavior in utero: A comparison of chemomyelotomy and spinal transection. Developmental Psychobiology 17:661–674.

Smotherman, W. P., S. R. Robinson, and B. J. Miller. 1986a. A reversible preparation for observing behavior of fetal rats in utero: Spinal anesthesia with lidocaine. Physiology and Behavior 37:57–60.

Smotherman, W. P., W. P. Woodruff, S. R. Robinson, C. Del Real, S. Barron, and E. P. Riley. 1986b. Spontaneous fetal behavior after maternal exposure to ethanol. Pharmacology Biochemistry and Behavior 24:165–170.

Solomon, P. R., and K. E. Gottfried. 1981. The septohippocampal cholinergic system and classical conditioning of the rabbit's nictitating membrane response. Journal of Comparative and Physiological Psychology 95:322–330.

Solomon, P. R., and W. W. Pendlebury. 1988. A model systems approach to age related memory disorders. Neurotoxicology 9:443–462.

Solomon, P. R., S. D. Solomon, E. R. Van der Schaff, and H. E. Perry. 1983. Altered activity in hippocampus is more detrimental to classical conditioning than removing the structure. Science 220:329–331.

Solomon, P. R., D. Pomerleau, L. Bennett, J. James, and D. L. Morse. 1989. Acquisition of the classically conditioned eyeblink response in humans over the life span. Psychology & Aging 4(1):34–41.

Spelt, D. K. 1948. The conditioning of the human fetus in utero. Journal of Experimental Psychology 38:338–344.

Stickrod, G. 1981. In utero injection of rat fetuses. Physiology and Behavior 28:5–7.

Stickrod, G., D. P. Kimble, and W. P. Smotherman. 1982a. In utero taste/odor aversion conditioning in the rat. Physiology and Behavior 28:5–7.
Stickrod, G., D. P. Kimble, and W. P. Smotherman. 1982b. Met-5-enkephalin effects on associations formed in utero. Peptides 3:881–883.
Thompson, T. 1986. Issues in developmental behavioral pharmacology. In Advances in Behavioral Pharmacology, Vol. 5, Developmental Behavioral Pharmacology, N. A. Krasnegor, D. B. Gray, and T. Thompson, eds. Hillsdale, N.J.: Lawrence Erlbaum Associates.
Voorhees, C. V. 1985. Fetal anticonvulsant syndrome in rats: Effects on postnatal behavior and brain aminoacid content. Neurobehavioral Toxicology and Teratology 7:471–482.
Yanai, J. 1984. An animal model for the effect of barbiturate on the central nervous system. In Neurobehavioral Teratology: Drugs of Use and Abuse, J. Yanai, ed. New York: Elsevier.
Yanai, J., A. Bergman, and R. Shafer. 1981. Audiogenic seizures and neuronal deficits following early exposure to barbiturate. Developmental Neuroscience 4:345–350.
Yokel, R. A. 1983. Repeated systematic aluminum exposure effects on classical conditioning in the rabbit. Neurobehavioral Toxicology and Teratology 5:41–46.

Neurobehavioral Time Bombs:
Their Nature and Their Mechanisms

Roger W. Russell

Basic to a complete discipline of neurobehavioral toxicology is the recognition that behavior is but one of the properties of living organisms that are affected by the chemical environments in which they live. Behavior does not exist independently of dynamic biochemical and electrophysiological processes taking place constantly in various structural (morphological) sites within the body. The major objective of this chapter is to place behavior in its proper perspective within the "integrated organism" (Russell, 1979). This is done by discussing the nature and mechanisms involved in three examples of what are generally referred to as "progressive degenerative dementias" (PDDs). It is well to begin with consideration of a general framework within which the trilogy may be analyzed.

Relations Between Behavior and Chemical Environment

It is well recognized that the biological effects of chemicals in the external physical environment can only be a result of physiochemical interactions between molecules of the agent and receptor sites on particular molecules present in the body (Doull, 1980). Technological advances have now made it possible to study some of the "cascade" of biological events that follows such an interaction, although there is still much more to learn. The events may be viewed as progressing from the molecular level through morphological sites—synapses, neurons, nerve networks, nuclei, and systems—eventually to exert an effect on

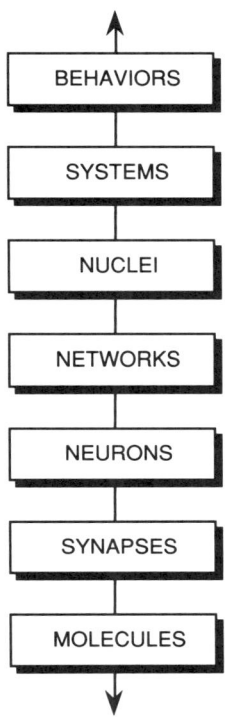

FIGURE 1 Sites of action of neurochemical events in the nervous system. The diagram represents levels of increasing complexity extending from molecules to endpoints measurable as changes in behavior.

behavior (Figure 1). Although the matter is not pursued here, it should be remembered that research in psychosomatics has shown that behavior may produce consequent changes even at the molecular level. The sequence of the neurochemical events taking place in these sites is shown diagrammatically in Figure 2. Events a_1 to a_n preceding the formation of the chemical–receptor complex, AR, are involved in the processes by which the chemical, A, reaches its site of action. The transport of A to its receptor may involve progress through several different membranes and chemical milieu, during which A may undergo biotransformation. In some cases the resulting molecule may be much more potent in terms of its biological effects than the parent substance. Conversion of the organophosphorus pesticide parathion to paraoxon, the active form, is an example. Binding of A to its receptor site may be reversible or irreversible. In the latter instance, receptor molecules must be synthesized de novo; hence, recovery from the effects produced by A may be delayed.

Effects e_1 to e_n following formation of the AR complex are independent of those preceding it, but are influenced by the state of the organ-

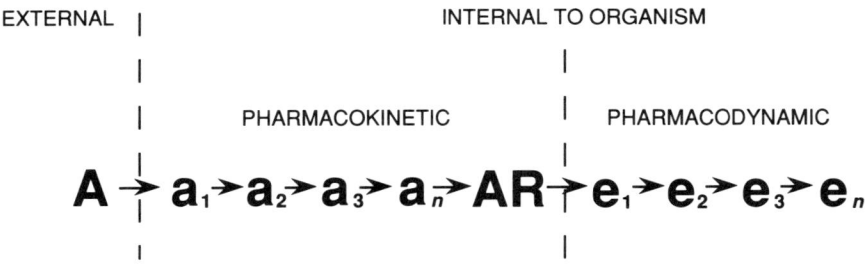

FIGURE 2 Sequence of neurochemical events taking place between the entry of an exogenous chemical into the body and its effects on behavior. The chemical, A, is transported to its site of action, a_1–a_n, where it binds to its specific receptor, R, initiating an extensive series of events, e_1–e_n, leading to effects on behavior.

ism and by other processes that interact with those stimulated by AR. It is logical that the further an endpoint is from its receptor activation, the greater is the possibility that other events may influence the nature of the effect. Some forms of behavior are linked more directly to their biochemical correlates than others. Where the linkage is direct (e.g., in sensory-reflexive responses), changes in biochemical events are reflected in specific changes in behavior, but where the linkage is diffuse, as it is in cognitive behaviors, changes in biochemical events may affect a variety of behavior patterns.

The nature of the relation between the magnitude of exposure to an environmental chemical and its effects on behavior is familiar to psychologists, as well as to pharmacologists and toxicologists. It takes the general form of an ogival or cumulative normal population curve. Very low levels of exposure produce no effects. As exposure increases, a level is reached where behavioral effects begin to appear, the "basal threshold." Further increases induce proportional changes in behavior until a level is reached, the "terminal threshold," at which behavioral malfunctions begin to occur. Activity prior to the terminal threshold is characteristic of self-regulatory, self-correcting biological processes, to which the term "homeostasis" has been applied. However, there are limits to this plasticity. A premise on which the concepts of terminal threshold and of behavioral plasticity depend is that some low magnitude of exposure exists for all chemical substances which will not produce an effect no matter how long the exposure. Coupled with this is the corollary that *all* substances will produce an effect at some higher level of exposure. It follows from this that any chemical introduced into the physical environment may set an "ecological trap" for the behavior of living organisms.

An example is presented next which illustrates major points drawn from studies of both human and animal models designed to provide information about chemicals affecting the cholinergic neurotransmitter system. These chemicals are involved in what may well be the widest diversity of purposes of any substances known today. They are applied therapeutically, a new anticholinesterase (antiChE) presently undergoing mass clinical trials for potential treatment of Alzheimer's disease. They appear in both indoor and outdoor environments as pesticides. Some were developed but not used during World War II as the so-called nerve gases. They have a basic neurochemical mechanism in common.

CHOLINESTERASE IN PROGRESSIVE DEGENERATIVE DEMENTIAS

The role of acetylcholinesterase (AChE) is normally associated with the inactivation (hydrolysis) of the neurotransmitter, acetylcholine (ACh) once the transmitter has been released into the synaptic cleft and, therefore, beyond the presynaptic side of the cholinergic synapse (Figure 3). However, evidence has been accumulating that AChE also appears to be involved presynaptically. Results of recent experiments have demonstrated that presynaptic AChE and the high-affin-

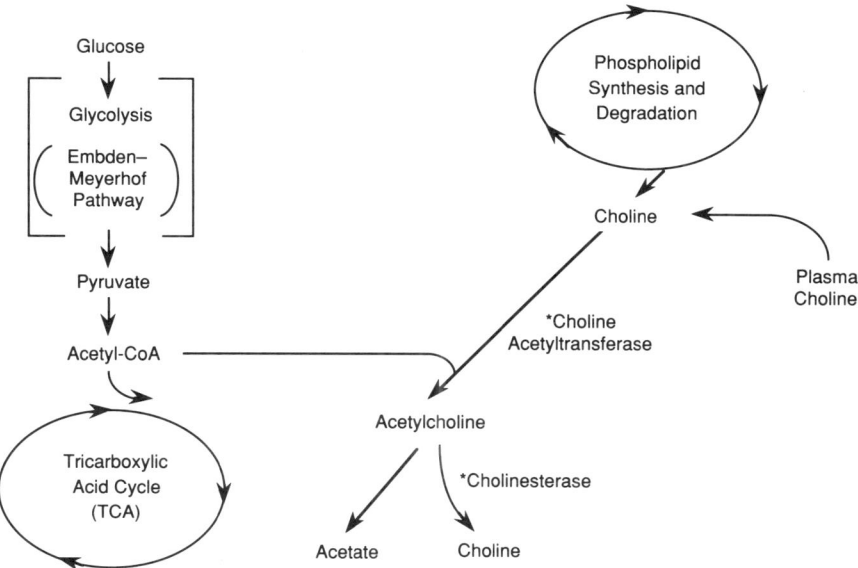

FIGURE 3 Metabolic pathways involved in the normal biosynthesis and hydrolysis of the neurotransmitter acetylcholine.

ity transport (HAChT) of the acetylcholine precursor, choline, are localized very close to each other on the cholinergic terminal membrane, suggesting some functional relationship between the two mechanisms (Raiteri et al., 1986). Furthermore, the existence of molecular heterogeneity (Atack et al., 1983; Michalek et al., 1981) and the different cellular distributions of AChE support the view that the enzyme may have more functions than hydrolyzing ACh postsynaptically (Greenfield, 1984).

Preclinical Studies of Acetylcholinesterase and Behavior

The possibility that manipulation of presynaptic AChE independently from the postsynaptic enzyme might produce differential effects on behavior awaits the invention of techniques for varying the one without the other in the intact organism. Meanwhile, it is relevant to the development of the present theme to summarize briefly the nature of behavioral effects when available antiChEs are employed as pharmacological tools.

Early experiments using animal models concluded that exposure to antiChEs produced differential effects on behavior, some behavior patterns being affected and others not. Behaviors affected involved the extinction of old responses that were no longer appropriate in coping with new environmental demands (Russell, 1958). More recent research has made it quite clear that cognitive behaviors (learning and memory) are particularly sensitive to manipulation of AChE activity. Dose-effect relations indicate that, behaviorally, the cholinergic system is capable of adaptive changes only within limits. Behavioral subsensitivity characterizes levels of AChE activity below this "normal" range and supersensitivity, levels above it. Activity at both extremes is associated initially with nonadaptive responses. However, prolonged changes in AChE activity at these extremes may initiate compensatory mechanisms within the cholinergic system [e.g., downregulation of muscarinic receptors (mAChRs), differential recovery of AChE isoenzymes] that are paralleled by the return of behavior to normal. Such "tolerance development" is an important form of behavioral homeostasis and also has significant implications for the use of antiChEs (e.g., physostigmine) as therapeutic agents in neurodegenerative disorders involving hypofunctioning of the cholinergic system [e.g., Alzheimer's disease (DAT)].

Acetylcholinesterase in Behavioral Disorders

Major neurochemical, morphological, and behavioral symptoms of DAT are summarized in Figure 4. Available evidence indicates that

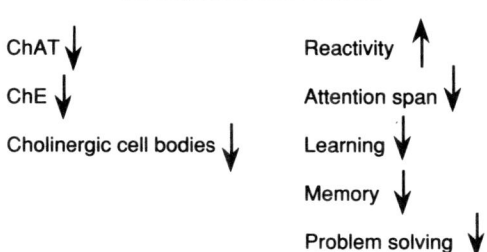

FIGURE 4 Major symptoms of Alzheimer's disease (DAT). Levels of activity of the synthesizing choline acetyltransferase (ChAT) and inactivating cholinesterase (ChE) enzymes are decreased. Relatively selective loss of cholinergic neurons occurs in certain regions of the brain, particularly the projections from large cholinergic cell bodies in the basal forebrain (nucleus basalis of Meynert) to the neocortex and projections from the medial septum to the hippocampus. Behavioral effects are characterized by hyperreactivity and by decreases in attention span, learning, memory, and other cognitive functions.

AChE activity is decreased in DAT: ". . . the first report of an altered distribution of acetylcholinesterase molecular forms in a disease of the central nervous system" distinguished three such forms in postmortem tissues from both the normal and the DAT neocortex (Atack et al., 1983). Losses in activity levels appeared selectively in the intermediate form assayed in DAT samples. Involvement of presynaptic AChE in human behavioral disorders is further suggested by the fact that deteriorative neuronal changes found in DAT are rich in AChE. As these changes increase, the AChE activity decreases. The results of such investigations are interpreted as indicating that changes in cortical cholinergic innervation are an important feature in pathogenesis and progressive development (PDD) (Struble et al., 1982).

The possibility that antiChEs, alone or in conjunction with other means for manipulating the cholinergic system when it is hypofunctional, might serve a therapeutic purpose has been under consideration for several years. Indeed, physostigmine continues to be a therapeutic strategy by which the half-life of ACh in the synaptic cleft is prolonged by decreasing its hydrolysis (inactivation). The varied success obtained, as well as the difficulties (i.e., short half-life, peripheral side effects, very narrow therapeutic window) involved in the clinical application of this particular compound, are reflected in a number of reports during the past decade (Davis and Mohs, 1982).

The PDDs considered in this example have characteristic behavioral sequelae that include significant deterioration of memory and other cognitive functions such as language, spatial or temporal orientation, judgment, and abstract thought. These behavioral changes are not readily discernible during early stages of the disorders and, even later, are difficult to differentiate from senile dementia accompanying the aging processes. It is generally accepted that final confirmation of DAT depends upon evidence of cellular degeneration in a particular area of the brain. The possibility that at least some PDD may result from exposures to chemicals in the external environment has been recognized in a recent publication on dementing diseases by the U.S. National Institutes of Health (1987).

The major points emphasized here are that detailed analyses of neurochemical mechanisms of action underlying behavior can provide (1) knowledge about differential behavioral effects of different toxicants upon which differential diagnostic criteria may be established; (2) information necessary for regulating exposures; (3) rational bases for patient management in neurodegenerative disorders, thereby eliminating procedures that have high probabilities of being unsuccessful.

THE TRUTH ABOUT A FALSE TRANSMITTER

Some half-century ago the possibility that a false precursor leading to the synthesis of a false transmitter might serve as a means for examining neurotransmitter systems at a molecular level began to receive attention. In vivo studies of choline analogues began to appear in scientific journals. Criteria that must be satisfied if a compound is to be accepted as a false transmitter were established. The first direct demonstration that a choline analogue, triethylcholine, could be acetylated and released in cholinergic synthesis of false transmitters, might provide valid animal models for studying clinical states involving dysfunctions in the central nervous system (CNS). A few experiments were designed to use such behaviors. The relevance of the early studies to PDD was recognized when it became apparent that the lack of natural precursor chemicals in the diet or the presence of a false precursor could significantly alter normal functioning of the cholinergic system and thus affect behavior (Jenden et al., 1987).

A major difficulty faced in the earlier studies was to demonstrate that chronic dietary administration of a choline analogue did in fact result in functionally significant replacement of choline in the synthesis of an endogenous analogue of ACh. Ways in which the availability of quantitative analytical techniques eventually solved this problem are illustrated in a series of experiments in our laboratories reported during the past five years, involving the false precursor N-amino-N,N-dimethylaminoethanol (N-aminodeanol, NADe) (Newton and Jenden, 1985): NADe is taken up by the choline transport system in competi-

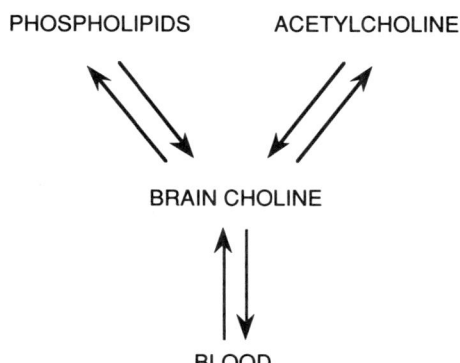

FIGURE 5 Competition for available choline. One hypothesis for the selective vulnerability of cholinergic neurons in Alzheimer's disease asserts that a competition for available choline occurs between biochemical pathways involved in maintaining the membrane integrity of cholinergic neurons (phospholipids) and in synthesizing the neurotransmitter (acetylcholine).

tion with choline. It is acetylated by choline acetyltransferase (ChAT), stored as O-acetyl-N-aminodeanol (ANADe) in vesicles, and released on stimulation. Stores of ACh are depleted as they are replaced with NADe. Upon release, ANADe interacts with both muscarinic and nicotinic receptors and is hydrolyzed by AChE. Because the potency of ANADe at these receptors is only 4 and 17 percent that of ACh, respectively, there occurs a profound interference with cholinergic transmission, particularly at muscarinic sites. Replacement of choline with NADe in the diet of weaning rats for periods of 60–120 days results in the replacement of 85–95 percent of free choline by free NADe in brain, plasma, and peripheral tissues; ChAT is reduced in the cortex, hippocampus, striatum, and ileum, suggesting the loss of cholinergic neurons. This evidence shows that NADe satisfies the neurochemical requirement of a false precursor, leading to the synthesis of a false cholinergic transmitter, and enables the study of these behavioral and physiological consequences.

We had two major objectives in mind. The first was to test a hypothesis about the etiology of DAT, and the second, to develop a useful animal model for the disorder. According to the hypothesis, competition for available choline is the central element (Jenden, 1986). Cholinergic neurons use choline both for the synthesis of the neurotransmitter ACh and, via phospholipid metabolism, as a structural component in cell membranes (Figure 5). The hypothesis states that when there exists a deficiency in the normal supply of choline for use in the two biochemical pathways involved, cholinergic neurons may break down membrane phosphatidylcholine to maintain the choline concentration required for ACh synthesis. This could lead to what has been imaginatively termed "autocannibalism" of the neuron (Wurtman et al., 1985). Conditions under which such degeneration could occur include (1) a relatively short supply of choline and (2) the overuse of choline for ACh synthesis. It has been suggested that "this competition could be precipitated by a number of inherited or acquired characteristics, is likely to be age-related and could result in failure of synaptic transmission, of mechanisms for cell membrane renewal or both" (Jenden, 1986). Although this hypothesis has not yet been put to an adequate test, it is consistent with a number of facts already available. Our approach to an experimental test involves the replacement of choline with its analogue NADe.

General observations during our early experiments failed to show any gross neurobehavioral toxicity, although the animals showed a discernible hypertonia when handled. This did not, of course, mean that the replacement of choline by NADe had no concomitant behavioral or physiological effects. For this reason an extensive series of experiments has now been carried out using quantitative assays of variables known to be cholinergically "coded." Initially these variables were measured only between 25 and 32 days on the NADe diet, yet even after this relatively short time, they showed some significant effects (Newton et al., 1986). A more extensive series of experiments (as yet unpublished) has now been completed, measuring many more variables periodically as the level of replacement of choline by NADe increased progressively over a much longer time. The results provide the data necessary to relate magnitudes of behavioral and physiological changes to levels of the false transmitter. There were no significant effects on total caloric intake, on the maintenance of body fluid balance, or on core body temperature, results which strongly suggest that the behavioral effects described below are unlikely to be attributable to imbalances in basic homeostatic mechanisms. The behavioral changes concomitant with increasing replacement of choline by NADe may be summarized in three major categories. The first includes measures that are primarily *sensory-reflexive* in nature, i.e., innate, appearing without the necessity for learning (e.g., reflexive responses to electric shock and acoustic stimuli). Such behaviors were affected significantly, being evidenced in sensory hypersensitivity and motoric hyperreactivity. Clearly apparent in DAT and other degenerative disorders of aging are symptoms of a *sensory-perceptual* nature, e.g., failure to "understand" events in the physical and psychosocial environments, spatial disorientation, and eventually a general loss of response to most stimuli. The effects were reminiscent of those observed in earlier experiments in which the transport of choline to the site of ACh synthesis in nerve endings was impaired by a specific inhibitor, hemicholinium-3 (HC-3), administered intracerebroventricularly (Russell and Macri, 1978), and of those found in septally lesioned animals with reduced cholinergic activity. The third category of behavioral effects included learning, memory, and other *cognitive processes*. Because these are among the early signs of dysfunction in degenerative disorders involving the cholinergic system, they received particular attention and were studied in several different test situations. Measures that are intrinsically dependent upon the functioning of the cholinergic system showed a consistent pattern of effects: NADe animals took more trials to learn, made more errors, were slower in their response times, and had poorer memories. Furthermore, these effects increased progressively as the available supplies of choline decreased.

Earlier in this chapter the hypothesis of autocannibalism was introduced as a possible mechanism underlying behavioral disorders associated with hypofunctioning of the cholinergic system, a condition that might be a consequence of competition for available choline supplies between the needs for it to maintain the membrane integrity of cholinergic neurons and to synthesize ACh. In the present experiments the supply of choline was progressively replaced by the much less efficient NADe. In the continuing competition, both free and lipid-bound choline were very significantly reduced. Paralleling these reductions were highly significant impairments of behavioral variables, the changes being analogous to those characteristic of such primary degenerative disorders as DAT. Research in our laboratory is now underway to study still further the neurochemical, behavioral, and histopathological properties of the model to determine whether it can be influenced by drugs intended to enhance cholinergic function and to discover the extent to which the various effects may be reversed when the diet is returned to normal. At the present time we believe that it is a potentially useful model for research on DAT and similar human disorders. We also believe that it may have significant implications for their pathogenesis.

The example discussed in this section of my trilogy illustrates mechanisms of action by which chemicals included (or not included) in the daily diet may induce malfunctions of normal behaviors. It also shows how the effects produced may be considerably delayed in their appearance. Earlier in this volume, Dr. Spencer provides some very striking examples of "long-latency neurotoxic disorders" as evidenced in humans and in animal models (see Spencer, 1987; Spencer et al., 1987). Since 1930 an increasing concern has been developing about such delayed neuropathies and their behavioral components. Particular attention has been directed to populations at special risk (i.e., pregnant women, young children, and workers exposed occupationally). However, concern generalized to populations at large has become evident in such regulations as those requiring the labeling of foodstuffs for their chemical compositions. Knowledge about the neurochemical modes of action underlying behavioral effects can contribute not only to regulation, but also to the invention of procedures for early detection of delayed neurotoxicities and to treatment of them once they are recognized.

NEUROTROPHIC FACTORS AND BEHAVIOR[1]

The third part of this trilogy involves hypotheses about interactions of brain "transplants," neurotrophic factors, and behavior. For reasons that will become apparent, this relatively new area of study is already at one of the more exciting frontiers in the biomedical sciences. In 1981, in a paper "A Unifying Hypothesis for the Cause of Amyotrophic Lateral Sclerosis, Parkinsonism, and Alzheimer Disease," Appel presented the thesis ". . . that each of these disorders is due to lack of a disorder-specific neurotrophic hormone" (Appel, 1981). The hypothesis postulated that the failure of target tissues to supply necessary neurotrophic factors is the primary manifestation of disorders: ". . . in each system, the lack of an appropriate hormone released from postsynaptic cells would impair the viability of the presynaptic cells" (Appel, 1981). More specifically, in DAT the failure would be in the production of nerve growth factor (NGF) by hippocampal and cortical cells, resulting in a gradual deterioration of septal and basal nuclei and associated decreases in ChAT activity and ACh synthesis (Appel et al., 1986; Hefti, 1983). Because knowledge of such processes derived primarily from research on animal models, a basic step toward testing this hypothesis was to establish the presence of NGF in the human brain. The evidence now indicates ". . . that NGF acts as a trophic factor for cholinergic neurons in the human brain in a similar way as has been established in recent years for the rat brain" (Hefti et al., 1986). Furthermore, "the similarity between the response in rodents to entorhinal cell loss and that in AD [Alzheimer's disease] patients indicates that studies using the rodent model may be directly applicable to AD" (Cotman et al., 1986).

Neurotrophic Factors

Neurotrophic Factors and Their Environments

The term "trophic" has been defined as ". . . any relatively long term influence that passes from one cell or tissue to another either during development or in the mature state" (Varon and Bunge, 1978). Interest in the remarkable ways in which neurons form synapses and become affilitated with their eventual target cells and tissues during embryonic, fetal, and neonatal growth raised questions about the dynamic processes involved. In the early 1950s, Nobel Laureate R.W. Sperry proposed that during maturation of the nervous system, genetic "... specification of the neurons makes possible the formation of selective synaptic linkages on the basis of a chemoaffinity" (Sperry, 1951). Although this concept was on a fruitful path, it was not until somewhat later that the current model took a definite shape, with the hypothesis that neurotrophic factors might be produced by other "target" tissues to provide a favorable cellular environment for axonal generation and regeneration. Initially, the physiological role and the distribution of NGF were well characterized peripherally as synthesized by target tissues innervated by sympathetic and certain sensory neurons and taken up by those neurons to be transported retrogradely to their cell bodies. More recent evidence, derived from studies of both human material and animal models, supports the involvement of NGF in an analogous role in the CNS. In the brain, as in the periphery, NGF is elaborated by target cells and binds to specific receptors on the innervating neurons.

Behavioral adjustments (including cognitive processes) to ever-changing physical and psychosocial environments require a multitude of dynamic biochemical events that take place in defined morphological sites within the CNS. There is a rapidly growing body of knowledge describing mechanisms by which these sites are themselves influenced by biochemical processes. It appears that neurotrophic factors determine which neurons will survive during ontogeny and thereby regulate the development of neural pathways. Present results also indicate that (1) there may be a large class of neurotrophic factors, each specific to particular neuronal populations; (2) they regulate neuronal survival during adulthood as well as during earlier development; and (3) inadequate neurotrophic activity may lead to neurological malfunctions arising from nerve cell death. It has been proposed that, phylogenetically, "... specific heritable, trophic interactions during development, which determine cell survival and pathway size, form a substrate for neural evolution" (Black, 1986). Clearly the growth of neuronal cell structures to synapse on other "target" tissues is a process of great importance for the normal functioning of the nervous system and hence of the ability of individuals to cope with the demands of their environments.

Award of the 1986 Nobel Prize for Physiology or Medicine to Rita Levi-Montalcini and Stanley Cohen for their discoveries of neurotrophic factors called special attention to the potential significance of these agents not only neurologically, but also behaviorally. Results of Levi-

Montalcini's early research (1964) had suggested the possibility that interactions might occur between behavior and biochemical events involved in the synthesis or release of nerve growth factors. "It seems a reasonable hypothesis that there may exist a mechanism by which during the coding of new behavioral patterns, the effects of information input on protein synthesis increase the production of protein molecules capable of modifying the structure of nerve or glial cells. The structural modifications would then serve as engrams for long-term memory storage" (Russell, 1966). This prediction came to have special meaning when it was established that deficiencies in memory were about the most prominent characteristics of DAT. More recent research makes it clear that some special relationship does in fact exist between NGF and the cholinergic neurotransmitter system, with behavior as the third among three partners in the relationship.

Roles During Neuronal Development and Beyond

Early development of the nervous system is characterized by a very significant overproduction of cells. During a circumscribed phase of embryonic life, neuronal degeneration occurs by which a high percentage of these cells die, despite the fact that they develop properties of mature neurons. The degeneration generally coincides with the arrival of surviving cells in their target area, indicating that developmental survival depends upon NGF derived from target cells. The development of neural pathways and connections is similarly dependent, with neurons whose axons have extended to an inappropriate target area being eliminated. It has been suggested that during early development of the mammalian brain the "transient cells" ". . . function as neurons in a synaptic circuitry that disappears by adulthood" (Chun et al., 1987).

In addition to its role during early development, NGF may be involved in at least two series of events vital to the behavior of living organisms: (1) the maintenance and viability of a normally functioning mature nervous system, and (2) the modification of neural structures during behavioral adjustments to changing physical and psychosocial environments (e.g., memory). Indeed, it has been suggested ". . . that NGF may function not only as a trophic agent, but also as a modulator of neurotransmission in the CNS" (Rennert and Heinrich, 1986).

Recovery from Morphological Lesions

Evidence that central neurons exhibit a capacity for axonal sprouting and generation of new connections in response to injury, not only in the developing but in the mature CNS, has been accumulating for some time. In peripheral models, axotomy is followed by a dramatic increase in the density of NGF receptors in Schwann cells through which axonal regeneration must occur. In the CNS, lesions of the septohippocampal pathway are followed by an increase in NGF content in the hippocampus and septum. Of major interest is the fact that, both centrally and peripherally, NGF-supported regeneration occurs following injury to the cholinergic system. That the regeneration results in "correct wiring" (i.e., in functional connections) is evidenced by electrophysiological activity and the recovery of behaviors disrupted by the original lesion. The use of immunohistochemical techniques has made it possible to examine the morphological effects of NGF on axotomized septal neurons (Gage et al., 1988): infusion of NGF protected most of the immunoreactive neurons from degeneration and prevented the appearance of plaque-like neurons.

Findings that suggest the occurrence of neuronal death following brain lesions are based primarily on the detectability of neurotransmitter-related enzymes. Very recently results of experiments using special staining techniques have demonstrated ". . . that the initial loss of ChAT-positive neurons following fimbria-fornix transection is due mainly to a reduction of the ChAT-stainability rather than actual neuronal death" (Hagg et al., 1988). Apparently, during the initial period, neurons may be in a state of reversible trauma, before irreversible damage sets in. The behavioral characteristics of tissue transplants may be consequences of the ". . . activation of 'silent' pathways by neurotrophic factors. . . " (Stein and Mufson, 1987).

Brain Transplants

Early conceptualization of the roles of NGF suggested that, if the deficiencies underlying behavioral changes and neuronal degeneration in disorders such as DAT are insufficient concentrations of specific neurotrophic factors, it should be possible to manipulate these factors by administering them directly or by transplanting cells that are rich trophic sources. Research could lead to the chemical isolation, purification, and eventually, synthesis of the trophic molecule. Important steps have already been taken to implement these actions.

Information about the survival of brain transplants has come from studies of their effects, as well as from evidence of their continued existence. The capacity to survive within a host has been demonstrated repeatedly. Neurochemical and histological evidence, supplemented by the fact that neural transplants can reverse behavioral impairments induced by brain lesions, supports the conclusion that transplants not only survive but also influence the functional capacity of newly regenerated neural connections. The evidence also indicates the importance of the relationship between transplants and NGF: survival of a large number of cells and a considerably larger amount of nerve fiber formation occur in the presence of NGF. The question of whether the probability of survival may be affected by untoward immunological accidents has been investigated. A review of the present state of affairs has come to the conclusion that the CNS is a site where transplants may enjoy a "... prolonged, but not always indefinite, survival" (Mason et al., 1985).

Nerve Growth Factor, Brain Implants, and Behavior

It is now time to relate the information about neurochemical and morphological factors to behavior, the property heavily involved in the diagnoses of PDDs, in monitoring their progression, and in evaluating treatment outcomes. Again, DAT will be used as the main example, with special attention given to memory because of the central role it plays in normal coping behavior as well as in disease states.

Behavioral Recovery from Brain Lesions

During the past few years, relations between neurochemical events and changes in behavior have encouraged a spate of experimentation designed to manipulate NGF and neuronal regeneration as means of compensating for adverse behavioral effects of experimental brain lesions. Two experimental approaches have been used: (1) repeated (chronic) injections or infusion of NGF and (2) transplantation of neuronal or target tissue. Both of these are reported to have produced at least partial compensation for the effects of brain damage, restoring the pattern of cholinergic innervation and producing concomitant improvements in some but not all behavioral patterns.

A broad range of behaviors have served as dependent variables in the search for evidence that cellular changes following neuronal regeneration stimulated by implants or by cell suspensions are functional. Interest in motoric abnormalities in Huntington's disease has led to the development of animal models involving lesions produced by neurotoxins. Bilateral lesions in the striatum followed by bilateral fetal striatal implants reversed the spontaneous motor abnormalities induced by the lesions. Severe striatal neuronal cell loss and shrinkage following ibotenic acid lesions of the caudate-putamen produced hyperactivity that was completely compensated by "neural grafting," i.e., implantation of a dissociated cell suspension from fetal rat striatum into the lesioned sites. Cognitive deficits have been reduced in animals with frontal cortex implants after bilateral damage to the medial frontal cortex. Particular attention has been given to learning and memory, by using a variety of different assays. Retention has been found to be very significantly improved in lesioned animals following implantation of cholinergically rich cells when measured by such well-established techniques as spatial alternation, inhibited (passive) avoidance, and learning and memory in water or radial maze situations.

These examples establish that exogenous NGF and tissue implants may produce neural regeneration and concomitant full or partial recovery of behaviors impaired by brain lesions. The behavioral effects appear to be differential in the sense that some behaviors may be affected and others, not. Recovery may be transient or long range depending upon the procedure used to induce increases in NGF activity.

As knowledge about the effects of NGF and of transplanting tissues into brain grew at the basic bioscience level, there began to appear an interest in its implications for human therapy. In 1985, reports appeared describing the "first clinical trials," apparently begun some three years earlier, involving transplantation of adrenal medullary tissue into the striatum of patients with severe Parkinson's disease. The transplantations were autologous ("autografts"), i.e., involving tissues within the same individual. The results of a series of "continuing clinical experiments" by this group of investigators are summarized in the statement: "The brief and limited response in our patients after the transplantation seems to indicate a limited survival of the graft" (Backlund et al., 1987). Greater success has been reported by other investigators: "Our results suggest that grafting chromaffin cells in direct contact with both the cerebrospinal fluid and the caudate nucleus produced excellent amelioration of most of the clinical signs of Parkinson's disease in our two patients" (Madrazo et al., 1987).

Results of research on animal models had indicated greater success with transplants of fetal tissue than with tissue taken later in development. In 1988, results were reported of transplants in two patients with Parkinson's disease of tissue from a spontaneously aborted fetus of 13 weeks. One patient received transplants from the fetal substantia nigra; the other, from the fetal adrenal medulla. The grafts were placed within a cavity of the right caudate nucleus and in contact with CSF. Significant improvement was recorded in both cases. The investigators concluded that". . . the use of fetal tissue as donor grafts may prove superior to autografting to treat Parkinson's disease" (Madrazo et al., 1988).

As happens with most "breakthroughs" at a frontier of science, this picture is not as clear as it must eventually become. Disparities between the findings reported by various investigators have cast a "cloud over Parkinson's therapy" (Lewin, 1988). A final evaluation of the effectiveness of procedures designed to use endogenous chemicals (e.g., NGF) in the treatment of neurodegenerative disorders awaits much fuller information than has yet appeared publicly. The rationale is clear and research on animal models has made it persuasive. Even if its validity is established, ethical and legal questions must be answered.

CONTRIBUTIONS IN THE FUTURE

About a decade ago a committee of the National Research Council, investigating "Decision Making for Regulating Chemicals in the Environment" reported: "All difficult decisions are characterized by inadequate information. . . . Problems of regulating chemicals in the environment are particularly beset with information characterized by a high degree of uncertainty. For some aspects of these problems there exists no information at all" (National Research Council, 1975). One of the major objectives of this trilogy has been to indicate how such uncertainties may be narrowed by basic and clinical research designed to understand the neurochemical mechanisms by which exposures to toxicants affect behavioral and other biological indicators. Procedures may be developed by which such processes can be observed and measured. The procedures can provide more precise information about the relation between the amount of a toxicant reaching receptors in its target tissues ("effective dose") and the consequent effects on biological endpoints. The study of neurodegenerative diseases can also provide opportunities to follow the progression of chemically induced malfunctions in the nervous system to which behavior has been shown to be especially sensitive. It can be hoped that by relating behavior to other biological properties of living organisms, those responsible for regulation can be convinced that behavioral endpoints should be introduced more fully into the regulatory arena (Buckholtz and Panem, 1986).

NOTE

1. A more extensive discussion of this subject, including a more complete list of references, appeard in a recent review (Russell, 1988).

REFERENCES

Appel, S. H. 1981. A unifying hypothesis for the cause of amyotrophic lateral sclerosis, Parkinsonism and Alzheimer disease. Annals of Neurology 10:449–505.

Appel, S. H., Y. Tomozawa, and R. Bostwick. 1986. Trophic factors and neurologic disease. Pp. 75–85 in Alzheimer's and Parkinson's Diseases, A. Fisher, I. Hanin, and C. Lackman, eds. New York: Plenum.
Atack, J. R., E. K. Perry, J. R. Bonham, R. H. Perry, B. E. Tomlinson, G. Blessed, and A. Fairbairn. 1983. Molecular forms of acetylcholinesterase in senile dementia of the Alzheimer type: Selective loss of intermediate (10s) form. Neuroscience Letters 40:199.
Backlund, E. O., L. Olson, A. Seiger, and O. Lindvall. 1987. Toward a transplantation therapy in Parkinson's disease: A progress report from continuing clinical experiments. Annals of the New York Academy of Sciences 495:658–673.
Black, I. B. 1986. Trophic molecules and evolution of the nervous system. . Proceedings of the National Academy of Sciences 83:8249–8252.
Buckholtz, N. S., and S. Panem. 1986. Regulation and evolving science: Neurobehavioral toxicology. Neurobehavioral Toxicology and Teratology 8:89–96.
Chun, J. J. M., M. J. Nakamura, and C. J. Shatz. 1987. Transient cells of the developing mammalian telencephalon are peptide-immunoreactive neurons. Nature 325:617–620.
Cotman, C. W., M. Nieto-Sampedro, and J. W. Geddes. 1986. Synaptic plasticity in the hippocampus: implications for Alzheimer's disease. Pp. 99–117 in Treatment Development Strategies for Alzheimer's Disease, T. Crook, R. T. Bartus, S. Ferris, and S. Gershon, eds. Madison, Conn.: Mark Powley Associates.
Davis, K. L., and R. C. Mohs. 1982. Enhancement of memory processes in Alzheimer's disease with multiple-dose intravenous physostigmine. American Journal of Psychiatry 139:1421–1424.
Doull, J. 1980. Factors influencing toxicology. Pp. 70–83 in Toxicology: The Basic Science of Poisons, J. Doull, C. D. Klaasen, and M. O. Amour, eds. New York: Macmillan.
Gage, F. H., D. M. Armstrong, L. R. Williams, and S. Varon. 1988. Morphological response of axotomized septal neurons to nerve growth factor. Journal of Comparative Neurology 269:147–155.
Greenfield, S. 1984. Acetylcholinesterase may have novel functions in the brain. TINS 7:364–368.
Hagg, T., M. Manthrope, H. L. Vahlsing, and S. Varon. 1988. Delayed treatment with nerve growth factor reverses the apparent loss of cholinergic neurons after acute brain damage. Experimental Neurology 101:303–312.
Hefti, F. 1983. Alzheimer's disease caused by a lack of nerve growth factor? Annals of Neurology 13:109–110.
Hefti F., J. Hartkka, A. Salvatierra, W. J. Weiner, and D. C. Mash. 1986. Localization of nerve growth factor receptors in cholinergic neurons of the human basal forebrain. Neuroscience Letters 69:37–41.
Jenden, D. J. 1986. The pharmacology of cholinergic mechanisms and senile brain disease. Pp. 205–215 in The Bioloqical Substrates of Alzheimer's Disease, A. B. Scheibel and A. P. Wechsler, eds. New York: Academic Press.
Jenden, D. J., R. W. Russell, R. A. Booth, S. D. Lauretz, B. J. Knusel, M. Roch, K. M. Rice, R. George, and J. J. Waite. 1987. A model hypocholinergic syndrome produced by a false choline analog, N-aminodeanol. Journal of Neural Transmission (Suppl.) 24:325–329.
Levi-Montalcini, R. 1964. Growth control of nerve cells by a protein factor and its anti-serum. Science 143:105–110.
Lewin, R. 1988. Cloud over Parkinson's therapy. Science 240:390–392.
Madrazo, I., R. Drucker-Colin, V. Diaz, J. Martinez-Mata, G. Torres, and J. J. Becerril. 1987. Open microsurgical autograft of adrenal medulla to the right caudate nucleus in two patients with intractable Parkinson's disease. New England Journal of Medicine 316:831–834.

Madrazo I., V. Leon, C. Torres, C. Aguilera, G. Varela, F. Alvarez, A. Fraga, R. Drucker-Colin, F. Ostrosky, M. Skurovich, and R. Franco. 1988. Transplantation of fetal substantia nigra and adrenal medulla to the caudate nucleus in two patients with Parkinson's disease. New England Journal of Medicine 318:51.

Mason, D. W., H. M. Charlton, A. Jones, D. M. Perry, and S. J. Simmons. 1985. Immunology of allograft rejection in mammals. Pp. 91–98 in Neural Grafting in the Mammalian CNS, A. Bjorklund and U. Stenevi, eds. Amsterdam: Elsevier.

Michalek, H., G. M. Bisso, and A. Meneguza. 1981. Comparative studies on rat brain acetylcholinesterase and its molecular forms during intoxication by DEP and paraoxon. Pp. 847–852 in Cholinergic Mechanisms: Phylogaretic Aspects, Central and Peripheral Synapses and Clinical Significance, G. Papeu and H. Ladinsky, eds. New York: Plenum.

National Research Council. 1975. Decision Making for Regulating Chemicals in the Environment. Washington, D.C.: National Academy Press.

Newton, M. W., and D. J. Jenden. 1985. Mechanism and subcellular distribution of N-amino-N,N-dimethylaminoethanol (N-aminodeanol) in rat striatal synaptosomes. Journal of Pharmacology and Experimental Therapeutics 235:135–146.

Newton, M. W., R. D. Crosland, and D. J. Jenden. 1986. Effects of chronic dietary administration of the cholinergic false precursor N-amino-N,N-dimethylaminoethanol on behavior and cholinergic parameters in rats. Brain Research 373:197–204.

Raiteri, M., M. Marchi, and A. M. Caviglia. 1986. Studies on a possible functional coupling between presynaptic acetylcholinesterase and high-affinity choline uptake in the rat brain. Journal of Neurochemistry 47:1696–1699.

Rennert, P. D., and G. Heinrich. 1986. Nerve growth factor mRNA in brain: Localization by in situ hybridization. Biochemistry and Biophysics Research Communications 138:813–818.

Russell, R. W. 1958. Effects of "biochemical lesions" on behavior. Acta Psychologica 14:281–294.

Russell, R. W. 1966. Biochemical substrates of behavior. Pp. 185–246 in Frontiers in Physiological Psychology, R.W. Russell, ed. New York: Academic Press.

Russell, R. W. 1979. Neurotoxins: A systems approach. Pp. 1–7 in Neurotoxins: Fundamental and Clinical Advantages, I. Chubb and L.B. Geffen, eds. Adelaide, South Australia: University of Adelaide Press.

Russell, R. W. 1988. Brain "transplants," neurotrophic factors and behavior. Alzheimer Disease and Associated Disorders 2:77–95.

Russell, R. W., and J. S. Macri. 1978. Some behavioral effects of suppressing choline transport by cerebroventricular injection of hemicholinium-3. Biochemistry and Pharmacology Behavior 8:399–403.

Spencer, P. S. 1987. Guam ALS/Parkinsonism-dementia: A long-latency neurotoxic disorder caused by "slow toxin(s)" in food? Canadian Journal of Neurological Science 14:347–357.

Spencer P. S., P. B. Nunn, J. Hugon, A. C. Ludolph, S. M. Ross, D. N. Roy, and R.C. Robertson. 1987. Guam amyotrophic lateral sclerosis-parkinsonism-dementia linked to a plant excitant neurotoxin. Science 237:517–522.

Sperry, R. W. 1951. Mechanisms of neural maturation. Pp. 236–280 in Handbook of Experimental Psychology, S. S. Stevens, ed. New York: John Wiley & Sons.

Stein, D. G., and E. J. Mufson. 1987. Morphological and behavioral characteristics of embryonic brain tissue transplants in adult, brain damaged subjects. Annals of the New York Academy of Sciences 495:444–471.

Struble, R. G., L. C. Cork, P. J. Whitehouse, and D. L. Price. 1982. Cholinergic innervation in neuritic plaques. Science 216:413–415.

U.S. National Institutes of Health. 1987. Differential diagnosis of dementing diseases. Washington, D.C.: National Institutes of Health.

Varon, S. S., and R. P. Bunge. 1978. Trophic mechanisms in the peripheral nervous system. Annual Review of Neuroscience 1:327–361.

Wurtman, R. J., J. K. Blusztajn, and J. C. Maire. 1985. "Autocannibalism" of choline-containing membrane phospholipids in the pathogenesis of Alzheimer's disease—A hypothesis. Neurochemistry International 7:369–372.

Neurobehavioral Toxicity of Selected Environmental Chemicals: Clinical and Subclinical Aspects

Gerhard Winneke

The present chapter discusses established as well as controversial associations between human exposure to environmental chemicals and nervous system dysfunction and, in particular, examines if and to what extent psychological theory and methods may contribute to an early detection and evaluation of chemically induced neurotoxicity. Because animal models used in neurotoxicity testing for purposes of screening and of clarifying mechanisms of action, are covered in other parts of this volume, this chapter is restricted largely to findings in human populations which describe the outcome of exposure in terms of neurological or psychological dysfunction.

Many chemicals are known or suspected to affect nervous system functioning. Anger and Johnson (1985) point out that over 850 workplace chemicals can be classified as neurotoxic, and for 65 of them, the exposed population is estimated to exceed one million (Anger, 1986); it should be noted, however, that the nervous system may not be the primary target for some of these chemicals.

This review must be selective; therefore, only a few chemicals will be dealt with for which human environmental exposure is likely to occur, and for which neurotoxic effects in humans have been reported by using signs and symptoms of neurological or psychological deterioration as the endpoint. This is true for some metals (e.g., aluminum, lead, and mercury), as well as for some organic compounds such as specific solvents and solvent mixtures as well as polychlorinated biphenyls (PCBs). Secondary neurotoxicity due to carbon monoxide-induced hypoxia and delayed neurotoxicity of organophosphorus (OP) compounds is not covered in this chapter.

CLINICAL VERSUS SUBCLINICAL NEUROTOXICITY

Nervous system (NS) diseases may roughly be grouped into the categories of focal or nonfocal syndromes (Schaumberg and Spencer, 1987). This distinction is based on pathology. Typical examples or focal syndromes are the neurodegenerative diseases associated with the process of aging, namely, Parkinson's (PD), Alzheimer's (AD), and motoneuron (MD) diseases (Calne et al., 1986). In PD the primary areas of cell death are the dopaminergic neurons in the zona compacta of the substantia nigra. Typical of AD is the loss of cholinergic neurons in the medial basal forebrain, and the primary focus of MD pathology is the loss of upper (higher brain areas) and lower (spinal cord brain stem) motor neurons.

Such circumscribed pathology is typically absent in clinical syndromes induced by many neurotoxic chemicals, which therefore may be called nonfocal syndromes (Schaumberg and Spencer, 1987). Clinical manifestations of toxicant-induced damage to the central nervous system (CNS) usually take the form of toxic encrephalopathies, whereas insult to the peripheral nervous system (PNS) may give rise to different types of neuropathy or polyneuropathy. Such full-blown clinical syndromes are rare in occupational and environmental exposure today, due to increased hygienic awareness, improved preventive countermeasures, and improved early diagnosis.

It is recognized however that subtle, insidious, and rather nonspecific alterations of NS functioning, often classified as subclinical effects indicative of asymptomatic neurotoxic disease, do occur, may even be widespread under certain circumstances (Schaumberg and Spencer, 1987), and thus constitute a real hazard at the workplace and in the environment—particularly with respect to sensitive subgroups of the population.

The solvent literature is a rich source of signs and symptoms often carrying the label "subclinical" effects. These may be grouped into the broad categories of subjective complaints, neurophysiological changes, and psychological test results. If the size of the observed effect is sufficiently large, or if manifestations from different outcome categories occur together and thus support each other, clinical syndromes well known from the solvent literature can be diagnosed. Examples, particularly from Scandinavian researchers, are the "psycho-organic syndrome," the "neurasthenic syndrome," or especially in Denmark, "presenile dementia," if the symptomatology is indicative of CNS involvement or, in case of predominant PNS involvement, neuropathy or polyneuropathy.

If, however, one is dealing with rather isolated effects instead, which can only be extracted from background noise by elaborate statistical procedures, the label subclinical is attached to them. What, now, are the main features of subclinical effects? First, and above all, they are usually weak, with individual values typically within the normal range of fluctuation. This creates diagnostic problems and usually means that statistical group comparisions rather than single cases form the source of information. Second, they are essentially nonspecific which, together with their small size, creates problems of interpretation in terms of causality. If, for example, age, alcohol consumption, or deprived social status is also associated with the exposure to neurotoxic agents, it is by no means a trivial task—and frequently an impossible one—to disentangle those cause-effect strings in which occupational or environmental neurotoxicology is interested. Third, although the observed effects must resemble aspects of the pathological condition (e.g., general slowing of mental speed) to justify their classification as subclinical, they never fulfill all the diagnostic attributes of the clinical condition. Fourth, for this very reason, it is often difficult to identify their implications for health/adjustment. Just two examples: How relevant is an average slowing of nerve conduction velocity by 4 m/s if this is still within the range of normal variability? How important is an average drop of IQ by 4 points, if it is still well above the expected mean value of 100? These are the types of questions being asked not only in the scientific community but also by health administrators and, above all, by the affected individuals who serve as subjects (Ss) in such studies.

After these more general remarks, specific examples illustrate how neurotoxic effects of some environmental chemicals are being characterized in terms of neurobehavioral deterioration and the extent to which psychological test methods can be used to detect early, subclinical signs of NS involvement before irreversible damage occurs. Although clinical conditions are described to some extent, the majority of findings deal with subclinical neurobehavioral effects in asymptomatic subjects.

NEUROBEHAVIORAL EFFECTS OF SELECTED INDUSTRIAL CHEMICALS

Chemicals of more widespread environmental impact, for which neurological and psychological effects have been described, are considered here. These include metals—namely, aluminum, lead, and mercury—and organic chemicals—namely, polychlorinated biphenyls and solvents. The emphasis here is on general population exposure, although experience from workplace exposure is taken into account whenever it helps to clarify the issue at hand. Potent biological neurotoxins from edible plants and animals, which have given rise to endemic neurotoxicity in developing countries (e.g., Spencer et al., 1986, 1987), are beyond the scope of this chapter.

Effects in the Developing Nervous System

For some chemicals, namely, lead and mercury, children have been shown to be more vulnerable than adults in terms of neurotoxicity. These chemicals are considered first; in addition, PCBs are discussed under this heading because recent evidence suggests that young children may also be at particular risk for environmental PCB exposure.

Lead

Lead has been used by man since antiquity, and its detrimental health effects have been well known for centuries. It is probably the best studied neurotoxic compound, and comprehensive reviews covering chemical, environmental, and biological aspects in great detail—e.g., the recent Environmental Protection Agency (EPA, 1986) report—should be consulted for more in-depth information. Chemically speaking, this metal occurs in inorganic form, namely in the form of lead salts of widely different water solubility, as well as in organic form. Although the organometallic compounds have been found to be highly neurotoxic in acute occupational exposure (Grandjean, 1984), chronic low-level exposure to inorganic lead constitutes a more important public health issue (Lansdown and Yule, 1986).

Inorganic lead enters the body by way of inhalation and ingestion; absorption is better in infants than in adults. Blood lead concentration (PbB) is an representative marker of current lead exposure, whereas tooth lead concentrations have been used as markers of past exposure (Needleman et al., 1979). Both placental transfer and blood-brain transfer of lead occur, so that prenatal exposure and CNS involvement are possible. Lead is considered a nonessential metal. Its toxicity may be explained largely by interference with different enzyme systems: lead inactivates these enzymes by binding to sulfydryl (SH) groups or by competitive interaction with other essential metal ions. Therefore, almost all organs or organ systems can be considered potential targets for lead: depending on duration and degree of exposure, a wide range of biological effects has been documented, the more critical of which are those on heme biosynthesis, erythropoiesis, and the nervous system. For a number of reasons related both to exposure and to CNS vulnerability, children between the ages of 9 months and 6 years are particularly at risk according to the Centers for Disease Control (CDC, 1985).

Both the peripheral and the central nervous systems may be involved in lead neurotoxicity, although PNS effects seem to be more prominent in occupational lead exposure of adults, whereas CNS involvement is more characteristic of childhood lead exposure. Acute symptomatic lead poisoning has been linked to the swallowing of lead-based paint and is often associated with encephalopathy at PbB exceeding 100 µg/dL (Chisolm, 1971). Lead encephalopathy is clinically characterized by some or all of the following symptoms (CDC, 1985): coma, seizures, ataxia, apathy, incoordination, vomiting, clouded consciousness, and loss of previously acquired skills. Children surviving lead encephalopathy typically present with neurological and psychological sequelae, including focal EEG abnormality, cramps, intelligence deficit, hyperactivity, distractibility, and reduced impulse control (Byers and Lord, 1943; Perlstein and Attala, 1966; Smith et al., 1963).

Such clinical findings have led to the hypothesis that long-term low-level childhood exposure to lead might be associated with subclinical neurobehavioral deficit in asymptomatic children as well, which due to its subtlety may often go undetected.

Since the early 1970s this hypothesis has been tested in about 30 cross-sectional studies using different psychological tests as well as behavior ratings to assess the degree of CNS involvement, and PbB or tooth lead levels as markers of current or past exposure. The variety of psychological functions and tests covered in these studies in different combinations may be given roughly as follows: psychometric intelligence in most studies was assessed by means of the Wechsler Scales (WISC, WPPSI), although other tests were used in some studies (e.g., the McCarthy Scales of children's ability, the Stanford-Binet, or the British Ability Scales). The following have been added to IQ measures in several of these studies: perceptual motor integration using the Bender Gestalt Test, the Benton Test, or the Frostig Scales; gross or fine motor coordination using the Purdue Pegboard; finger-wrist tapping and the Osertesky Motor Scales; reaction and attentional performance using delayed and serial choice reaction times; vigilance performance; behavior ratings by means of the Connor, Wherry-Weiss-Peters, and Rutter Scales, and measures of educational attainment.

Inconsistency of outcome, differences in study design, and confounder structure interfere with any simple, straightforward conclusion as to whether neurobehavioral deficit in asymptomatic children is truly associated with, or even caused by, low-level childhood lead exposure. A comprehensive review (EPA, 1986, p. 145) arrived at the following careful conclusions: "As for CNS-effects, none of the available studies on the subject, individually, can be said to prove conclusively that significant cognitive (IQ) or behavioral effects occur in children at blood lead levels <30 µg/dL." However, the most recent neurobehavioral studies of CNS cognitive (IQ) effects collectively demonstrate associations between neuropsychological deficits and low-level lead exposures in young children resulting in blood lead levels ranging to below 30 µg/dL. Some more recent cross-sectional studies (Fulton et al., 1987; Hatzakis et al., 1987; Norby-Hansen et al., 1989), not covered in the above mentioned review, generally agree with this conclusion, in that small but significant cognitive and attentional deficit was observed at low blood or tooth lead levels.

In addition to these cross-sectional approaches, typically lacking precise exposure histories, first results from several prospective studies are now beginning to be published, four of which have recently been reviewed (Davis and Svendsgaard, 1987). In all of these studies, repeated blood lead sampling starting at birth (cord blood) was done to describe the early exposure history, and repeated outcome assessment was done at regular intervals using the Bayley Scales of Infant Development as the instrument. In their review, Davis and Svendsgaard (1987, p. 299) conclude: "There can now be little doubt that exposure to lead, even at blood-levels as low as 10–15 g/dL, and possibly lower, is linked with undesirable developmental outcomes in human fetuses and children. These effects include impaired neurobehavioral development, reduced gestational age, lowered birth weight, and other possible effects on early development and growth."

In view of some divergent findings between studies, in the present author's opinion, this must be qualified as a bold statement, although it is certainly true that some of these results do raise concern about persistent neurobehavioral effects of low-level lead exposure at early stages of brain maturation. Such concern is supported by animal studies showing long-lasting neurobehavioral deficit in different species after perinatal lead exposure associated with blood levels below 30 µg/dL (Winneke, 1986). It should be added, however, that no convincing mechanism has yet been proposed to account for such deficit.

Mercury

Mercury also belongs to those metals known to, and used by, man since ancient times; for centuries it has been employed primarily for therapeutic purposes. For more detailed information on chemical, environmental, and biomedical aspects, comprehensive reviews should be consulted (Berlin, 1986; Clarkson et al., 1984).

There are different physical and chemical forms of mercury—namely, metallic, mercurous, and mercuric mercury. Vapors of metallic mercury are primarily an occupational problem and are not considered here; instead, the emphasis is on those stable organometallic compounds known as methylmercury (MeHg). It is through these compounds that increased environmental exposure may occur in segments of the general population because inorganic mercury released into the environment from a variety of sources is methylated by microorganisms present in bodies of fresh and ocean waters and thus enters the aquatic food chain. The highest MeHg concentrations have been measured in large predatory fish, such as shark or tuna. Consequently, populations with high fish consumption must be considered at risk from MeHg exposure.

Absorption of MeHg in the gastrointestinal tract is almost complete, and its distribution in the body is rather uniform. There is no placental barrier for MeHg, which is found in all fetal tissue; MeHg concentrations in fetal blood are typically higher than those of the mother. Methylmercury also enters the hair as soon as it is formed; thus, MeHg concentration in hair is an excellent noninvasive marker of exposure.

Knowledge about the neurological and psychological sequelae of high MeHg exposure was gained primarily in two catastrophic incidents of mass poisoning, namely, the Minamata Bay tragedy in Japan in 1950 (Harada, 1966) and an outbreak in Iraq 20 years later (Bakir et al., 1973). Population exposure in the Minamata incident was through contaminated fish from Minamata Bay, which had been polluted for years by metallic mercury from industrial sources; this was then methylated by marine microorganisms and thus introduced into the food chain. Thousands of people were exposed, and hundreds of cases of MeHg poisoning have been documented. Methylmercury exposure in Iraq was through ingestion of seed grain treated with an MeHg fungicide; the grain had been ground into flour to make bread. About 7,000 people were hospitalized with signs and symptoms of poisoning, and more than 400 of them died.

The clinical picture of MeHg poisoning is characterized by sensory, motor, and cognitive deficit. The earliest effects are nonspecific symptoms (e.g., complaints of paresthesia, general malaise, and blurred vision). Later on, signs of neurotoxicity appear, such as constricted visual field ("tunnel vision"), deafness, dysarthria, and ataxia. Mental disturbances and alterations of the chemical senses may occur as well. Clarkson et al. (1984) mention three important features of MeHg effects, namely, their irreversibility, their selective neurotoxic character with predominant CNS involvement, and the long latency between cessation of exposure and onset of symptoms, which may extend from a few weeks to several months or even years.

Whereas irreversibility of CNS effects is most likely due to loss of neurons, the reason for the long latency period is not known. Clarkson et al. (1984, p. 302) speculate that latency periods of several years "may be partially explained by psychogenic overlay which modifies the symptoms or subclinical lesions which may be revealed by the aging factor."

Another important feature of MeHg neurotoxicity is the particular vulnerability of the developing CNS, which has been observed in human cases as well as in animal models. In both the Minamata and the Iraq outbreaks, pregnant women with only minor symptoms of MeHg poisoning occasionally gave birth to children with severe CNS damage. The clinical picture was dose-dependent. At high maternal MeHg blood levels, microcephaly, hyperreflexia, and severe motor and mental impairment were prominent. For lower degrees of exposure, subtle deficits were difficult to diagnose shortly after birth but became increasingly pronounced later on. Psychomotor impairment and persistent abnormal reflexes were found at hair levels exceeding 50 mg/kg. The mildest cases presented with signs of the minimal brain dysfunction syndrome, characterized by hyperactivity and attention deficit (Amin-Zaki et al., 1974). The likelihood of mental retardation increased with increasing maternal MeHg hair levels. In the Minamata case, follow-up studies revealed strong associations between cord blood MeHg levels and mental retardation in 20-year-old victims of prenatal exposure (Harada et al., 1977).

Thus far, only two studies deal with subclinical signs of MeHg exposure at environmentally elevated levels. In one such study (McGill Group, 1980) in Cree Indians exposed to MeHg from fish, some associations were found between tone and reflexes of Cree boys and MeHg hair levels of their mothers during pregnancy. Such effects occurred at much lower MeHg levels than those previously found to be associated with neurotoxicity. Because these effects were mild and somewhat isolated, however, doubts as to their substantive nature have been raised (Clarkson et al., 1984). A more recent, ongoing study in New Zealand (Kjellstrom et al., 1986) used a pair-matching approach to study developmental retardation due to low-level in utero exposure to MeHg from fish at hair levels exceeding 6 mg/kg. Developmental status was assessed by means of the Denver Developmental Screening Test. From a basic cohort of 11,000 mother-child pairs, 31 with elevated MeHg levels between 6 and 20 mg/kg were compared with pair-matched controls. Significant dose-related developmental delay was found at age 4. Results from subsequent psychometric testing at age 6 have not yet been published.

Polychlorinated Biphenyls

Polychlorinated biphenyls typically are mixtures of several compounds differing in terms of number and position of chlorine substituents. In most industrialized countries, PCBs are no longer used in "open" systems but continue to be used in "closed" systems such as hydraulic pumps, transformers, or heat exchangers.

The PCB compounds are biologically persistent and, therefore, accumulate in the food chain. Marine mammals are a particular target: average PCB concentrations of 160 mg/kg of fat have been measured in marine mammals in the North Sea. In human fat tissue, average PCB concentrations are between 1 and 2 mg/kg of fat. There is an age-related increase of PCBs in fat tissue.

Toxicological effects of PCB exposure in man were first observed in the context of mass poisoning in Japan in 1968 (Kuratsune, 1972). The first and most obvious signs were skin affections resembling chloracne in about 1,000 persons. The cause of poisoning was found to be PCB-contaminated rice oil. Besides skin affections, the clinical picture of the disease, which soon became known as *yusho* (oil disease), was characterized among others by pigmentation of fingernails, alopecia, porphyria, and decreased concentrations of immunoglobulin M (IgM). An increased number of stillbirths and of small-for-age babies was observed.

Apart from such outbreaks of acute poisoning, low-level chronic PCB exposure occurs through foodstuffs. Human breast milk contains elevated PCB concentrations: average values of 1–2.5 mg of PCB per kilogram of fat were measured in West Germany (DFG, 1984). The PCBs are known to cross the placenta. Neonatal effects of transplacental PCB exposure have been studied recently by using psychological and neurological criteria (Jacobson et al., 1985; Rogan et al., 1986). In one of these studies (Rogan et al., 1986) the Brazelton Neonatal Assessment Scale was used to assess reflexive and motor behavior and to track the state of about 900 neonates regularly from the first three weeks after birth to 24 months of age. The PCB level was measured in cord and maternal serum at term. Multiple regression analysis revealed significant PCB associations only for hypotonicity and hyporeflexia; birth weight and head circumference were not related to PCB serum levels.

The second study examined the effects of transplacental and neonatal PCB from fish in 242 children born to women who ate fish from Lake Michigan and in children form non-fish-eating mothers (Fein et al., 1984; Jacobson et al., 1985). The PCB level in cord serum predicted lower birth weight and smaller head circumference (Fein et al., 1984). In addition, cognitive performance in the Visual Recognition Memory Test (Fagan and McGrath, 1981) at 7 months of age exhibited significant association with cord serum PCB levels but not with breast milk PCB levels after control for confounding (Jacobson et al., 1985). Thus, there is some evidence for the ability of prenatal PCB exposure to affect cognitive and neuromuscular development in the neonate, although unfortunately, possible simultaneous MeHg exposure was not taken into account.

EFFECTS IN THE ADULT NERVOUS SYSTEM

The effects of aluminum and solvents are considered here, the neurotoxicity of which has been studied only in adults so far.

Aluminum

Aluminum is abundant in the earth's crust. Its toxicity for humans has been rated low in the past because it was considered to be almost nonabsorbable from the gastrointestinal tract. It is now clear, however, that both inhaled and ingested aluminum is absorbable to some extent. For a detailed review of chemical, environmental, toxicological, and biomedical aspects, the reader is referred to Elinder and Sjogren (1986). Aluminum exists in organic and inorganic form. The inorganic aluminum salts have different water solubilities.

The neurotoxicity of aluminum was first detected in animal studies: Epileptic cramps and neurofibrillary degeneration were observed after direct brain injection or parenteral application of different aluminum salts (De Boni et al., 1976; Sorensen, 1974). Information about aluminum-induced neurotoxicity in humans was gained when dialysis patients developed a progressive dementing illness, which often proved fatal if untreated by chelation. Clinical signs of this type of brain damage include speech and motor disturbances, memory deficit, personality changes, dementia, and seizure disorders. Although there was some debate as to the etiological contribution of aluminum to this disease, which has become known as dialysis dementia, it is now accepted that the use of aluminum-containing phosphate-binding gels or of water with high aluminum content was the cause of this dementing illness. Reduced renal function may also lead to a significant accumulation of aluminum in the body, associated with dialysis dementia in patients who had never undergone dialysis treatment, but who had taken large doses of aluminum hydroxide.

A possible role of aluminum in the pathogenesis of Alzheimer's disease is being discussed (Crapper and De Boni, 1980). This hypothesis rests on partial similarity of dialysis dementia to presenile and senile features of Alzheimer's disease, in both clinical and pathological terms. It has been shown, for example, that in Alzheimer's patients, aluminum selectively accumulates in the nucleus of the brain cells that form the neurofibrillary tangles, typical of the Alzheimer condition; neurofibillary tangles have also been observed subsequent to injection of aluminum salts in cats and rabbits. In some studies, elevated aluminum levels in the gray matter of Alzheimer's patients with normal kidney function were found as well. There are, however, several contradictory findings, so that the evidence supporting an association between environmental aluminum exposure and Alzheimer's disease must still be considered circumstantial (Elinder and Sjogren, 1986).

Organic Solvents

Organic solvents represent a large, chemically heterogeneous group of chemicals which are liquids between 0 and 250°C. Traditionally they are used for the extraction, solution, or suspension of water-insoluble materials—namely, fats, lipids, resins, and polymers. Solvents may be grouped into aliphatic hydrocarbons (e.g., hexane), aromatic hydrocarbons (e.g., toluene), halogenated hydrocarbons (e.g., trichloroethylene), alcohols, ketones (e.g., methyl ethyl ketone), esters (e.g., butyl acetate), and mixtures (e.g., white spirit).

Due to the wide variety of applications, occupational exposure and, to a lesser degree, general population exposure are frequent. The nervous system is the primary target for inhaled solvents because of their lipophilic characteristics. Whereas narcotic action is the predominant biological effect in the CNS, functional and structural effects ranging from neurophysiological changes to severe polyneuropathies have been reported to occur in the PNS. Comprehensive reviews covering relevant aspects of chemistry, exposure, and biomedical effects are available [Riihimaki and Ulfvarson, 1986; World Health Organization (WHO), 1985].

The PNS neurotoxicity of hexacarbons, carbon disulfide, and acrylamide is well established both clinically and experimentally. The range of effects covers subclinical neurophysiological alterations and full-blown polyneuropathies with slow recovery, depending upon the degree and duration of exposure, as well as potentiation by "innocent bystander" chemicals (Schaumberg and Spencer, 1987).

Much of the CNS neurotoxicity of organic solvents is explainable in terms of narcotic action. Short-term human exposure under experimental conditions has resulted in prenarcotic reversible effects such as psychomotor slowing or vigilance decrement at low levels of exposure (Dick and Johnson, 1986). As yet it is not clear whether repeated prenarcotic exposure over years may eventually give rise to irreversible brain damage. It has, however, been shown that for some compounds such as trichloroethylene, styrene, and carbon disulfide, as well as solvent mixtures, chronic low-level exposure is associated with perceptual and motor retardation which, from the very design of the different studies, could not be explained as an acute reversible effect. In summarizing several such studies the conclusion was drawn (Gamberale, 1986, p. 217) "...that the measurement of behavioral performance has been demonstrated to possess more general applicability in human studies than other methods."

Case control studies from Scandinavian countries, using records from disability pensions, generally support psychological findings from cross-sectional studies in workers chronically exposed to organic solvents, in that a higher prevalence of neuropsychiatric disorders or toxic encephalopathies was found after long-term occupational exposure to solvents or solvent mixtures (Hogstedt and Axelson, 1986). Syndromes in such cases have become known as psycho-organic syndromes in Finland and Sweden or, primarily in Denmark and Norway, as presenile dementia. Despite a number of methodological drawbacks the conclusion has been drawn (Hernberg, 1984) that solvent-induced toxic encephalopathies do exist, although they may present with considerable problems of differential diagnosis in individual cases.

CONCLUSION

The preceding examples have been selected to illustrate how psychological tests have been used to detect subtle psychological deficit resulting from neurotoxic insult of environmental/occupational chemicals. In some instances, functional deficit has been reported to occur at exposure levels that generally do not induce alterations of neurological or neurophysiological functions. Despite this apparent sensitivity, such findings have been met with considerable skepticism, in the scientific world as well as among administrators. This appears to be due mainly to the following shortcomings: (1) lack of consistency, (2) lack of theory, and (3) lack of significance for health or adjustment.

Lack of Consistency

It is true that neurobehavioral findings reported from one laboratory often do not prove replicable in other laboratories. This is very obvious in neuropsychological lead research and interferes with efforts to establish dose-response modeling. Apart from analytical differences and differences in study design or confounder structure, part of this inconsistency is due to lack of standardized testing. This refers both to instrumental and to procedural aspects of psychological testing. Two examples are given for the instrumental aspect to illustrate this point: (1) The Bender Gestalt Test, which has been used in several studies, is a well-established clinical tool for the early detection of brain damage, with well-standardized stimulus material and standardized instructions, but different scoring systems (e.g., Koppitz, GFT system). These different systems have in fact been shown to produce different exposure-related effects for the same lead levels (Trillingsgaard et al., 1985). (2) Reaction time (RT) is another case in point. The paradigm is apparently simple but actually covers a wide range of cognitive demands (e.g., simple RT, choice RT, delayed RT, serial choice RT). The Shakov-derived delayed RT paradigm of Needleman has produced different lead-related outcomes depending on similarity to the original procedure (Hunter et al., 1985; Winneke et al., 1985).

Conclusion There is a need for more rigorous standardization of test procedures, and computerized testing could be an important step forward. Whereas for adults in occupational exposure settings such developments, based on the WHO core battery, have proved promising, very little has been done at the lower end of the age continuum. In addition to standardization efforts, strategies of quality assurance for psychological data must be developed to be able to compare outcomes from different studies. In this context the advantages of parametric variation (Weiss, 1978), e.g., increase of task difficulty, should also be studied and exploited in a more systematic manner to increase the sensitivity and the comparability of psychological outcome measures and, possibly, to clarify their validity. One such example is the interaction of task difficulty and lead-induced neurobehavioral deficit in serial choice reaction performance (Winneke et al., 1989): It was shown that lead-induced deficit occurred for high but not for low signal rates. The type of deficit resembles clinical observations in children presenting with attention deficit disorders.

Lack of Developmental Perspective

It is true that the selection of psychological tests in many neurotoxicity studies has been guided primarily by availability and convenience, rather than by considerations based on experience from developmental psychology. Just two examples illustrate the point: Psychometric intelligence has been a preferred outcome measure in many studies. Even in prospective studies of early developmental lead exposure, the Bayley Scales have been used, which are known to possess poor predictive validity. Instead, the work of Fagan and coworkers on visual recognition-memory (Fagan and McGrath, 1981) offers more promising features, in terms of across-age continuity, because its predictive validity for later cognitive development has been shown to be higher than that of the Bayley Scales. Another important aspect of developmental continuity is extrapolation of neurobehavioral effects across species. Cognitive performance is a case in point. Much of the abundant animal literature on agent-induced learning and memory deficits is difficult to extrapolate to the human level, because the preferred cognitive models developed and used in clinical or basic contexts are almost incompatible with the typical models used in behavioral pharmacology and toxicology.

Conclusion There is a need for the development and use of more specific measures with known functional significance and greater validity across different age groups. In addition, closer collaboration between those engaged in animal and human research in the field of neurobehavioral toxicology is necessary to develop cognitive paradigms that allow for a more direct extrapolation across species.

Uncertain Health Significance

Subclinical effects are necessarily difficult to evaluate in terms of their implication for health and adjustment. One such example is the ongoing discussion about the relevance of an average exposure-related IQ drop in later academic achievement. For psychology to contribute more successfully to the solution of environmental health problems it is necessary to systematically exploit biological or disease states as frames of reference for the interpretation of results from psychological toxicity studies. Examples for studies in infants and children are small-for-age or low-birthweight babies, perinatal hypoxia, infectious or traumatic brain injuries, minimal brain dysfunction, dyslexia, and epilepsy. Examples for studies in adults are the effects of normal aging and of age-related neurological disorders (e.g., Alzheimer's or Parkinson's diseases and other dementing illnesses), particularly in their early degenerative stages.

Conclusion Systematic validation of functional tests that use biological processes or neurological disease entities as frames of reference is needed for the interpretation of neurobehavioral findings in terms of health adjustment.

REFERENCES

Amin-Zaki, L. S. Elhassani, M. A. Mjeed, T. W. Clarkson, R. A. Doherty, and M. R. Greenwood. 1974. Intrauterine methylmercury poisoning in Iraq. Pediatrics 54:587–595.

Anger, W. K. 1985. Neurobehavioral tests used in NIOSH-supported worksite studies, 1973–1983. Neurobehavioral Toxicology 7:357–368.

Anger, W. K. 1986. Workplace exposures. In Neurobehavioral Toxicology, A. Annau, ed. Baltimore: The Johns Hopkins Press.

Anger, W. K., and B. L. Johnson. 1985. Chemicals affecting behavior. Pp. 51–148 in Neurotoxicity of Industrial and Commercial Chemicals, J. L. O'Donoghue, ed. Boca Raton, Fla.: CRC Press.

Bakir, F., F. Damlougi, L. Amin-Zaki, M. Murtadha, A. Khadlidi, N. Y. Al-Rawi, S. Tikriti, H. I. Dhahir, T. W. Clarkson, J. Smith, and R. A. Doherty. 1973. Methylmercury poisoning in Iraq. Science 181:230–241.

Berlin, M. 1986. Mercury. Chap. 16 in Handbook on the Toxicology of Metals, second edition, L. Fridberg, G. F. Nordberg, and V. Vouk, eds. Amsterdam: Elsevier.

Byers, R. K., and E. E. Lord. 1943. Late effects of lead poisoning on mental development. American Journal of Diseases of Children 66:471–494.

Calne, D. B., A. Eisen, E. McGeer, and P. Spencer. 1986. Alzheimer's disease, Parkinson's disease, and motoneuron disease: Abiotropic interaction between ageing and environment? Lancet 2:1067–1070.

Centers for Disease Control. 1985. Preventing Lead Poisoning in Young Children. Document 99-2230. U.S. Department of Health and Human Services, Atlanta, Georgia.

Chisolm, Jr., J. J. 1971. Lead poisoning. Scientific American 224:15–23.

Clarkson, T. W., R. Hamada, and L. Amin-Zaki. 1983. Mercury. Pp. 326–389 in Changing Metal Cycles and Human Health, J. O. Nriagu, ed. Berlin: Springer.

Crapper, D. R., and U. De Boni. 1980. Aluminium. Pp. 326–335 in Experimental and Clinical Neurotoxicity, P. S. Spencer and H. H. Schaumburg, eds. Baltimore: Williams and Wilkins.

Davis, J. M., and D. J. Svendsgaard. 1987. Lead and child development. Nature 329:297–300.

De Boni, U., A. Otvos, J. W. Scott, and D. R. Crapper. 1976. Neurofibrillary degeneration induced by systemic aluminium. Acta Neuropathologica 35:285–288.

DFG—Deutsche Forschungsgemeinschaft. 1984. Rueckstaende und Verunreinigungen in Frauenmilch. Mitteilung VII der Kommission zur Pruefung von Rueckstaenden in Lebensmitteln. Weinheim: Verlag Chemie.

Dick, R. B., and B. L. Johnson. 1986. Human experimental studies. Pp. 348–387 in Neurobehavioral Toxicology, Z. Annau, ed. Baltimore: Johns Hopkins University Press.

Elinder, C. G., and B. Sjogren. 1986. Aluminium. Chap. 1 in Handbook on the Toxicology of Metals, second edition, L. Friberg, G. F. Nordberg, and V. Vouk, eds. Amsterdam: Elsevier.

Fagan, J. F., and S. K. McGrath. 1981. Infant recognition memory and later intelligence. Intelligence 5:121–130.

Fein, G. G., J. L. Jacobson, S. W. Jacobson, P. M. Schwartz, and J. K. Dowler. 1984. Prenatal exposure to polychlorinated biphenyls: Effect on birth size and gestational age. Journal of Pediatrics 105:315–320.

Fulton, M., G. M. Raab, G. Thomson, D. Laxen, R. Hunter, and W. Hepburn. 1987. Influence of blood lead on the ability and the attainment of children in Edinburgh. Lancet 1:1221–1226.
Gamberale, F. 1986. Application of psychometric techniques in the assessment of solvent toxicity. Pp. 203–224 in Safety and Health Aspects of Organic Solvents, V. Riihimaki and U. Ulfvarson, eds. New York: Liss.
Grandjean, P., ed. 1984. Biological Effects of Organolead Compounds. Boca Raton, Fla.: CRC Press.
Harada, Y. 1966. Study group on Minamata disease. In Minamata Disease, M. Matsuma, ed. Humamato: Humamato University Press.
Harada, M., T. Fujino, and K. Kabashima. 1977. A study of methylmercury concentration in the umbilical cords of the inhabitants born in the Minamata area. Brain and Development 9:79–84.
Hatzakis, A., A. Kokkevi, C. Maravelias, K. Katsouyanni, F. Salaminos, A. Kalandidi, A. Koutselinis, C. Stefanis, and D. Trichopoulos. 1987. Psychometric intelligence deficits in lead-exposed children. In Lead Exposure and Child Development, M. A. Smith, L. D. Grant, A. I. Sors, eds. Boston: Kluwer Academic Publishers.
Hernberg, S. 1984. Die subklinische Wirkung von Losungsmitteln and Loseungsmittelgemischen auf das Nervensystem. Pp. 47–56 in Verhandlungen der Deutschen Gesellschaft fur Arbeitsmedizin, H. Konietzko and F. Schuckmann, eds. Stuttgart: Gentner Verlag.
Hogstedt, L., and O. Axelson. 1986. Longterm health effects of industrial solvents—A critical review of the epidemiological research. Medicina del Lavorno 77:11–22.
Hunter, J., M. A. Urbanowicz, W. Yule, and R. Lansdown. 1985. Automated testing for reaction time and its association with lead in children. International Archives of Occupational and Environmental Health 57:27–34.
Jacobson, S. W., G. G. Fein, J. L. Jacobson, P. M. Schwartz, and J. K. Dowler. 1985. The effect of intrauterine PCB exposure on visual recognition memory. Child Development 56:853–860.
Kjellstrom, T., P. Kenenedy, S. Wallis, and C. Mantell. 1986. Physical and mental development of children with prenatal exposure to mercury from fish. Stage 1: Preliminary tests at age 4. National Swedish Environmental Protection Board, Report 3080.
Kuratsune, M. 1972. Epidemiologic study on yusho, a poisoning caused by ingestion of rice oil contaminated with a commerical brand of polychlorinated biphenyls. Environmental Health and Perspectives 1:119–128.
Landsdown, R., and W. Yule, eds. 1986. The Lead Debate: The Environment, Toxicology and Child Health. London: Croom Helm.
Needleman, H. L., C. Gunnoe, A. Leviton, R. Reed, H. Peresie, C. Maher, and P. Barrett. 1979. Deficits in psychologic and classroom performance of children with elevated dentine lead levels. New England Journal of Medicine 300:689–695.
Norby-Hansen, O., A. Trillingsgaard, I. Beese, T. Lyngbye, and P. Grandjean. 1989. Neuropsychological profile of children in relation to dentine lead level and socioeconomic group. In Lead Exposure and Child Development, M. A. Smith, L. D. Grant, A. I. Sors, eds. Boston: Kluwer Academic Publishers.
Perlstein, M. A., and R. Attala. 1966. Neurologic sequelae of plumbism in children. Clinical Pediatrics 5:292.
Riihimaki, V., and U. Ulfvarson, eds. 1986. Safety and Health Aspects of Organic Solvents. New York: Liss.
Rogan, W. J., B. C. Gladen, J. D. McKinney, N. Carreras, P. Hardy, J. Thullen, J. Tingelstad, and M. Tully. 1986. Neonatal effects of transplacental exposure to PCBs and DDE. Journal of Pediatrics 109:335–341.
Schaumburg, H. H., and P. S. Spencer. 1987. Recognizing neurotoxic disease. Neurology 37:276–278.

Smith, H. D., R. L. Baehner, T. Carney, and W. J. Majors. 1963. The sequelae of pica with and without lead poisoning. A comparison of the sequelae of five or more years later. I. Clinical and laboratory investigations. American Journal of Diseases of Children 105:609–616.

Spencer, P. S., P. B. Nunn, J. Hugon, A. Ludolph, and D. N. Roy. 1986. Motoneuron disease on Guam: Possible role of a food neurotoxin. Lancet 1:965.

Spencer, P. S., J. Hugon, A. Ludoph, P. B. Nunn, D.N. Ross, and H. H. Schaumburg. 1987. Discovery and partial characterization of primate motor-system toxins. Pp. 221–231 in Selective Neuronal Death, G. Bock and M. O'Connor, eds. Ciba Foundation Symposium 126.

Trillingsgaard, A., O. Norby-Hansen, and I. Beese. 1985. The Bender Gestalt Test as a neurobehavioral measure of preclinical visual-motor integration deficits in children with low-level lead exposure. Pp. 189–198 in Neurobehavioral Methods in Occupational and Environmental Health. Copenhagen: World Health Organization.

U.S. Environmental Protection Agency. 1986. Air Quality Criteria for Lead, Vol. I–IV. EPA Report No. 60/8-83/028. Springfield, Va.: National Technical Information System.

Weiss, B. 1978. The behavioral toxicology of metals. Federation Proceedings 37:22–27.

Winneke, G. 1986. Animal Studies. Pp. 217–234 in The Environment, Toxicology and Child Health, R. Lansdown and W. Yule, eds. London: Croom Helm.

Winneke, G., U. Beginn, T. Ewert, C. Havestadt, U. Kramer, C. Krause, H. L. Thron, and H. M. Wagner. 1985. Comparing the effects of perinatal and later childhood lead-exposure on neuropsychological outcome. Environmental Research 38:155–167.

Winneke, G., A. Brockhaus, W. Collet, and U. Kramer. 1989. Modulation of lead-induced performance deficit in children by varying signal rate in a serial choice reaction-task. Neurotoxicology and Teratology 11(6).

World Health Organization. 1985. Chronic Effects of Organic Solvents on the Central Nervous System and Diagnostic Criteria. Copenhagen.

The Health Effects of Environmental Lead Exposure: Closing Pandora's Box

Deborah C. Rice

Lead has been recognized as a poison from ancient times to the present (Cantarow and Trumper, 1944; Oliver, 1914). Over the last decade or so, attention has focused on the subtle effects of environmental exposure at levels presently considered "normal" in our industrialized age [Landsdown and Yule, 1986; Mahaffey, 1985; National Academy of Sciences (NAS), 1980; Needleman, 1980; Rutter and Russell Jones, 1983]. The resultant research has inspired lively and sometimes heated debate concerning the nature and possible threshold of these effects. The recognition that lead produces intellectual impairment in children, as well as other health effects, has resulted in the progressive tightening of regulation of lead in the United States and other countries, including the phaseout of lead from gasoline [Environmental Protection Agency (EPA), 1984; see Johnson and Mason, 1984, for review of U.S. lead regulations].

Exploration over the last decade of the effects of environmental exposure to lead, neuropsychological and otherwise, represents a case study in the scientific and political procedures, problems, and failures inherent in such an endeavor. As such it can be used as a model for discussion in light of the theme of Part III of this volume—chemical time bombs. All individuals in industrialized societies carry a significant body burden of lead, a situation that will take at least a generation to change even if lead were removed from the environment instantaneously tomorrow. What have been, and what can we anticipate will be, the consequences of this mass exposure? What are the issues critical to this evaluation, and what should be done differently in the future?

HISTORY OF THE PRESENT PERSPECTIVE ON THE ENVIRONMENTAL NEUROTOXICITY OF LEAD

Although lead is a common element in the earth's crust, its ubiquitous presence in bioavailable forms in the environment is due largely to the activities of humans (see Lin-Fu, 1985; Smith, 1986 for reviews). Lead has been used in metalworking and in pottery glazes for millennia. The Romans used lead for plumbing as well as a sweetening agent in wine and other foods. The industrial revolution and the addition of lead to gasoline in the 1920s have resulted in dramatic increases in environmental lead levels (Elias et al., 1975). Present blood levels of industrial populations are highly correlated with the amount of leaded gasoline in use (Hunter, 1986). Present environmental levels are several orders of magnitude above preindustrial levels (NAS, 1980; see Table 1). The body burden of lead in human bones is presently 500-fold greater than in prehistoric times, and the present diet of Americans contains 100 times more lead than prehistoric diets (NAS, 1980).

The inclusion of tetraethyllead as a gasoline additive in the 1920s was a landmark event that resulted in a steep increase in lead emitted into the environment (Elias et al., 1975). The overtly toxic effects of lead were already recognized at that time; the use of lead as a

TABLE 1 Comparison of Estimated Natural Levels of Lead in the Environment with Typical Present-Day Levels

Medium	Natural Concentration	Present Concentration	Approximate Ratio, Present/Natural
Air			
Rural/remote	0.01–0.1 ng/m^3	0.1–100 ng/m^3	10–1,000
Inhabited	0.1–1.0 ng/m^3	0.1–10 g/m^3	100–10,000
Soil			
Rural/remote	5–25 g/g	5–50 g/g	1–2
Inhabited	5–25 g/g	10–5,000 g/g	2–200
Water			
Fresh	0.005–10 g/L	0.005–10 g/L	1
Ocean	0.001 g/L	0.005–0.015 g/L	10
Food	0.0001–0.1 g/g	0.01–10 g/g	100

SOURCE: NAS (1980).

gasoline additive engendered grave warnings by health professionals concerning the potential threat to the general health as a result of lead exposure (Rosner and Markowitz, 1985). This concern was based on occurrences of mortality and severe neurologic and psychiatric signs in workers exposed during manufacture of this additive. A committee convened by the Surgeon General warned in 1926 (in Rosner and Markowitz, 1985): "It remains possible that if the use of leaded gasoline becomes widespread, conditions may arise very different from those studied by us. . . . Longer experience may show that even such slight storage of lead as was observed in these studies may lead eventually in susceptible individuals to recognizable or to chronic degenerative diseases of a less obvious character." Despite the recommendation by the committee that the matter be studied further, the interests of the automotive and oil industries won out, lead remained in gasoline, and no further data were collected.

In the 1940s, it was recognized by astute physicians that children who had been treated for lead poisoning suffered permanent sequelae in the form of neurological damage (Byers and Lord, 1943). High-level lead exposure in children at that time was via lead-based paint. Byers and Lord reported poor school performance, impulsive behavior, short attention span, restlessness, and occasional neurological signs in these children. These observations were later replicated by other investigators (Jenkins and Mellins, 1957; Perlstein and Attala, 1966; Thurston et al., 1955).

In the 1970s, concern arose in the United States and elsewhere that the tons of lead being introduced into the environment every year by the use of leaded gasoline, as well as other industrial processes, were producing significant health effects, particularly in children. The new concern was that common environmental levels of lead were producing intellectual impairment in children that had no overt signs of lead poisoning. Early in the decade, attention focused on children who had ingested lead-based paint (de le Burde and Choate, 1972; Lin-Fu, 1972). Deficits in IQ, fine motor performance, and behavioral disorders such as distractibility and constant need for attention were observed in children who had never exhibited overt signs of lead intoxication.

A new understanding of the insidious effects of lead on the intellectual capacity of a large number of children arose with the landmark study of Needleman et al. in 1979. These investigators reported decreased IQ and increased incidence of distractibility and inattention in middle-class children with no exposure to lead from paint. The conclusion to be drawn from this research was that environmental sources were responsible for the increased lead burden in these children, and that this environmental contamination at levels that had come to be regarded as "normal" could be insidiously robbing children in industrialized countries of their intellectual birthright. Largely as a result of that study, the last decade has witnessed intense research into the health effects of lead and the sources of exposure of the general population. The issue has generated a great deal of political as well as scientific controversy. Involved have been physicians, epidemiologists, chemists, geologists, animal researchers, representatives of the lead industry, and members of a host of government agencies in a number of countries. The result of this intense scrutiny is that probably more is known about the health effects of lead than any other noncarcinogenic environmental contaminant.

OVERVIEW OF MODERN STUDIES

There have been a number of cross-sectional (retrospective) studies since 1979 concerning the effects of lead on intellectual and other behavioral functions in children. The general trend has been to study children with increasingly lower body burdens of lead and to focus on middle class rather than disadvantaged children. These studies have been extensively reviewed (cf. Mahaffey, 1985; Rutter and Russell Jones, 1983), although new important studies have been published very recently. There are also several prospective studies going on, in which the mothers are recruited before the birth of their infants, and the infants are followed in a longitudinal manner. This design is stronger than a cross-sectional design, and these studies will undoubtedly continue to provide important information over the next several years. This section reviews recently published data from the prospective studies, as well as the new cross-sectional studies. Alternate methods of testing for nervous system effects, as well as other recently reported health effects of lead, are described briefly.

Prospective Studies

Reproductive Effects

It has long been recognized that industrial exposure to high lead levels produced an increased incidence of miscarriages and stillbirths, and that infants that did survive failed to thrive and exhibited neurological abnormalities. Although the situation is less clear for lower exposure to lead, recent studies provide evidence that low-level lead exposure causes reproductive problems. Increased maternal blood lead levels are associated with increased incidence of preterm delivery (McMichael et al., 1986) and decreased gestational age (Dietrich et al., 1987; Moore et al., 1982). Blood lead has also been found to be associated with increased spontaneous abortion (McMichael et al., 1986). Higher lead burden may also be associated with minor but not major physical abnormalities (Needleman et al., 1984), although this is not a universal finding (Ernhart et al., 1985, 1986; McMichael et al., 1986). Increased maternal blood lead level is also associated with abnormal reflexes, poor muscle tone, and neurological soft signs such as jitteriness, hypersensitivity, and abnormal cry in the infant (Ernhart et al., 1985, 1986). It must be stressed that the maternal and infant blood lead levels in these studies were in the range considered normal or average for people in industrialized societies (2–15 µg/dL in most cases).

It is well established that premature or small-for-date infants are at greater risk for a variety of behavioral and other health problems. Such children have more trouble in school and require special help more often than other children (Schraeger et al., 1966; Weiner et al., 1968). Thus, these individuals are likely to represent an ongoing cost to society over and above any special medical intervention that might be associated with the neonatal period.

Early Behavioral Effects of Perinatal Lead Exposure

Obviously the functional effects of perinatal lead exposure are not separate from the effects discussed in the preceding section, but will in part be related to them. (In fact, controlling for effects such as gestational age in evaluation of behavioral effects may underestimate the effects of lead.) There are at least three prospective studies at present in which women are recruited during pregnancy and the offspring are monitored at specified ages. In one study by Bellinger and colleagues (Bellinger et al., 1987a) performance on the Bayley Mental Development Index (MDI) at 6, 12, and 24 months of age was associated with cord but not postnatal blood lead levels (Figure 1). The difference between the high (mean 14.6 µg/dL) and low (mean 1.8 µg/dL) blood lead groups was 4–7 points. Assessment of these children at 57 months of age (Bellinger et al., 1987b) revealed that performance on the General Cognitive Index of the McCarthy Scales was associated with blood lead levels at 24 but not 57 months of age (after adjusting for possible confounders). Blood levels averaged 6.8 µg/dL at 24 months and 6.4 µg/dL at 57 months.

In the study by Dietrich and colleagues (Dietrich et al., 1987), it was found that each log unit increment in blood lead was associated with a covariate-adjusted reduction of 5.7 points on the MDI; the reduction was 8.0 points if the effects on gestational age and birth weight were included. One year after birth, prenatal blood lead lev-

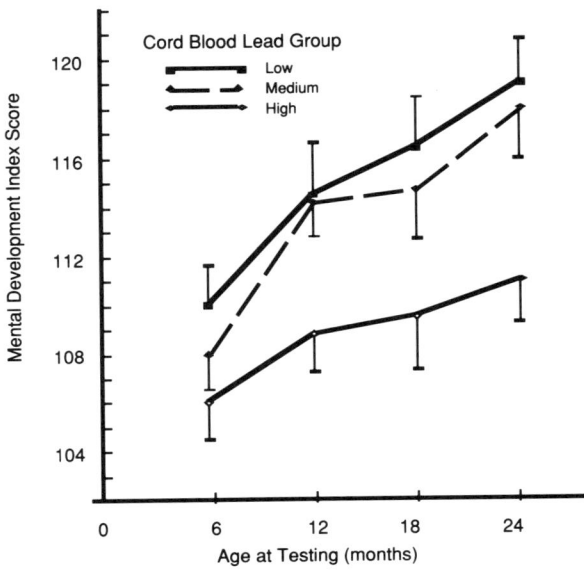

FIGURE 1 Mean Mental Development Index scores at four ages in infants according to lead level in umbilical cord blood.

SOURCE: Bellinger et al. (1987a).

els were negatively correlated with MDI, Bayley Psychomotor Development Index (PDI), and Bayley Infant Behavioral Record (IBR). The IBR revealed higher activity levels and more negative social-emotional response. In the third prospective study, in Port Pirie, South Australia (McMichael et al., 1986), a decrease of 2 points in the MDI scale for every 10 µg/dL increase in blood lead levels was observed at 24 months of age. Performance was found to be more related to postnatal than prenatal blood lead levels; however, no assessment was performed before 2 years of age. It is possible that early testing would have revealed significant prenatal exposure effects.

In a retrospective study, Winneke et al. (1985a,b) found that performance on a variety of neurobehavioral and intellectual tasks at 6–7 years of age was attributable approximately equally to maternal levels at birth (average 9 µg/dL) and to current blood levels in the children.

Retrospective Studies on Correlation of Lead Body Burden and Behavior in Grade School Children

Since the study by Needleman and colleagues (Needleman et al., 1979), a number of investigators have examined the effects of moderate-

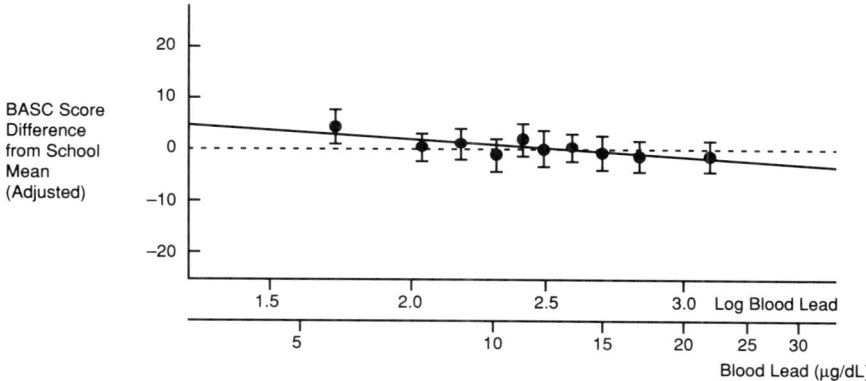

FIGURE 2 British Ability Scales Combined (BASC) score (mean and 95 percent confidence intervals) for groups of children ordered by blood lead.
SOURCE: Fulton et al. (1987).

level exposure on intellectual functioning in children. Such studies have usually included some measure of intelligence (IQ), school functioning, teachers' rating of classroom behavior, or specific measures of attentional mechanisms. Most recent studies have utilized populations with lower body burdens of lead than the children assessed by Needleman. For example, Fulton et al. (1987) reported a linear relationship between intellectual functioning and log blood lead concentration for blood lead values between approximately 5 and 25 µg/dL (mean about 10 µg/dL) in children living in Edinburgh, with no indication of a threshold for lead effect (Figure 2). Results were significant after adjusting for potential confounders. Another study of middle-class children in New Zealand (Silva et al., 1988) reported high correlations between log blood lead (mean 11 µg/dL) and measures of inattention and hyperactivity, after adjusting for confounding variables. A number of other studies published since 1980 have also reported a negative association between lead body burden and performance (Hansen et al. 1987; Hatzakis et al., 1987; Hawk et al., 1986; Schroeder et al., 1985; Winneke and Kraemer, 1984; Winneke et al., 1983; Yule et al., 1981), although this finding has not been universal. (This issue is discussed in a later section.)

Later Behavioral Concomitants of Increased Lead Burden

The consequences of early poor performance as a result of lead exposure in terms of grade retention or need for special education have been little investigated. In a follow-up of children from the

TABLE 2 Indices of Academic Failure

Dentine Lead Level	Academic Aid[a]	Grade Retention[b]
Low	17.0% (8/47)	4.3% (2/47)
Midrange	18.6% (13/70)	11.6% (8/69)
Elevated	36.4% (8/22)	22.7% (5/22)
Total	20.9% (29/139)	10.9% (15/138)

[a] $\chi^2(2) = 3.84, p < 0.20$.
[b] $\chi^2(2) = 5.61, p < 0.10$.
SOURCE: Bellinger et al. (1984).

Needleman et al. (1979) study, Bellinger et al. (1984) reported a fivefold increase in grade retention and a twofold increase in the need for academic aid in teenagers, based on tooth lead levels as 5 and 6 year olds (Table 2). Barrett (1978) reported a dose-related increase in unsatisfactory school performance as a function of increased free erythrocyte protoporphyrin (FEP) levels (a measure of lead exposure). These results are not surprising in view of the effects of lead on classroom performance in the early grades. Needleman et al. (1979) reported dose-dependent disordered classroom behavior as measured by a teacher's rating scale (Figure 3). These results were replicated by Yule et al. (1981) and Lansdown et al. (1983) in British children and Hatzakis et al. (1987) in Greek children. Yule also reported that children with high lead levels exhibited more deviant performance on tests of conduct problems, inattentive-passive, and hyperactivity scales. Such early attentional deficits and their associate behaviors place children at risk for academic failure and behavior problems (Horn and Packard, 1985). It is hoped that investigators will continue to follow the children tested initially in the early grades, collecting data on school performance, special needs, and antisocial behavior.

Effects of Lead on Other Neuropsychological Endpoints

Lead is associated with increased reaction time (Figure 4) and increased errors on various performance tasks (Hatzakis et al., 1987; Needleman, 1983; Winneke and Kraemer, 1984; Winneke et al., 1983, 1985a,b). Using the Second National Health and Nutrition Examination Survey (NHANES II) data base, Schwartz and colleagues found an association between lead and increased hearing threshold in children with blood levels between 5 and 45 µg/dL, with no threshold for

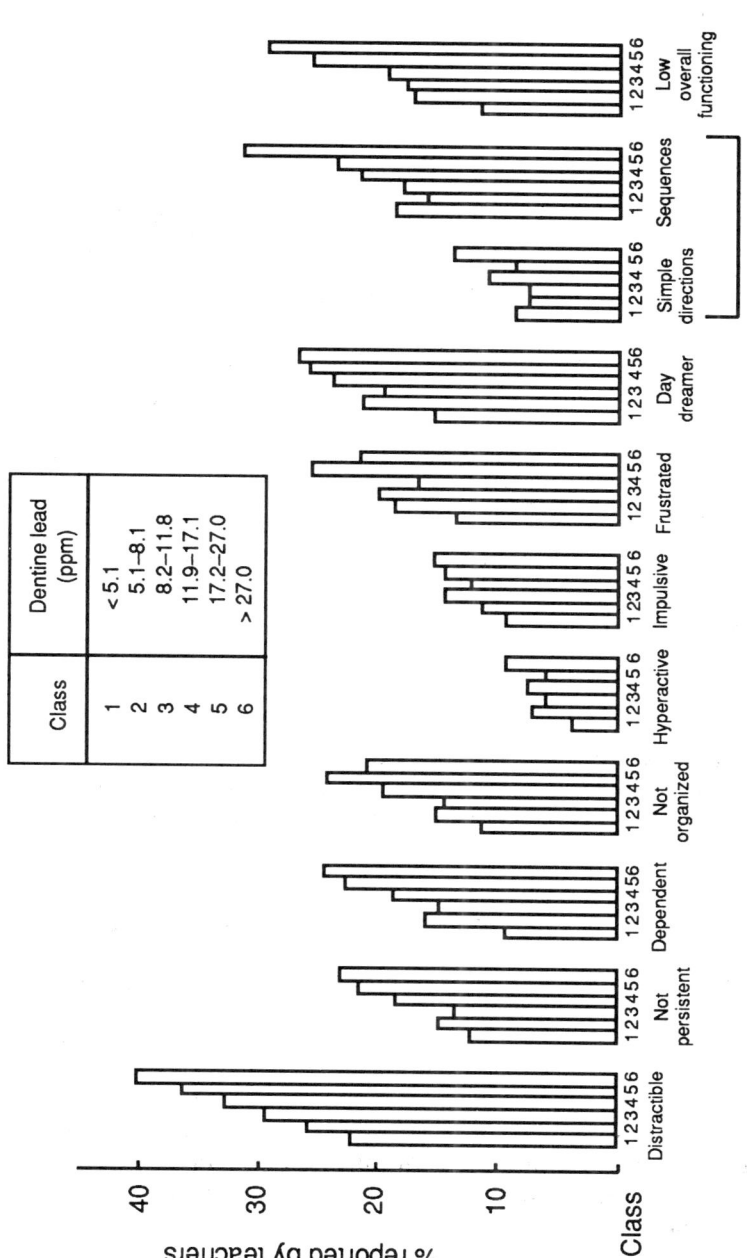

FIGURE 3 Distribution of negative ratings by teachers on 11 classroom behaviors in relation to dentine (tooth) lead concentrations.
SOURCE: Needleman et al. (1979).

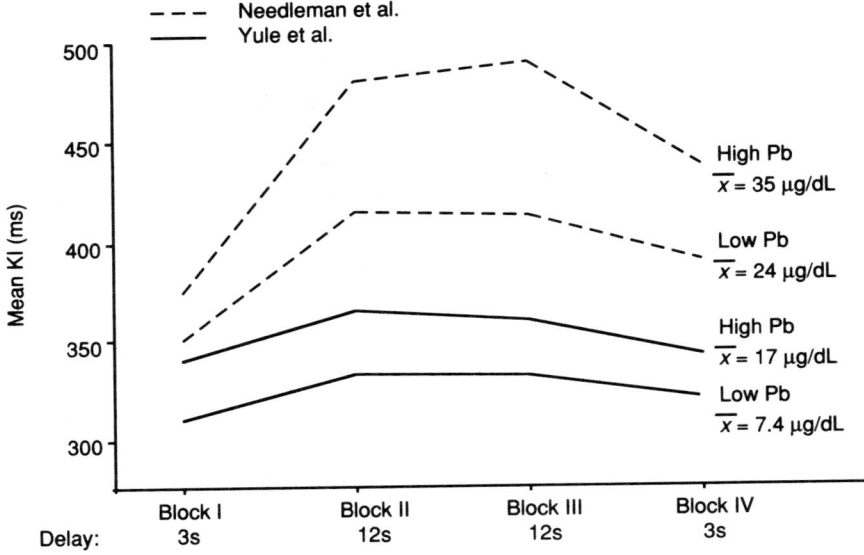

FIGURE 4 Performance of children on a simple reaction time task as function of blood lead levels (K1 = time on task).

SOURCE: Needleman (1987b).

effect (Schwartz and Otto, 1987) (Figure 5), as well as slowed nerve conduction velocity at blood lead levels above 20-30 µg/dL (Schwartz et al., 1988). Blood lead levels of 15 µg/dL and below are also associated with changes in EEG pattern and auditory evoked potentials (Otto, 1987; Otto et al., 1981).

Other Health Effects of Environmental Lead Exposure

In addition to effects on the nervous system, low-level exposure to lead affects a number of important metabolic processes. The most widely recognized of these are changes in the hematopoietic system. High intake of lead produces anemia, an effect influenced by iron status. Lead also inhibits a number of enzymes involved in heme biosynthesis (see Moore and Goldberg, 1985 for review). The result is a buildup of some of the precursors involved in heme synthesis. Inhibition occurs at somewhat different levels for different enzymes, but some are reliably affected at blood lead values of 10–15 µg/dL, which are observed routinely in the general population. The buildup of blood protoporphyrins is the basis of the screening program for undue lead exposure of children [Centers for Disease Control (CDC), 1985], because these precursors are easier to measure than blood lead itself. Aside from frank anemia observed only at relatively high blood lead levels, the significance of these biochemical changes is in their potential contribution to lead neuropathology and as markers of exposure.

Lead interferes with vitamin D synthesis at low (environmental) levels (Rosen, 1985). Such an effect has important health implications in terms of calcium homeostasis, cell differentiation, and immunoregulatory capacity. Results from the NHANES II survey indicate that increased lead burden is associated with decreased stat-

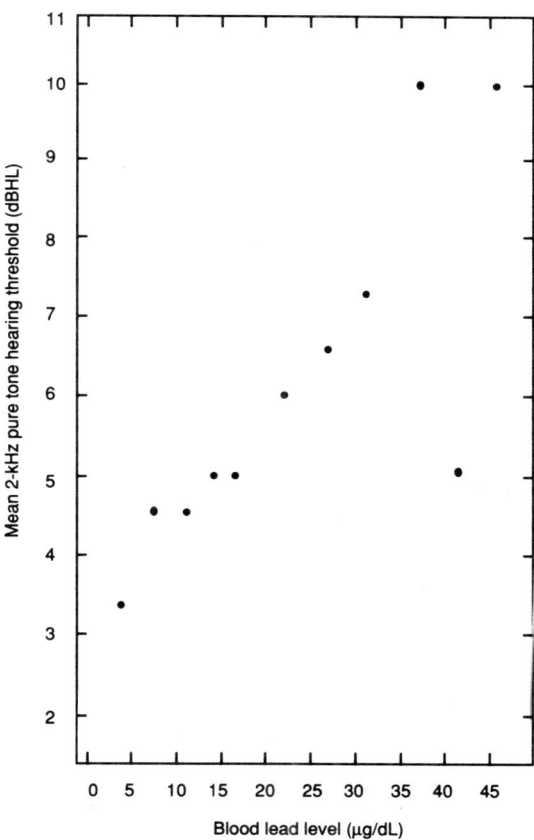

FIGURE 5 Relationship of 2-kHz pure tone hearing thresholds and blood lead levels in 4,519 NHANES II subjects aged 14–19 years.

SOURCE: Schwartz and Otto (1987).

ure in children (Schwartz et al., 1986). Increased lead burden has also been linked to increased blood pressure in males, again by using the NHANES II data base (Schwartz, 1986). Both effects may be mediated by effects on calcium homeostasis.

CRITICAL ISSUES

Markers of Lead Exposure

Blood is a poor marker of lead exposure, because it indicates only recent exposure. Bone represents a record of total past exposure, but not the pattern over time, which may be of critical importance to the type and degree of neuropsychological damage. Bone obviously cannot be utilized in human studies as a measure of lead exposure, although the recent ability to determine in vivo bone lead by use of x-ray fluorescence represents an important contribution that it is hoped will be utilized in future research. Deciduous teeth are probably a reasonable substitute for certain ages and have been utilized in some retrospective studies. Prospective studies provide the opportunity to measure lead exposure, currently performed by means of blood lead, at regular intervals from before birth onward. Such a design ensures that the history of lead exposure is known at least approximately and provides the opportunity to correlate behavior against current and past blood levels. Retrospective studies, on the other hand, suffer from the inability to determine both the pattern and the degree of past exposure to lead. Retrospective studies sometimes use only one blood lead measure as a marker of lead exposure, an important limitation. It is extremely relevant, for example, that two of three prospective studies found that performance at 2 years of age was correlated with in utero, but not postnatal, blood level. In the study that has reported results past 2 years of age, performance at 57 months was correlated with blood levels at 24 months, but not 57 months, of age. If these studies had been performed retrospectively, the results would have been negative.

The positive results being obtained in prospective studies deserve attention in a different vein. The implication of these findings is that the blood lead body burden of women, reflected by maternal and cord blood lead levels, is important to at least the early well-being of children. The women described in these studies, and their offspring, had blood lead levels that are typical in our environment—the result of simply living in a present-day industrialized society. One important unanswered (and unaddressed) question is the contribution of total maternal body burden, rather than blood level, to the risk to the infant. Bone is the most significant level compartment in the body, and it is established that lead increases in bone throughout the life span of humans (Barry, 1975). Because a large number of women are at present delaying childbearing until relatively late, they may be exposing their fetuses to an increased burden of lead as a result of mobilization from bone. The calcium turnover from bone increases during pregnancy and lactation, and there is some evidence that bone lead may be mobilized as well during pregnancy and lactation, long after exposure has ceased (Thompson et al., 1985). The amount of lead in milk in humans is correlated with having lived for at least five years in a high-traffic area, regardless of whether this occurred during childhood or adulthood (Debeka et al., 1986). Because women who were born in the 1950s and 1960s have, in general, a significantly higher total body burden of lead than previous generations (due to exposure by leaded gasoline), this represents a potentially important problem for years to come, despite the fact that lead in the environment is presently decreasing.

Statistical Interpretation

The interpretation of these human studies, perhaps particularly the retrospective one, depends to a great degree on the statistical methods employed. Statisticians involved in these studies often disagree among themselves on the most appropriate way in which to analyze the data. Although the resolution of such controversies is best left to the statisticians, there are matters of scientific interpretation that may be addressed here. An issue of critical importance is the potential for overcontrolling potential confounding variables. An example of overcontrolling mentioned above is control for gestational age, which itself may be affected by lead burden. Another example is the factoring out of mother's IQ and measures of socioeconomic status and maternal care scores, which may be influenced by the mother's lead burden, and may in turn influence the child's lead burden at birth as well as during childhood. A potential partial solution may be the choice of highly homogeneous populations for study, or even populations in which lead and the typical confounders vary inversely. For example, in the Boston prospective study (Bellinger et al., 1987b), families with higher socioeconomic standing tended to live in a more urban area, with the result that lead body burdens tended to be higher in this group. Consequently, the association between lead and infant performance became stronger when potential confounders were included in the analysis.

An issue of extreme importance is the power of any study to de-

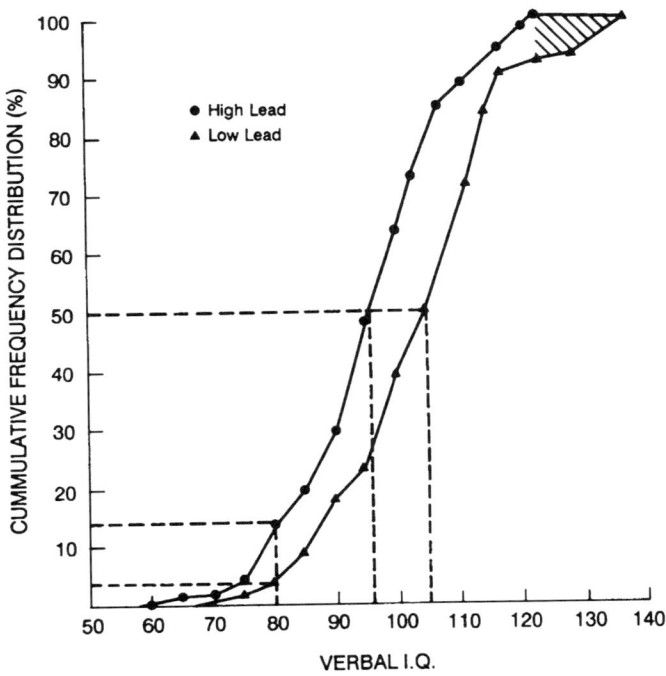

FIGURE 6 Cumulative frequency distribution of verbal IQ scores in high- and low-level lead subjects: A shift in the median of 6 points is associated with a fourfold increase in the risk of IQ below 80.

SOURCE: Needleman (1983).

tect an effect (reject the null hypothesis) if an effect is present. For example, in an analysis by Needleman (1987a) performed on 14 recent studies, for which power to detect an effect could be ascertained, the power varied from 0 to 0.52. Nine studies reported a significant effect ($p < 0.05$), two were in the right direction ($p = 0.12$–0.15), and two showed no effect (the study with power = 0 was not analyzed). As pointed out by Needleman, the results taken together are indicative of a consistent effect of lead on intellectual performance of children.

One issue mentioned routinely by both primary investigators and reviewers in relation to causality is the possibility of a phantom that is covariant with lead and is the actual causative agent of the intellectual impairment, as discussed by Needleman and Bellinger (1986). This covariate would have to be in effect for inner city children exposed to paint and for middle-class children exposed directly or indirectly via tap water (as in Edinburgh) or fallout from gasoline. In addition, the covariant hypothesis ignores the large body of data from animals, particularly monkeys, that report behavioral deficits analogous to those found in children (i.e., deficits in attention and information processing).

Another statement often made by reviewers with respect to the effects of lead is that they are small, representing only a few percent of the total variance. However, in general a 10-µg/dL increase in blood lead results in about a 4–6 point decrease in intelligence. That degree of deficit represents 0.30–0.45 standard deviation of the normal distribution and results in a significant shift of the population. For example, Needleman (1983) reported cumulative IQs in children with high and low lead levels for which the mean differed by 6 points. The resulting distributions revealed that the number of children with IQs below 80 increased by a factor of four in the high-lead group, whereas the number of children with IQs over 120 decreased by an equal amount (Figure 6). If it is in fact the case that increasing blood lead levels from 5 to 15 µg/dL (or from 0 to 10 µg/dL) shifts the population IQ by approximately 0.4 standard deviation, this represents a profound and calamitous health effect for any society.

Adequacy of Behavioral Methodology

One problem with the use of intelligence scales is that results are very heavily environmentally determined. This has often resulted in the effect of lead apparently decreasing when environmental factors such as socioeconomic status and scores of home care were included in the analyses. It therefore would be highly desirable to develop tests that were less environmentally influenced. Tests such as vigilance tasks or reaction time may be less environmentally determined than tests of IQ and seem to be sensitive to impairment by lead, as mentioned previously. A fruitful approach may be to adapt procedures proven to be sensitive to the effects of lead in animals, particularly monkeys, for use with children. Research from the University of Wisconsin, as well as from our laboratory, has shown consistent effects on tests of attention and distractibility that could easily be adapted for children.

Assessment of sensory system function, particularly by psychophysical means, seems a promising avenue of research and should reflect the diffuse damage produced by toxicants such as lead. For example, the decrement in hearing threshold as a function of increased blood lead (Schwartz and Otto, 1987), although small (and therefore requiring a large study to detect), undoubtedly represents subtle neuronal damage. It would be extremely interesting to test frequency or amplitude difference thresholds in addition to absolute frequency thresholds, for two reasons. First, there is evidence that difference thresholds may degrade before absolute frequency thresholds (Stebbins, 1982) and would, therefore, provide a more sensitive indicator of lead neurotoxicity. Second, the ability to detect changes in frequency and amplitude is extremely important to the understanding of speech. Such testing may be especially relevant in view of the reported deficits on auditory processing as a result of lead exposure, reported by some investigators.

Testing of visual system function may also prove a fruitful avenue of research. A number of investigations in animals as well as in workers occupationally exposed to lead report visual deficits, particularly at low luminance, as a result of lead exposure. Such deficits probably could be detected only by psychophysical techniques, because electrophysiological techniques cannot be used for low-luminance assessment of functions other than purely retinal. Detection of subtle effects may require testing of a large number of subjects.

Does Lead Contribute to Aging?

It is well established from research on animals that lead produces neuronal degeneration, resulting in decreased numbers of nerve cells in various brain areas. Developmental lead exposure also results in a decrease in the amount of dendritic branching from nerve cells, representing a decrease in the ability of the nerve to communicate with its neighbors. It is also established that aging can produce these same effects. In fact, the brain areas affected most by lead and those that degenerate most quickly as a result of aging overlap to a great extent, at least in rodents and monkeys. Although there is presently no evidence for or against the hypothesis, it is conceivable that the effect of a lifetime of exposure to low levels of lead results in an acceleration of the normal process of degeneration of neural structures. Weiss (1980) discussed the consequences on functional mental age of a very slight acceleration in loss of functional capacity beginning at age 25. One-tenth of one percent acceleration would result in a "brain age" of 95 at 40 years of age (Figure 7). The actual situation could in fact be worse for certain individuals, because the deleterious effect of lead does not begin at age 25 but before birth. Additionally, the effects of lead on the nervous system of the developing organism may be more severe and different in kind than accelerated attrition of neurons.

The effects of lead on blood pressure may also contribute to mani-

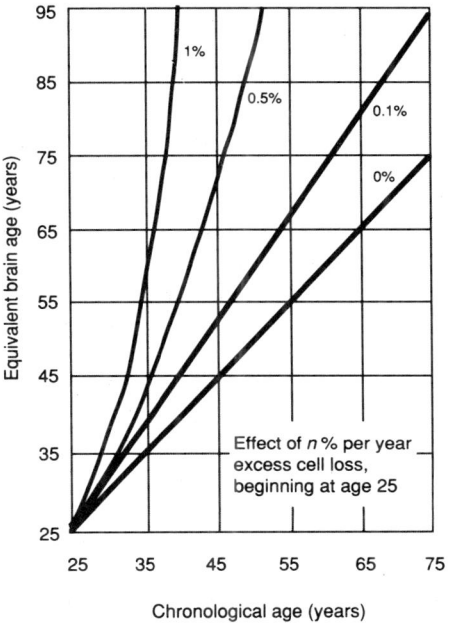

FIGURE 7 The "brain age" associated with different degrees of acceleration in the decline of brain functional capacity, if that decline begins at age 25.

SOURCE: Weiss (1980).

festation of diseases of aging, including heart and kidney disease, and stroke. The effects of lead on vitamin D metabolism may also have important lifetime repercussions. As pointed out by Grant (1986), "this effect is significant on two counts: (i) altered levels of 1,25-$(OH)_2$-vitamin D not only affects calcium homeostasis (affecting mineral metabolism, calcium as a second messenger and calcium as a mediator of cyclic nucleotide metabolism), but also likely affects its role in immunoregulation and mediation of tumerogenesis, and (ii) the effect of lead on 1,25-$(OH)_2$-vitamin D is a particularly robust one, with blood levels of 30–50 µg/dL resulting in decreases in the hormone that overlap comparable degrees of decrease seen in severe kidney injury or certain genetic diseases." Thus lead is implicated in the compromise of the body's ability to repair tissue, fight disease, regulate growth of abnormal cells, and maintain bone, among other effects. Such functions are often compromised in old age; lead may be contributing to these effects, particularly as a result of lifetime exposure.

Overview of Cost to Society

Much of the cost of lead exposure is invisible. The consequences to society of decreasing the IQ of thousands of individuals from 130 to 125 points, or of a few from 160 to 155, are of great significance but cannot be measured either in monetary terms or in terms of human suffering. What can be measured monetarily is the cost of individuals who require special services as a result of undue lead exposure. Such services may include institutional care, special education, lost wages, hospitalization, and various forms of treatment. Also included should be the cost of monitoring children for lead exposure as well as various abatement programs. An analysis performed a decade ago (Provenzano, 1980) for the United States estimated the cost at $0.4 to $1.0 billion (1978 dollars) annually. Since that estimate was made, the criteria for considering a child at potential risk for lead exposure have been made more conservative (CDC, 1985). It is generally recognized that blood lead levels above 15 µg/dL are undesirable for children (cf. EPA, 1984; Grant, 1986). About 15 percent of U.S. children have blood lead values above 15 µg/dL; the proportion is higher for poor and black children. This figure translates to 3 to 4 million U.S. children (Mushak and Crocchetti, 1987). The Committee on Environmental Hazards (1987) recommends that all children at risk be tested for undue exposure by erythrocyte protoporphyrin levels at 12 months of age, with later follow-ups for high-risk children. It also recommends vigorous lead abatement programs. In addition, the 1978 analysis was performed before the relationships of lead to blood pressure, vitamin D metabolism, and calcium homeostasis were known. Thus a modern estimate of the cost would undoubtedly be substantially higher.

Have Regulatory Agencies Been Slow to Act?

It has been clear for centuries that lead is toxic. It has been clear for 15 years that lead is irreversibly neurotoxic to children, and for several years at least that lead at levels observed routinely contributes to suboptimal behavioral functioning. Various U.S. federal agencies have been criticized for failing to act to protect the health of the public (Schoenbrod, 1980; Stein, 1980). Regulation of lead in the United States (and other countries) is controlled by a number of different agencies (Billick, 1981) (Figure 8). As pointed out by Schoenbrod (1980), this has provided an avenue for avoidance of action by blaming lead exposure on sources under the control of another agency. Various agencies have also blamed natural sources, despite the fact

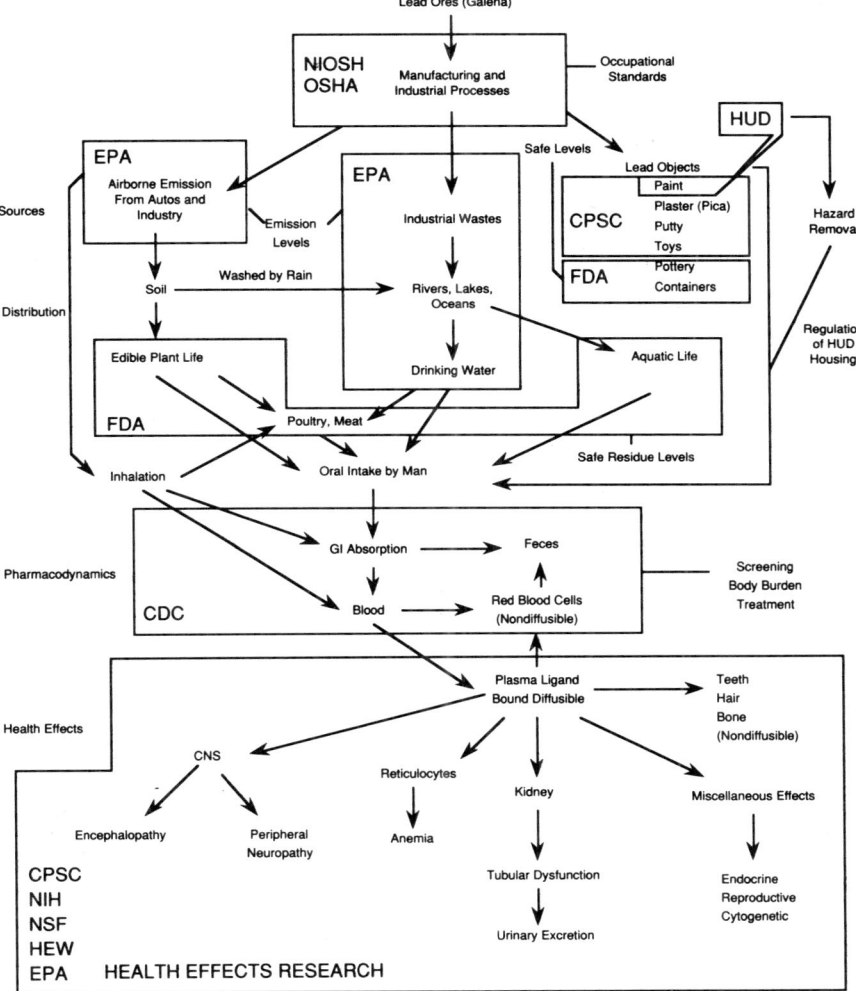

FIGURE 8 Ecodiagram showing movement of lead in the environment and areas of U.S. federal agency responsibility for control of exposure.

SOURCE: Billick (1981).

that they contribute considerably less than 1 percent of the human lead burden (Settle and Patterson, 1980).

The perceived slowness of regulatory agencies to act is probably the result of a number of factors. One is the disagreement among scientists and statisticians working in the field. As discussed above, the reasons for the controversy are due at least in part to the following factors:

- Methodological limitations including inadequate markers of lead exposure, environment-influenced instruments of neuropsychological function, and choice of populations in which these environmental factors and lead exposure are highly correlated
- Evaluating the data by simply counting studies as positive or negative, without looking at direction of effect across studies or power of individual studies to find an effect, and failure to perform meta-analyses
- Failure to utilize the animal literature in interpreting data from human studies
- Failure to recognize that a "small" effect (i.e., 2 to 3 percent of variance) does not translate to "insignificant"

In addition, there may be a reluctance on the part of regulatory agencies to regulate on the basis of psychological test data. For example, the EPA has focused on the effects of lead on heme synthesis and the hematopoietic system (EPA, 1984) and, more recently, on vitamin D metabolism (Grant, 1986). This may represent a real reluctance to regulate on the basis of behavioral data; conversely, regulators may be sensitive to the potential legal ramifications of the controversy over low-level effects on psychological functioning. It is hoped that recent studies demonstrating effects on behavior and intellectual functioning in middle-class children at blood lead levels typical in our society, as well as the analyses based on the NHANES data, will persuade regulatory agencies in their decision-making processes.

CLOSING PANDORA'S BOX

The level of lead in the environment in bioavailable form, distributed over the entire earth, has been increasing for thousands of years as a result of human activity. The industrial revolution accelerated the release of lead into the environment. The use of lead in paint still produces undue lead exposure in children, as a result of old paint in old houses and the present use of lead in paint for application to metals. The decision in the 1920s to add lead to gasoline has resulted in virtually universal exposure of the entire populations of industrialized countries. Humans now carry a lead burden 500 times that before lead mining began, and present intake is 100 times greater.

The United States has legislated the phaseout of lead from gasoline, and other countries are doing the same or contemplating such a move. Lead abatement and monitoring programs are in place in the inner cities, and children are being "treated" (by chelation) if exposure to lead reaches a specified level. The average level of lead in the blood of children as a result of these actions has decreased since the mid-1970s and will undoubtedly decrease even further. Can we therefore congratulate ourselves on our success in alleviating the problems produced by lead exposure and feel confident that lead toxicity will soon no longer be an issue to contend with? The answer is, unfortunately, no.

The adverse effects of in utero lead exposure are being characterized. The women presently in their childbearing years, and those that will be for the next 20 years, have been exposed during childhood to the highest lead levels since certain ancient civilizations. These individuals presumably carry a high level of lead in bone, which is available for mobilization into the fetus. In addition, there are millions of individuals who have had undue lead exposure as young children. Even if their exposure to lead decreases substantially, permanent damage has already been done. These individuals will be part of our society for another 60 to 70 years.

The effects of lead on various biochemical functions, particularly calcium homeostasis, may stress a wide variety of functions over the course of the life span, resulting in premature breakdown of these functions and accelerated aging. The ongoing insult of lead to the brain may also result in an accelerated decrease in mental functioning. These deleterious effects would presumably manifest themselves for a considerable period of time in our populations even if lead exposure ceased tomorrow.

It is clear, then, that environmental lead indeed represents a chemical time bomb in two senses: First, exposure to lead over years or a lifetime may result in health effects late in life, as well as very early in life. (We cannot know this yet, because individuals exposed in the 1940s and 1950s will not reach old age until after the turn of the century.) Second, decisions made by industrialists and governments in the 1920s and before will have *unavoidable* effects on individuals not yet born. It will be at least one more generation before this particular Pandora's box can be closed.

REFERENCES

Barrett, R. 1978. Auditory-Visual Integration, Intelligence, and Reading in Normal and Lead-Poisoned Children. Ph.D. thesis, University of Pittsburgh.

Barry, P. S. I. 1975. A comparison of concentrations of lead in human tissues. Brit. J. Indust. Med. 32:119–139.

Bellinger, D., H. L. Needleman, R. Bromfield, and M. Mintz. 1984. A followup study of the academic attainment and classroom behavior of children with elevated dentine lead levels. Biol. Trace Elem. Res. 6:207–223.

Bellinger, D., A. Leviton, C. Waternaux, H. L Needleman, and M. Rabinowitz. 1987a. Longitudinal analysis of prenatal and postnatal lead exposure and early cognitive development. New Engl. J. Med. 316:1037–1043.

Bellinger, D., J. Sloman, A. Leviton, C. Waternaux, H. Needleman, and M. Rabinowitz. 1987b. Low level lead exposure and child development: Assessment at age 5 of a cohort followed from birth. Pp. 49-53 in Proc. 6th Intern. Conf. Heavy Metals in the Environ., S. E. Lindberg and T. C. Hutchinson, eds. Edinburgh: CEP Consultants.

Billick, I. H. 1981. Lead: A case study in interagency policy-making Environ. Health Perspect. 42:73–79.

Byers, R. K., and E. E. Lord. 1943. Late effects of lead poisoning on mental development. Amer. J. Dis. Child. 66:471–494.

Cantarow, A., and M. Trumper. 1944. Lead Poisoning. Baltimore: Williams and Wilkins.

Centers for Disease Control. 1985. Preventing lead poisoning in young children—United States. Morbid. Mortal. Weekly Rep. 34:66–73.

Committee on Environmental Hazards. 1987. Statement on childhood lead poisoning. Pediat. 79:457–465.

Dabeka, R. W., K. F. Karpinski, A. D. McKenzie, and C. D. Bajdik, 1986. Survey of lead, cadmium and fluoride in human milk and correlation of levels with environmental and food factors. Fd. Chem. Toxic. 24:913–921.

de la Burde, B., and M. S. Choate. 1972. Early asymptomatic lead exposure and development at school age. J. Pediat. 87:638–642.

Dietrich, K. N., K. M. Krafft, R. L. Bornschein, P. B. Hammond, O. Berger, P. A. Succop, and M. Bier. 1987. Low-level lead exposure: Effects on neurobehavioral development in early infancy. Pediat. 80: 721–730.

Elias, R., Y. Hirao, and C. Patterson. 1975. Impact of present levels of aerosol Pb concentrations on both natural ecosystems and humans. Pp. 257-271 in International Conference on Heavy Metals in the Environment, Vol. II.

Environmental Protection Agency. 1984. Regulation of fuels and fuel additives: Lead phasedown. Fed. Reg. 49(150):31032–31050.

Ernhart, C. B., A. W. Wolf, M. J. Kennard, H. F. Filipovich, R. J. Sokol, and P. Erhard. 1985. Intrauterine lead exposure and the status of the neonate. Pp. 35–37 in Int. Conf. Heavy Metals in the Environment, Vol. 1, T. D. Lekkas, ed. Edinburgh: CEP Consultants Ltd.

Ernhart, C. B., A. W. Wolf, M. J. Kennard, and P. Erhard. 1986. Intrauterine exposure to low levels of lead: The status of the neonate. Arch. Envir. Health 41:287–298.

Fulton, M., G. Raab, G. Thomson, D. Laxen, R. Hunter, and W. Hepburn. 1987. Influence of blood lead on the ability and attainment of children in Edinburgh. Lancet 1(8544):1221–1226.

Grant, L. D. 1986. Assessment and management of risk associated with airborne lead in the United States: Key issues in the 1970s and 1980s. Pp. 467–502 in Health Effects of Lead, M. C. B. Hotz, ed. Ottawa: The Royal Society of Canada, Commission on Lead in the Environment.

Hansen, O. N., A. Trillingsgaard, I. Beese, T. Lyngbye, and P. Grandjean. 1987. A neuropsychological study of children with elevated dentine lead level. Pp. 54–56 in Proc. 6th Intern. Conf. Heavy Metals in the Environ., S. E. Lindberg and T. C. Hutchinson, eds. Edinburgh: CEP Consultants.

Hatzakis, A., A. Kokkevi, K. Katsouyanni, K. Maravelias, J. K. Salaminios, A. Kalandidi, A. Koutselinis, K. Stefanis, and D. Trichopoulis. 1987. Psychometric intelligence and attentional performance deficits in lead-exposed children. Pp. 204–209 in Proc. 6th Intern. Conf. Heavy Metals in the Environment, S. E. Lindberg and T. C. Hutchinson, eds. Edinburgh: CEP Consultants.

Hawk, B. A., S. R. Schroeder, G. Robinson, D. Otto, P. Mushak, D. Kleinbaum, and G. Dawson. 1986. Relation of lead and social factors to IQ of low-SES children: A partial replication. Amer. J. Mental Defic. 91:178–183.

Horn, W. F., and T. Packard. 1985. Early identification of learning problems: A meta-analysis. J. Educat. Psychol. 77:597–607.

Hunter, J. 1986. The distribution of lead. Pp. 96–126 in Lead Toxicity: History and Environmental Impact, R. Lansdown and W. Yule, eds. Baltimore: The Johns Hopkins University Press.

Jenkins, C. D., and R. B. Mellins. 1957. Lead poisoning in children. AMA Arch. Neurol. Psychiat. 77:70–78.

Johnson B. L., and R. W. Mason. 1984. A review of public health regulations on lead. Neuro. Toxicol. 5:1–22.

Landsdown, R., and W. Yule, eds. 1986. Lead Toxicity: History and Environmental Impact. Baltimore: The Johns Hopkins University Press.

Lansdown, R., W. Yule, M. Urbanowicz, and I. B. Millar. 1983. Blood lead, intelligence attainment and behavior in school children: Overview of a pilot study. Pp. 267–296 in Lead Versus Health: Sources and Effects of Low Level Lead Exposure, M. Rutter and R. Russell Jones, eds. Chichester: Wiley & Sons.

Lin-Fu, J. S. 1972. Undue absorption of lead among children—A new look at an old problem. New Engl. J. Med. 286:702–710.

Lin-Fu, J.S. 1985. Historical perspective on health effects of lead. Pp. 43–64 in Dietary and Environmental Lead: Human Health Effects, K. R. Mahaffey, ed. New York: Elsevier.

Mahaffey, K. R., ed. 1985. Dietary and Environmental Lead: Human Health Effects. New York: Elsevier.

McMichael, A. J., G. V. Vimpani, E. F. Robertson, P. A. Baghurst, and P. D. Clark. 1986. The Port Pirie cohort study: Maternal blood lead and pregnancy outcome. J. Epidemiol. Commun. Health 40:18–25.

Moore, M. R., and A. Goldberg. 1985. Health implications of the hematopoietic effects of lead. Pp. 261–314 in Dietary and Environmental Lead: Human Health Effects, K. R. Mahaffey, ed. New York: Elsevier.

Moore, M. R., A. Goldberg, S. J. Pocock, A. Meredith, I. M. Stewart, H. MacAnespie, R. Lees, and A. Low. 1982. Some studies of maternal and infant lead exposure in Glasgow. Scott. Med. J. 27:113–122.

Mushak, P., and A. Crocchetti. 1987. The Nature and Extent of Lead Poisoning in Children in the United States: A Report to Congress (draft). U.S. Dept. of Health and Human Services, Agency for Toxic Substances and Disease Registry.

National Academy of Sciences, Committee on Lead in the Human Environment. 1980. Lead in the Human Environment. Washington, D.C.: National Academy Press.

Needleman, H. L., ed. 1980. Low Level Lead Exposure: The Clinical Implications of Current Research. New York: Raven Press.

Needleman, H. L. 1983. Lead at low doses and behavior of children. Neurotoxicol. 4:121–133.

Needleman, H. L. 1987a. Low level exposure and children's intelligence: A quantitative and critical review of modern studies. Pp. 1–8 in Proc. 6th Internat. Conf. Heavy Metals in the Environment, S. E. Lindberg and T. C. Hutchinson, eds. Edinburgh: CEP Consultants.

Needleman, H. L. 1987b. Introduction: Biomarkers in neurodevelopmental toxicology. Environ. Health Perspect. 74:149–152.

Needleman, H. L., and D. J. Bellinger. 1986. Type II fallacies in the study of childhood exposure to lead at low dose: A critical and quantitative review. Proc. Internatl. Workshop on the Effects of Lead Exposure on Neurobehavioral Development, Edinburgh.

Needleman, H. L., C. Gunnoe, A. Leviton, R. Reed, H. Peresie, C. Maher, and P. Barrett. 1979. Deficit in psychologic and classroom performance of children with elevated dentine lead levels. New Engl. J. Med. 300:689–695.

Needleman, H. L., M. Rabinowitz, A. Leviton, S. Linn, and S. Schoenbaum. 1984. The relation between prenatal exposure to lead and congenital anomalies. J. Amer. Med. Assoc. 251:2956–2959.

Oliver, T. 1914. Lead Poisoning from the Industrial Medical and Social Points of View. New York: Paul B. Hoeber.

Otto, D. 1987. The assessment of neurotoxicity in children: Electrophysiological methods. Pp. 139–158 in Toxic Substances and Mental Retardation. S. R. Schroeder, ed. Washington, D.C.: Amer. Assoc. on Mental Defic.

Otto, D., V. Benignus, K. Muller, and C. Barton. 1981. Effects of age and body lead burden in young children. I. Slow cortical potentials. Electroencephalogr. Clin. Neurophysiol. 52:229–239.

Perlstein, M. A., and R. Attala. 1966. Neurological sequelae of plumbism in children. Clin. Pediat. 5:292–298.

Provenzano, G. 1980. The social costs of excessive lead exposure during childhood. Pp. 299–315 in Low Level Lead Exposure: The Clinical Implications of Current Research, H. L. Needleman, ed. New York: Raven Press.

Rosen, J. F. 1985. Metabolic and cellular effects of lead: A guide to low level toxicity in children. Pp. 157–186 in Dietary and Environmental Lead: Human Health Effects, K. R. Mahaffey, ed. New York: Elsevier.

Rosner, D. and G. Markowitz. 1985. A "gift of God"? The public health controversy over leaded gasoline during the 1920s. Amer. J. Public Health 75:344–352.

Rutter, M., and R. Russell Jones. 1983. Lead vs. Health: Sources and Effects of Low Level Exposure. Chichester: Wiley & Sons.

Schoenbrod, D. 1980. Why regulation of lead has failed. Pp. 259–266 in Low Level Lead Exposure: The Clinical Implications of Current Research, H. L. Needleman, ed. New York: Raven Press.

Schrager, J., J. Lindy, S. Harrison, J. McDermott, and E. Killins. 1966. The hyperkinetic child: Some consensually validated behavioral correlates. Exceptional Children 32:665–667.

Schroeder, S. R., B. Hawk, D. Otto, P. Mushak, and R. E. Hicks. 1985. Separating the effects of lead and social factors on I.Q. Environ. Res. 38:144–154.

Schwartz, J. 1986. The relationship between blood lead levels and blood pressure. Pp. 379–414 in Health Effects of Lead, M. C. B. Hotz, ed. Ottawa: The Royal Society of Canada, Commission on Lead in the Environment.

Schwartz, J., and D. Otto. 1987. Blood lead, hearing thresholds, and neurobehavioral development in children and youth. Arch. Environ. Health 42:153–160.

Schwartz, J., C. Angle, and H. Pitcher. 1986. The relationship between childhood blood lead levels and stature. Pediat. 77:281–288.

Schwartz, J., P. J. Landrigan, R. G. Feldman, E. K. Silbergeld, and E. L. Baker. 1988. Threshold effect in lead-induced peripheral neuropathy. J. Pediat. 112:12–17.

Settle, D. M., and C. C. Patterson. 1980. Lead in albacore: Guide to lead pollution in America. Science 207:1167–1175.
Silva, P. A., P. Hughes, S. Williams, and J. M. Faed. 1988. Blood lead, intelligence, reading attainment, and behavior in eleven year old children in Dunedin, New Zealand. J. Child Psychol. Psychiat. 29:43–52.
Smith, M. 1986. Lead in history. Pp. 7–24 in Lead Toxicity: History and Environmental Impact, R. Lansdown and W. Yule, eds. Baltimore: The Johns Hopkins University Press.
Stebbins, W. C. 1982. Concerning the need for more sophisticated animal models in sensory behavioral toxicology. Environ. Health Perspect. 44:77–85.
Stein, J. M. 1980. An overview of the lead abatement program response to the silent epidemic. Pp. 279–284 in Low Level Lead Exposure: The Clinical Implications of Current Research, H. L. Needleman, ed. New York: Raven Press.
Thompson, G. N., E. F. Robertson, and S. Fitzgerald. 1985. Lead mobilization during pregnancy. Med. J. Aust. 143:131.
Thurston, D. L., J. N. Middelkamp, and E. Mason. 1955. The late effects of lead poisoning. J. Pediat. 47:413–423.
Weiner, G., R. V. Rider, W. C. Oppel, and P. A. Harper. 1968. Correlates of low birth weight, psychological status at eight to ten years of age. Pediat. Res. 2:110–118.
Weiss, B. 1980. Conceptual issues in the assessment of lead toxicity. Pp. 127–134 in Low Level Lead Exposure: The Chemical Implications of Current Research, H. L. Needleman, ed. New York: Raven Press.
Winneke, G., and Kraemer, V. 1984. Neuropsychological effects of lead in children: Interaction with social background variables. Neuropsychobiol. 11:195–202.
Winneke, G., V. Kraemer, A. Brockhaus, U. Ewers, H. Kujanek, H. Lechner, and W. Janke. 1983. Neuropsychologic studies in children with elevated tooth-lead concentrations. II. Extended study. Int. Arch. Occup. Envir. Health 108:231–252.
Winneke, G., U. Beginn, T. Ewert, C. Havestadt, U. Kraemer, C. Krause, H. L. Thron, and H. M. Wagner. 1985a. Comparing the effects of perinatal and later childhood lead exposure on neuropsychologic outcome. Environ. Res. 38:155–167.
Winneke, G., A. Brockhaus, W. Collet, U. Kraemer, C. Krause, H. L. Thron, and H. M. Wagner. 1985b. Predictive value of different markers of lead-exposure for neuropsychological performance. Pp. 44–47 in Int. Conf. Heavy Metals in the Environment, Vol. 1, T. D. Lekkas, ed. Edinburgh: CEP Consultants.
Yule, W., R. Lansdown, I. Millar, and M. Urbanowicz. 1981. The relationship between blood lead concentration, intelligence, and attainment in a school population: A pilot study. Dev. Med. Child. Neurol. 23:567–576.

Chemical Time Bombs: Environmental Causes of Neurodegenerative Diseases

Peter S. Spencer

The known adverse effects of chemical substances on the human nervous system and special sense organs probably cover the waterfront of the combined disciplines of neurology, psychiatry, ophthalmology, and otorhinology, not to mention significant chunks of internal medicine. If we are truly interested in developing neurotoxicity tests with wide applicability and utility, it will be essential to consider the wealth of noncognitive, adverse behavioral effects induced by chemical agents acting on the nervous system and its target organs. Thus, in setting this goal, we must recognize, as does Roger Russell, that agents with potential for neurotoxic effects are widely deployed in the environment and are not limited to the well-known workplace chemicals and environmental pollutants. Indeed, some of the most widespread and troublesome substances are found as natural toxins of plants and animals; witness the crippling effects of *Lathyrus sativus* and *Manihot esculenta* (cassava) in Africa and Asia, the extraordinarily high incidence of ciguatera neurotoxicity in the Pacific Islands, and the unsolved worldwide problem of tetanus associated with the neurotoxin of *Clostridium* bacilli. Of course, although these are everyday realities for vast numbers of our fellow beings, they are of little consequence to those in developed countries who generally enjoy a varied diet and excellent hygiene, and who are understandably more concerned about avoiding contact with potentially harmful man-made substances. Even here, however, our professional sights are often limited to a rather small number of well-known substances such as lead salts, carbon disulfide, and *n*-hexane; and from this well-trodden ground, we sometimes commit the egregious error of denouncing all metals and solvents as neurotoxicants. Why do we seem to ignore other important subjects, such as the often well-characterized neurotoxic potential of therapeutic agents, which probably accounts for the majority of recognized cases of chemical neurotoxicity in developed countries? Why do we generally fail to consider the neurotoxic potential of food additives and fragrance raw materials when several such agents have been shown to produce neurobehavioral toxicity in animals or humans. Why was the discovery of methylphenyltetrahydropyridine (MPTP)—one of the most important neurotoxins of recent times—for many years essentially ignored by behavioral toxicologists? Because it was perceived to be only a contaminant of a street drug and therefore of no significance for the "real world" of occupational and environmental neurotoxicology? Are we really prepared to be this shortsighted, and is our vision of neurotoxicology and test method development limited to workplace chemicals and environmental pollutants? Are we willing to pass by the many specialties of neurobehavioral toxicology? Fortunately, as Norman Krasnegor demonstrates, the developmental behavioral toxicologist is better acquainted than most with the need to protect against the adverse effects of therapeutic drugs such as thalidomide, as well as those associated with chemicals, environmental pollutants, and contaminants of the food supply.

ESTABLISHING CRITERIA FOR CHEMICAL NEUROTOXICITY

At one time or another, we have all been guilty of making sweeping statements implying that certain groups of substances (e.g., metals, solvents, pesticides) have inherent neurotoxic properties. Such statements are not only inaccurate but also patently misleading. We need to develop much more critical criteria for neurotoxicity. The problem is illustrated immediately when a chemical is labeled a neurotoxin without any consideration of dose. Obviously, anticholinesterase chemicals have the potential for inducing dramatic neurobehavioral changes but, as Russell points out, the very same pharmacological properties have been exploited (without success, it is widely acknowledged) as symptomatic therapy for Alzheimer's disease. Thus, anticholinesterase chemicals are not neurotoxins but, at certain doses, exert neurotoxic effects. An even better example of the importance of dose is provided by the vitamin pyridoxine, well known to be an absolute requirement for normal metabolism. In this case, neuropathy follows both inadequate and megavitamin intake of Vitamin B_6. Obviously, pyridoxine is not a neurotoxin but, in sufficiently large dosage, the chemical does have neurotoxic potential. Similarly, lead, mercury, and acrylamide are not neurotoxins per se, but they are able to cause neurobehavioral changes at certain levels of exposure.

Once this point is understood, we are in a position to proceed to the question of connecting a specific chemical to one or more neurobehavioral effects. Our lack of precision here is equally as misleading. Consider the statement that acrylamide is a peripheral nervous system (PNS) neurotoxicant. Although it is certainly true that acrylamide has the potential to induce peripheral neuropathy in humans and animals, it is also apparent that this is a function of dose and duration of exposure. At certain doses, acrylamide is able to induce encephalopathy, with confusion, disorientation, memory disturbances, and hallucinations; at other doses, ataxia and dysarthria predominate. Thus, acrylamide is a chemical with neurotoxic potential capable of inducing a number of neurobehavioral effects that vary in nature with dosage and duration of exposure. An analogous situation exists with thallium, carbon disulfide, *n*-hexane, and a host of other chemicals with neurotoxic potential. Such considerations are more, much more, than semantic niceties; our use of language in describing chemical neurotoxicity is often so imprecise that it serves to mislead the public rather than to inform.

Our most dismal performance, however, is reserved for our failure to develop solid criteria that must be met before a chemical is identified as having neurotoxic potential in human subjects or is linked with a human neurological or developmental disorder. Stunning examples are provided by the diatribe over solvents and aluminum. In the case of solvents, where specific chemical substances are clearly associated etiologically with various types of neurological deficit, some have been prepared to indict and convict all chemicals labeled as solvents. Are we unaware that there are many different classes of chemicals which make up this heterogenous group of substances, and that the large majority of solvent chemicals has never been tested for neurotoxic potential? Apparently, this paucity of information has not prevented some from claiming solvent chemicals as causal of human dementia. Equally as disturbing is the rationale for continuing to link aluminum with the etiology of Alzheimer's disease. This idea began when neurofilamentous accumulations induced by aluminum were observed by light microscopy to be superficially similar to those seen in Alzheimer's disease. Somehow, the idea continued even though neurofilaments of experimental aluminosis were found by transmission electron microscopy to be identical to those induced by a host of other substances and quite distinct from the characteristic paired-helical filaments of Alzheimer's disease. More fuel was added when dialysis dementia was linked to aluminum toxicosis, although the clinical features and neuropathology of dialysis dementia and Alzheimer's disease are distinctly different. Nevertheless, the idea of a link between Alzheimer's disease and aluminum toxicosis has become so entrenched that the public is encouraged to believe it. The (nonspecific) observation of aluminum (and other metals) in neurons and senile plaques in Alzheimer's disease and related disorders has further encouraged those faithful to the aluminum hypothesis. However, as Gerhard Winneke observes, the evidence supporting an association is (at best) circumstantial.

This discussion, of course, is intended neither to defend solvents—many of which may well prove to have neurotoxic potential once they are tested—nor to imply that aluminum has no etiologic relationship with Alzheimer's disease. Rather, it is a plea for the exercise of extreme caution in the best tradition of scientific conservatism before statements are made about cause-and-effect relationships between chemical substances and human neurobehavioral effects. More specifically, it is a call for the development of a set of guidelines that must be met before a chemical is accepted as causal of human neurobehavioral dysfunction. Identification of those chemicals that unequivocally fulfill criteria for a potential human neurotoxin is an essential first step in the development and validation of neurobehavioral test methods for chemical neurotoxicity. Such tests will require reference compounds linked to the various types of neurobehavioral deficit under study.

The Minamata tragedy, reviewed by Winneke, provides a graphic illustration of the steps that should be taken before a substance is accepted as causally responsible for human neurotoxic disease. Simply stated, the neurological illness seen in humans and cats was reproduced in the latter both by feeding methylmercury-contaminated fish and by administering authentic samples of the suspect chemical agent. Because pure methylmercury was able experimentally to induce in cats a disorder indistinguishable from that seen in the affected feline population of Minamata Bay, and the nature of the feline disorder clearly paralleled the attendant human disease, it was possible to make strong statements about cause-and-effect relationships between methylmercury and neurodegenerative disease. Thus, in the absence of human experimentation, the strongest foundation on which we can rest our case for chemical neurotoxicity is the experimental induction in animals of a disorder equivalent to that seen in human subjects exposed to the compound under scrutiny. Such substances, of which there are relatively few, constitute the list of chemicals with proven human and animal neurotoxic potential (class 1 chemicals). A much larger number of chemicals has been associated with neurobehavioral effects in exposed humans (class 4 chemicals), but these reports are often poorly documented exposures to mixtures of ill-defined substances that have not been subjected to experimental scrutiny either in isolation or as a mixture. Important exceptions are the recurring reports of consistent neurobehavioral deficits associated with certain therapeutic agents and mixtures, where the exact chemical compositions as well as the dose and duration of exposure are often well documented (class 3 chemicals). Our task, of course, is to help devise experimental methods that will detect neurotoxic effects of substances which can then be prevented from entering the marketplace or removed therefrom. Chemicals with neurotoxic properties recognized in experimental animal studies constitute a group of agents suspected to possess human neurotoxic potential (class 2). By far the largest group of substances, however, contains that multitude of untested chemicals which is not known to be linked to any human neurobehavioral disorder (class 5). Thus, in summary, at least five categories may be recognized in our efforts to link chemical substances with potential adverse effects on the human nervous system, the strength of the association diminishing as one descends through the list:

Class 1. Experimentally proven in animals, with similar effects in humans
Class 2. Experimentally proven in animals, with unknown effects in humans
Class 3. Effects observed in humans but experimental evidence unavailable
Class 4. Possible effects in humans but experimentally unproven
Class 5. Effects unknown in humans and untested in animals

Important consequences flow from this conservative classification. For example, in the case of industrial solvents, the majority of which falls into classes 4 or 5, there is no justification at the present time for stating that solvents as a class have human neurotoxic potential. On the other hand, because solvents are so widely employed in industry and so little is known of their chronic neurobehavioral effects, there is every reason to improve knowledge in this area. Another example is the so-called Spanish Toxic Oil Syndrome, an epidemic of a new and remarkable disease that affected thousands of people in Madrid and its surrounds. Although epidemiological studies strongly link the disorder to the consumption of an illicit cooking oil, until a viable model of the disease is produced in a laboratory animal fed the suspect agent, there is always room for a small degree of doubt.

CLASSIFYING CHEMICALLY INDUCED HUMAN NEUROTOXIC DISORDERS

Russell argues convincingly that basic and clinical research designed to understand the neurochemical mechanisms by which exposures to toxicants affect behavioral indicators will lead us a long way toward a comprehension of the enormity of our task and the design of appropriate methods to detect, define, and even predict chemical neurotoxicity. In other words, we cannot simply devote our research time to the somewhat pedestrian task of developing test methods to fill the regulators' needs. With this in mind, therefore, it is appropriate to inquire how far we have come along the mechanistic path, where we need to concentrate research efforts to improve understanding, and how we should proceed in our attempts to identify the "atomic variety" of chemical time bomb. Our first chore, however, is to understand how to classify the adverse effects of chemical substances, first in the adult and then during brain development.

THE MATURE NERVOUS SYSTEM AS CHEMICAL TARGET

Although a satisfactory scientific nosology of chemicals with neurotoxic properties in adult human subjects remains an elusive goal, it is possible to offer a surprisingly useful framework for the eventual development of a comprehensive classification system. Ideally, this should be able to link the target of chemical attack to alterations in neural function that explain observed neurobehavioral changes and, in the clinical setting, provide a logical basis for prevention. One such attempt is a 1984 classification of human neurotoxic response that was based on cellular and subcellular targets of chemical agents. This recognizes the following sites of functional disruption or damage: the neuron, glial cells and myelin, nervous system vasculature, and muscle. Points pertinent to the present discussion are readily made by selectively considering agents that act on the mature nerve cell.

Three broad types of neuronal change induced by chemical substances are recognized, namely, functional perturbations of the excitable membrane, interference with neurotransmitter systems, and structural breakdown of dendrite, perikaryon, or axon.

The first type involves direct chemical interference with the excitable surface membrane of neurons, the consequent changes in electrical transmission, and the associated generation of neurobehavioral alterations. By and large, these effects appear and reverse rapidly, and the extent of dysfunction reflects the distribution of the chemical within the central or peripheral nervous system. The best examples are provided by agents that interfere with the normal passage of sodium ions across the nerve cell membrane. Some substances, such as tetrodotoxin (from the puffer fish) and saxitoxin (paralytic "shellfish" poison), act as channel blockers, whereas ciguatoxin, scorpion and anemone toxins, DDT, and pyrethroid insecticides, act to increase membrane permeability to sodium while the membrane is in either the resting or the active state, or both. Although the neurobehavioral effects of these channel agents range from discomfort (circumoral and distal-extremity paresthesias) to life-threatening dysfunction (respiratory paralysis), the salient point for the present discussion is that they are unlikely to have any long-lasting effects once the chemical has left the membrane receptor. Thus, although neurotoxicity may be fatal, agents of this type do not merit consideration as delayed-action chemical time bombs.

The second common locus of chemical attack upon the neuron is its neurotransmitter system, the proper regulation of which is often an absolute requirement for normal behavior. Chemicals may interfere with neurotransmitters at many levels, including their synthesis in the neuronal perikaryon, transport along the axon, packaging and release from synaptic vesicles, transport across the synaptic cleft, and reception by the target membrane, as well as the enzymatic breakdown or reuptake of excess transmitter in the nerve gap. Because these exigencies (points of vulnerability) exist for each and every neurotransmitter, the possible sites of chemical-induced perturbations are legion. Take, for example, agents active on cholinergic pathways, a subject discussed by Russell. Of course, the relationship between human behavioral changes and the selective actions of chemical agents on critical neurotransmitter systems extends far beyond the cholinergic system. Comparable examples may be drawn for chemicals active on central and peripheral catecholaminergic pathways. For example, monoamine oxidase inhibitors that increase the duration of action of synaptic catecholamines lead to mania, hyperreflexia, and involuntary movement. By contrast, the antihypertensive drugs tend to induce mental depression, weakness, and lethargy because they serve to deplete synaptic catecholamines. Other agents act on serotonergic pathways (lysergic acid diethylamide, LSD), gamma-aminobutyric acid (GABA) pathways (picrotoxin), glutamate pathways (beta-N-oxalylamino-L-alanine, BOAA), whereas some, such as the antipsychotics and opiates, influence a number of different pathways.

Most of the important adverse neurobehavioral effects of chemicals that perturb the function of neurotransmitter systems tend to appear rapidly, may be life threatening, and are generally considered to reverse without permanent sequelae. In this regard, many transmitter neurotoxins are as insignificant for our present thinking as compounds that only perturb excitable membranes. A notable exception is the persistent buccolingual-masticatory dyskinesia and choreiform movement of trunk and extremities (tardive dyskinesia) that often accompanies prolonged therapy with antipsychotic drugs. These involuntary movements are aggravated with emotional stress and concentration on motor tasks, and are intensified by the reduction or discontinuance of therapy, possibly because of the "unmasking" of supersensitive dopamine receptors. Chemically induced movement disorders also routinely accompany prolonged L-dopa (dihydroxyphenylalanine) therapy for parkinsonism. Although these two iatrogenic conditions probably account for a significant percentage of the disabling, chemical-induced neurological disease seen in developed countries, they have received little attention from neurobehavioral toxicologists concerned with environmental toxicity. However, for the neurologically and psychologically impaired patient who faithfully follows the doctor's prescription of L-DOPA or antipsychotics, the tardive appearance of these neurobehavioral effects may often produce an incapacitating or life-threatening state equal to or greater than the disease under treatment. A distinguished Spanish neurologist recently reported that fully 30 percent of his parkinsonian patients were found to have a drug-induced disorder that disappeared upon cessation of treatment. Thus, in addition to the mechanistic information available from such data, therapeutic agents also require much more rigorous neurobehavioral testing to predict and control the development of iatrogenic neurological and psychiatric states.

The different types of structural damage induced by chemical agents constitute for present purposes a very important third class of neuronal responses to chemical attack. Numerous chemicals and drugs are able to induce axonal degeneration without loss of the parent neurons; most require repetitive exposure. The neurobehavioral changes (usually peripheral neuropathy) develop rapidly or insidiously after weeks or months of exposure, and the disorders are reversible to the extent that damaged axons will regenerate and reconnect with sensory and motor targets in the peripheral nervous system. The clinical signs and symptoms develop in a distal, symmetrical, and temporally ascendant pattern in the extremities, with sensory loss and muscle weakness usually predominating in clinical significance over the commonly attendant autonomic dysfunction. Axonal neuropathy in humans, animals, or both, is known or (from the pattern of neurobehavioral dysfunction) suspected to occur with repetitive exposure to workplace chemicals (e.g., acrylamide, ethylene oxide, carbon disulfide, *n*-hexane, methyl *n*-butyl ketone, dimethylaminopropionitrile, certain organophosphates), therapeutic drugs (vincristine, chloramphenicol, thalidomide, disulfiram, isoniazid), household chemicals (thallium, arsenic, zinc pyridinethione), abused substances (ethanol), and natural toxins (buckthorn). Some are painful (thallium); others are associated with autonomic dysfunction (acrylamide), prominent pyramidal signs (leptophos, methyl bromide), or optic nerve changes (ethambutol). Although the molecular and cellular mechanisms underlying these toxic effects are unknown, the resulting pattern of neurobehavioral dysfunction is remarkably stereotyped and often readily studied in experimental species. Clinical experience demonstrates that recovery is usually slow, with sensory and motor deficits disappearing in the reverse order of their appearance. Recovery of sensation and strength is usually well advanced within months or years after cessation of exposure to the offending agent. Sometimes, when there has been extensive damage to ascending or descending spinal pathways, there may be permanent sensory loss, truncal ataxia, or pyramidal signs. The pathways involved in axonal neuropathies are also deleteriously affected with the normal advancement of age, but late-life appearance of neuropathy in previously recovered subjects is not recognized. In short, the fuse of this time bomb is short, and the limited damage associated with the explosion is largely reparable.

Perhaps the most important group of potential neurotoxins consists of those that trigger degeneration and loss of nerve cells. Examples include (1) thallium-induced lesions of the amygdala and periamygdaloid cortex, precipitating uncinate fits and peculiar affective disorders resulting from disruption of the limbic system; (2) inorganic mercury-induced cerebellar lesions associated with intention tremor, disordered speech, and ataxic gait; (3) lead-induced cerebral cortical damage leading to irreversible mental retardation; (4) organomercury-induced degeneration of dorsal root ganglion neurons and the calcarine and cerebellar cortex, with coincident sensory loss, tunnel vision, and ataxia; (5) MPTP-induced degeneration of nigrostriatal neurons causing parkinsonism; and (6) BOAA-induced loss of pyramidal neurons eliciting the spasticity of lathyrism. Because mature neurons are postmitotic and therefore irreplaceable, these types of neurobehavioral deficits are permanent. Moreover, because some neuronal groups at risk for toxic damage also normally undergo nerve cell attrition with advancing years, the combined effects of chemical-induced damage and age-related loss may lead to a permanent deficit that becomes relentlessly progressive in old age. Finally, and most significantly, because these regions of the brain are commonly endowed with a substantial functional reserve, the initial loss of neurons associated with chemical damage may be clinically silent and only unmasked years or decades later as the deleterious effects of age deliver the coup de grace. Here is a chemical time bomb with a very long fuse, and once it explodes, the damage accrues relentlessly until death supervenes. Long-latency neurotoxicity of this type has been associated with organomercurialism, MPTP-induced parkinsonism, and the cycad-associated neurodegenerative disorder of the western Pacific that displays facets of amyotrophic lateral sclerosis (ALS), parkinsonism, progressive supranuclear palsy, and senile dementia. As a result of these new observations, a search has recently begun to identify exogenous chemicals with neurotoxic properties that may play a key role in triggering some of the devastating neurodegenerative diseases of later life, notably ALS, Parkinson's and Alzheimer's diseases.

DEVELOPING NERVOUS SYSTEM AS CHEMICAL TARGET

This is the point at which we must turn our attention to the susceptibility of the nervous system during its formative stages, the subject of Krasnegor's incisive chapter. Because the developing nervous system is radically different in structure from its adult counterpart, a completely independent system must be deduced to classify the adverse effects of chemical substances. For example, whereas the adult neuron is postmitotic, static, and typically equipped with elaborate, branching dendrites for the receipt of electrochemically encoded information from neighboring nerve cells, the developing neuron divides, migrates, and has few cellular processes in contact with those of few other nerve cells. Similarly, during development, glial cells proliferate, their processes are mobile, and myelin formation is prominent. These and other factors, such as an absent blood-brain regulatory interface to control access of chemical substances to nervous tissue, radically alter the potential responses of the developing nervous system to chemical attack. Whereas specific or unique abnormalities are known to be caused by certain agents (thalidomide), structurally similar abnormalities may result from exposure to different chemicals. Conversely, different types of developmental anomalies may occur with a single noxious agent. Observations such as these suggest that there are factors besides the chemical nature of the agent which dictate the type of resulting damage. Genotype and species are important variables, but the overriding factor is developmental stage. Above all else, the timing of chemical attack appears to dictate the resulting effect.

It is well established that exposure to selected agents during development in utero may have dramatic consequences for behavior during postnatal maturation and young adulthood. Although behavioral abnormalities may be found in the absence of structural changes, either grossly or with a light microscope, it is a tenet of neurobiology that some (presently undetectable) alteration—for example, in synaptic organization or neurochemical anatomy—must underlie a change in neural function. That behavioral alterations induced by chemical agents have no structural basis whatsoever is untenable. What is needed is for the behavioral teratologist and the neurochemical anatomist to join forces to find out how to account for these behavioral abnormalities. As Krasnegor points out, the recent methodological breakthrough in studying the conceptus in the externalized uterus provides a remarkable new window of opportunity to research these phenomena. Substances of interest can now be studied for their neurochemical, structural, and behavioral effects at the time of gestation (when they have their putative action upon the developing brain), postnatally during neonatal development, at maturity, and even in old age.

Scientific Basis for Neurobehavioral Toxicity Testing

With this broad overview of the adverse effects of chemical substances on the human nerve cell during development and at maturity, we are in a position to assess our understanding of molecular and cellular mechanisms of neurotoxicity and to determine where there is a lack of information and which type of neurobehavioral effect constitutes the greatest threat to human health.

The mere fact that we have been able to construct a tenable basis for understanding the action of chemical substances on the human nerve cell is most encouraging. During development, the stage of cellular differentiation plays a major role in dictating the resultant structural (and probably neurobehavioral) alterations. Although a vast amount of work needs to be devoted to this subject, at least there is a logical basis for understanding why, for example, antimitotics rather than sodium channel agents have devastating effects. Similarly, in the adult, there is a satisfying mechanistic explanation for the rather similar signs and symptoms associated with chemicals that target one or another neurotransmitter (e.g., cholinergic) pathway even though the xenobiotics of interest (e.g., anticholinesterases, pesticides, and certain snake venoms) may have greatly disparate chemical structures. Far less satisfying is the current state of ignorance of the molecular and cellular mechanisms underlying neuronal and axonal degeneration. With few exceptions, such as the well-characterized action of diphtheria toxin on the Schwann cell, a similar state of ignorance exists for chemical toxins that target myelinating cells, muscle cells, the neuroendocrine system, and the intimate vasculature of the nervous system.

How can we improve understanding of the scientific basis of neurotoxicology and create a more solid foundation on which neurobehavioral toxicity testing can be developed? We have already noted the extraordinary new opportunity for a multidisciplinary attack on the effects of chemical substances on the nervous system during both in utero and postnatal development. Similarly, there are important opportunities for collaboration in understanding how chemicals may modify behavior in the adult. For example, the expertise of the neurophysiologist is required to explain the behavioral outcome of overexposure to membrane channel agents; neuropharmacologists are well equipped to discuss the functional consequences of neurotransmitter disruption; and neuropathologists are needed to offer a rational basis for neurobehavioral alterations associated with structural breakdown of the nervous system. Taken in concert, therefore, behavioral toxicologists will greatly strengthen their science if they join forces with multidisciplinary teams with expertise in many areas of the neurological sciences.

An additional collaborative opportunity for the behavioral toxicologist has been opened up by the advent of positron emission tomography (PET), a noninvasive imaging system that permits assessment of the functional status of the human and primate brain in real time. By careful selection of appropriate radioactive labels and their precise localization in the brain following systemic administration, the PET specialist is able to estimate the integrity of a particular brain region and sometimes detect lesions that are clinically silent at time of analysis but predictive of impending disease. The best example is the ability to detect (with a fluorodopa probe) subclinical lesions in the substantia nigra that predict the likely onset of parkinsonism later in life. Because subtle behavioral differences have been reported in individuals prior to the onset of clinical parkinsonism, there is an important opportunity for collaborative research using the methodology of both behavioral toxicology and PET. The obvious place to start is with MPTP-lesioned primates with fluorodopa PET evidence of nigrostriatal damage but no clinical signs of parkinsonism. Such animals would then be candidates for testing the behavioral effects of brain implants designed to restore normal levels of dopamine neurotransmitter, as discussed by Russell. Unfortunately, even though a bona fide primate model of human parkinsonism was available to test the efficacy of brain transplants in restoring normal behavior, the international neurosurgical community saw fit to proceed directly with human experimentation on Parkinson patients. Although a few have been helped, the results overall have been disappointing.

Our final task is to decide where the "atomic" variety of chemical time bomb is likely to be deployed in the broad environment. Implicit in this question is the notion of a long fuse, a surprisingly large explosion, and a devastating, irreversible outcome. Thus, we are not concerned here with the short-latency effects of certain chemicals on excitable membranes or the reversible consequences of pharmacological disruption of a neurotransmitter system. Certainly, disorders such as drug-induced tardive dyskinesia are of considerable relevance because of their poor reversibility. Degeneration of axons and myelin is also of some concern, because these disorders are usually either slowly reversible (peripheral neuropathy) or irreversible (spasticity). Yet none of these conditions can be compared to the new and frightening concept of long-latency neuronal toxicity, in which the chemical exposure purportedly occurs decades prior to the clinical appearance of a neurodegenerative disease that is not only irreversible, but also relentlessly progressive, totally incapacitating, and even when treated, inevitably fatal.

Just as the drama of the potential fetotoxicity of chemical substances unfolded as a consequence of the effects of a therapeutic drug, discovery of the principle of long-tendency neurotoxicity has come not from testing either workplace chemicals or widespread environmental pollutants but rather from systematic study of high-incidence neurodegenerative disease that in one case (MPTP) was linked to a contaminant of a street drug and, in another, a neurotoxic plant (cycad). Experimental animal studies confirmed the suspicion that MPTP was responsible for inducing parkinsonism in a group of California drug addicts. More importantly, certain individuals who were exposed to MPTP, but who remained clinically intact, were shown by fluorodopa PET to have sustained damage to the substantia nigra. Because this pathway is highly susceptible to age-related neuronal attrition, these subjects are currently being followed in the expectation that the combined effects of toxic neuronal damage and age-related cell loss will eventually overcome the considerable functional reserve of this pathway, whereupon they too will develop progressive clinical parkinsonism.

An even more troubling possibility has emerged from study of a prototypical neurodegenerative disease known as western Pacific ALS and parkinsonism-dementia (P-D) complex. Decades of research on this disappearing familial disease have ruled out inherited and viral factors as causal and have clearly indicated a nontransmissible environmental trigger. The vast majority of evidence incriminates seed of the neurotoxic cycad plant, formerly an important source of medicine or food in all three areas where high-incidence ALS/P-D has been found. Epidemiological studies of populations migrating to and from Guam have clearly established that the disorder may be acquired during the first 20 years of life but may remain clinically silent for up to 35 years or more. Two ideas have been advanced to explain this phenomenon: one (discussed above in relation to MPTP) proposes the additive effects of subclinical, chemical-induced neuronal depletion at the time of exposure, coupled with age-related attrition of the same neuronal population; the other proposes the existence of one or more chemical substances in the cycad plant that act as a "slow toxin." The latter idea evolved from the intensive study of individual patients with documented heavy cycad exposure in the first or second decades of life, who developed clinical ALS less than 15 years later. Because age-related neuronal attrition cannot possibly be involved in the etiology of ALS in subjects who develop aggressive disease prior to age 30, some other explanation is needed. The slow-toxin hypothesis proposes the existence (in cycad seed) of an agent which, after single or multiple exposure, establishes an irreversible sequence of molecular and cellular events that lead progressively to changes in neuronal integrity and eventually to degeneration. Once sufficient target neurons have undergone degeneration (perhaps 50 percent of the anterior horn cells in the case of ALS), the previously covert disease becomes clinically apparent. A somewhat analogous situation exists with delayed peripheral neuropathies induced by organophosphates, except in this case the time to onset of clinical disease is measured in weeks and, once established, the disease is not progressive. The best analogy, however, may be with cancer, and this consideration may provide clues as to where to look for molecular mechanisms underlying the action of a putative slow neurotoxin. The analogy should also be read as an indication of the massive level of funding that is urgently needed to begin to determine whether disorders such as Alzheimer's disease are triggered by early exposure to exogenous chemicals.

How do these new concepts of long-latency neurotoxicity influence current concerns over workplace chemicals and environmental pollutants? First and foremost, we must rigorously explore the concept of slow neurotoxins in the hope of identifying the chemical nature of substances that exhibit this property. Once these critical pieces of information are in hand, we should be able to identify comparable chemical factors throughout the human environment and recommend steps for the prevention of exposure. Just as the Guamanian public has been warned of the long-term hazards associated with use of cycad seed for food and medicine, one day we may be able to advise industry of comparable slow toxins in the workplace. Although intensive laboratory research is mandatory to reach this goal, it is also well worthwhile subjecting patients who develop neurodegenerative disease at a young age to the most intensive exploration of their chemical exposure history. Here is a very special research opportunity for the behavioral toxicologist to work in cooperation with the neurologist; the latter has access to the patients, whereas the former should be uniquely equipped with a broad knowledge of the potential adverse effects of chemical substances in all environmental loci.

The second corollary to be drawn from the new concern over long-latency neurotoxicity is the need for collaborative behavioral and neuroanatomical studies to identify which populations of nerve cells are most susceptible to the aging process, and how such changes influence behavior. Additionally, there is an absolute requirement for research focus on chemicals that have the ability to destroy age-sensitive neuronal populations. Compounds such as trimethyltin and the glutamate excitotoxins are of very great interest in this regard, but does the list also include the classical environmental pollutants that have figured so centrally in the chapters presented in Part III of this volume? Space permits consideration of just three: solvents, lead, and mercury.

The neurobehavioral effects of solvents have occupied an extraordinary amount of space in the recent literature although, with few exceptions, it has been difficult to obtain consistent, clear-cut evidence of the inherent neurotoxicity of chemicals that fall within the many classes making up this heterogeneous collection of substances. Although some have sought to link dementing states to occupational exposure of often ill-defined mixtures of solvents, there is no clinical or neuropathological evidence to suggest that these substances have the capacity to induce long-latency neurodegenerative diseases. Certainly, we can justifiably include certain individual solvents (e.g., *n*-hexane, methyl *n*-butyl ketone) in class 1 substances that have established animal and human neurotoxic potential, but the associated disorders are largely reversible (neuropathy) or persistent (ototoxicity). Because the structure and function of the critical cellular components of peripheral nerves and inner ear also decline with age, elderly subjects with preexisting neurotoxic damage to these structures may be more markedly affected than their unexposed peers. There is also special concern over the occupational solvent carbon disulfide because, in addition to its ability to trigger psychosis and neuropathy, there are several reports suggesting the tardive onset of a form of parkinsonism. With this notable exception, there is no evidence to suspect that solvents represent the types of chemical time bombs that concern us here.

In a well-written advocacy for the rigorous control of environmental levels of lead, Rice proposes that this potential neurotoxicant merits special consideration because the metal has a very long half-life in the body. Less convincing is her stand that this may result in long-latency effects, such as a neurodegenerative disorder appearing late in life. It is far from clear, as she proposes, that lead has a propensity to attack age-sensitive populations of neurons. Nevertheless, because lead is likely to be mobilized from bone in advancing age (during the process of demineralization), the suggestion of tardive toxic and neurotoxic effects needs to be considered seriously. A few authors have linked lead with amyotrophic lateral sclerosis, but the case is far from proved and the leading current proponent has recently discarded the idea. Another solitary investigator has extrapolated experimental neuropathologic observations to propose a link between lead and Alzheimer's disease, but the idea in general is given no credence. Needleman (1980) pointed out that the mobilization of lead from the bones of the elderly is synchronous in some subjects with restricted intake of proteins, calories, and other trace elements, raising the possibility that some of the cognitive changes in older people are an effect of lead. He concluded: "The behavioral and biochemical status of older subjects with respect to both lead exposure and lead mobilization could well be a fertile area for investigation." Perhaps, one might dare to add, the potential importance of this subject merits diverting some of the current interest in lead neurotoxicity from the heroic use of statistics to uncover minor changes of dubious significance in the intellectual performance of young subjects with modest blood levels. Isn't the possibility of relentlessly progressive late-life decline in intellectual performance at least as important as the possibility of being robbed of a few IQ points during early development?

Winneke's review of Minamata disease is also of special relevance to long-latency neurotoxicity because of the recognition by some Japanese authorities of clinical variants in which manifestations of toxicity worsened after contamination had ceased or in whom the signs and symptoms of methylmercury poisoning appeared after a delay of some years. This was initially reported in the 1970s as *Minamata disease of late onset*, and some alleged cases have been verified at autopsy. Several explanations have been advanced to account for this phenomenon, including (1) the psychological condition of people who are eager to be compensated (latent Minamata disease is not recognized in local legal circles), (2) long-lasting but slight damage due to a minimal amount of organic mercury remaining in the brain (unlikely in view of data on the accumulation and metabolism of ingested mercury), and (3) the effect of aging on latent Minamata disease. In the light of recent understanding about long-latency neurotoxic disorders, the latter proposal clearly merits close study.

CONCLUSION

My major goal in discussing the various points raised in the preceding four chapters is to propose a firm scientific foundation on which to accept substances as potential human neurotoxins and to place in perspective the relative severity of the adverse effects induced by chemical substances by analyzing their sites and mechanisms of action. I have argued strongly that a comprehensive understanding of behavioral neurotoxicology can only be achieved if we consider all types of chemical substances that attack the nervous system. Of course these compounds must be rigorously tested and regulated, but we will only understand the true magnitude and awfulness of their potential effects, and be able to devise appropriate test methods to detect such changes, if we are prepared to draw freely from the entire body of knowledge available to the science of neurotoxicology.

REFERENCE

Needleman, H. L., ed. 1980. Low Level Lead Exposure: The Clinical Implications of Current Research. New York: Raven Press.

PART IV
Behavioral Aspects of Neurotoxicity:
Regional Issues

Neurobehavioral Toxicity Testing in China

Liang You-Xin

INTRODUCTION

It is hardly an overstatement that "environment" has become one of the key words of our time. A great number of different chemicals are produced for use at home, in industry, in agriculture, and in the control of diseases. It is now estimated that the universe of chemical compounds exceeds 5 million and at least 80,000 chemicals are available in the open market. It is also estimated that 1,000–2,000 new chemicals enter the market yearly and consequently pass into the environment (Geyer et al., 1986). Therefore, environmental pollution due to various kinds of chemical waste, dust, fumes, smog, and vapor has become a tangible challenge to human society.

In China, along with the rapid growth of industrialization, occupational exposure to toxic, particularly neurotoxic, substances has expanded both in the extent of exposure and in the spectrum of toxicants. According to a nationwide investigation during 1979–1981, there were 1,031,775 workers in 51,574 enterprises exposed to lead, mercury, benzene, organophosphorus pesticides, or trinitrotoluene (TNT). The overall prevalence of occupational poisoning induced by these five chemicals reached 1.3 percent among a total population of 987,934 examined (Table 1) (Gu et al., 1985). The challenge has grown since the swift expansion of industrialization from urban areas to rural areas where numerous small-scale industries were built at town and village levels during recent years. By 1985, there were about 2,225,000 such

TABLE 1 Prevalence of Intoxication from Five Chemicals in Question

Chemicals	Workers Examined ($\times 10^4$)	Cases of Chronic Intoxication	
		Number ($\times 10^4$)	Prevalence (%)
Lead	35.50	0.627	1.766
Mercury	6.20	0.166	2.675
Benzene	50.90	0.268	0.526
Organophosphorus pesticides	1.70	0.058	3.429
Trinitrotoluene	4.20	0.110	2.619

SOURCE: Gu et al. (1985).

enterprises across the country. It was reported that 65 percent of the measured levels of chemical as well as some physical agents (e.g., noise) from 2,321 worksites exceeded the prescribed maximum allowable concentration (MAC) or permissible exposure level (PEL). The overall prevalence of lead, benzene, mercury, and chromium poisoning reached 6 percent, which is about fivefold greater than corresponding levels found in state-owned industries in urban areas (Yu and Gu, 1987).

From the viewpoint of protecting workers from irreversible damage by chemicals, particualarly neurotoxicants, approaches for early detection of reversible neurotoxicity at the subclinical stage of exposure are a growing concern. Over the years, scientific research on, and applications of, neurobehavioral toxicity testing have been extensively conducted in Finland, the United States, and some countries in Europe. Emphasis has been mainly on the development of a test battery for the purpose of early detection of neurobehavioral effects due to occupational exposure to neurotoxic agents in the working population. Helena Hanninen is regarded as one of the most respected pioneers in this field that she called "toxicopsychology." Her booklet "Behavioral Test Battery for Toxicopsychological Studies" (Hanninen and Lindstrom, 1979) has provided detailed information about test items, instruments, and procedures for researchers involved in behavioral studies. Other pioneering work contributed by international efforts has played a key role in the initiation of neurobehavioral toxicity testing in China. Growing numbers of Chinese researchers are showing interest in exploring the use of neurobehavioral tests and recognizing that a proper test battery may serve as one of the most useful tools in the early detection of adverse effects on the nervous system. They realize that (Weiss, 1983)

- many toxicants, including certain kinds of heavy metals, organic solvents, pesticides, and air pollutants act primarily on the nervous system;
- many poisonings, before they show overt clinical signs and symptoms, may be preceded by vague, subjective, and nonspecific psychological complaints; and
- finally, there are substances whose actions, although not mediated directly through nervous system mechanisms, produce biochemical changes that result in distinct behavioral effects.

As a result, a new era of using neurobehavioral toxicity tests for evaluating psychological effects of occupational hazards developed in China during the 1980s.

The objectives of the present chapter are to provide an overview of the development of neurobehavioral toxicity testing in China and to suggest roles that these methods can play as that nation faces up to its own challenges in occupational safety and health. Case studies of exposures to carbon disulfide, lead, video display terminal (VDT) operation, and static magnetic fields serve as examples to provide readers with a better understanding of the current status and future direction of the contributions of neurobehavioral testing to meeting these challenges.

OVERVIEW OF RESEARCH ACTIVITIES

Carbon Disulfide

In China, carbon disulfide (CS_2) is used mainly in the production of viscose rayon fiber and cellophane films (Liang et al., 1983). Approximately 50,000 workers are reportedly exposed to CS_2 in these industries. Because control measures have been relatively effective in lowering the CS_2 concentration at worksites, overt CS_2 poisoning due to high-level exposure is no longer a serious problem. However, the neuropsychological complaints related to long-term exposure to low levels of CS_2 are not uncommon. Our pilot study on CS_2 (Liang et al., 1983) was initiated by both the need for medical surveillance in the viscose rayon industry and the use of a neurobehavioral test battery for screening the subclinical effects of milder exposure. This was the first systematic use of a neurobehavioral test battery for occupational epidemiology in the region.

A battery of psychological tests involving mood states, intellectual activity, visual perception, short-term memory, and performance speed was administered to 98 male workers exposed to CS_2; all were from a viscose rayon factory in Shanghai. The sample was divided into

TABLE 2 Comparision of Behavioral Scores Between CS_2-Exposed and Nonexposed Groups

Test	Mean SD		P Value
	Exposed	Nonexposed	
Similarities	6.71 ± 1.59	6.40 ± 1.89	>0.05
Digit span	43.10 ± 7.91	44.49 ± 8.16	>0.05
Digit symbol	14.26 ± 8.17	17.23 ± 10.79	<0.05
Block design	39.15 ± 17.19	44.56 ± 14.51	<0.01
Reaction time	14.41 ± 3.48	10.75 ± 3.62	<0.05
Santa Ana	21.90 ± 4.42	25.49 ± 3.70	<0.01
Finger tapping	18.52 ± 2.48	20.14 ± 4.63	<0.01
Total scores	148.16 ± 32.08	173.65 ± 31.51	<0.001

NOTE: SD = standard deviation.
SOURCE: Liang et al. (1983).

three groups according to the level of esposure during the past 15 years: i.e., group I, a mean exposure level of CS_2 above 10 mg/m^3, the present maximum allowable concentration of CS_2 adopted in China; group II, approximately 10 mg/m^3; and group III, below 10 mg/m^3. As controls, 91 nonexposed workers from a textile machinery plant were used and matched for sex, age, and educational background.

There were statistically significant differences in test scores between the exposed and the control groups for most items used (Table 2). An exposure-effect relationship was found among the three groups with different exposure levels. For example, in group I, a total score of 134.04 ± 27.54 could be reliably discriminated from that of from the control group (172.83 ± 30.96); group II, a total score of 150.67 ± 30.46 showed differences of a lesser magnitude; whereas group III (155.27 ± 34.19) was no different from controls. These findings coincided with the results of neurophysiological testing in the same investigation. The motor conduction velocity (MCV) and conduction velocity of slow fiber (CVSF) of ulnar nerve in the CS_2-exposed group were found to be 57.2 and 44.8 m/s, respectively, values significantly lower than those of the comparison group (64.0 and 47 m/s, respectively; Liang et al., 1984).

The present study of CS_2 not only was exploratory work toward long-term research on neurobehavioral toxicity tests, but also provided information about effects of low-level CS_2 exposure on the nervous system that could be used to reevaluate the safety of the current

MAC adopted in China. The conclusions drawn from this pilot study may be summarized as follows:

- Neurobehavioral toxicity tests can be recommended as a useful adjunct to other conventional tools in occupational epidemiology.
- Levels of CS_2 of about 10 mg/m^3 may be close to the minimum effect level causing changes of neurobehavioral and neurophysiological functions in an occupationally exposed population.
- For purposes of safety, a proposal to amend the MAC of CS_2 from 10 mg/m^3 to a lower level is deemed worthy of consideration.

Lead

In recent years, occupational exposure to lead has been adequately controlled to prevent the occurence of overt lead poisoning; such controls have been quite successful in many countries including China. Yet milder exposure to lead in certain occupations has been associated with a series of nonspecific neurasthenic symptoms, slowed nerve conduction velocity, and impaired psychological performance indicative of effects on the central and peripheral nervous systems. Impaired neuropsychological performance in lead workers has also been found in the absence of peripheral nervous system effects even in workers with blood lead levels below that traditionally accepted as safe (He, 1987).

In China, occupational exposures to lead occur mainly in smelting and battery plants. According to a nationwide survey in 1979–1981, the prevalence of lead poisoning was 1.77 percent in 355,000 lead workers examined. Our research focused on clarification of the effects of low-level lead exposure on neurobehavioral function: 24 workers exposed to lead from a small battery plant and 24 controls from a food product manufacturer were investigated by using an expanded neurobehavioral test battery comprised of nine test items reflecting mood state, intelligence, memory, perception, vigilance, and psychomotor performance. The exposed group had an occupational exposure to lead for periods varying from 3 to 8 years at levels in air ranging from 0.06 to 0.34 mg/m^3, time weighted average (TWA) 8 hours. The average blood lead level, free erythrocyte level, and zinc protoporphyrin level were, respectively, 59.0 ± 31.1, 279.7 ± 178.6, and 326.6 ± 202.4 µg/dL. These values were significantly higher than those in the control group.

The results indicated that all of the neurobehavioral tests, except pursuit aiming II, showed significant reductions in test scores for the

TABLE 3 Scores in Asymptomatic Lead-Exposed and Nonexposed Groups

Test	Exposed (n = 24)	Nonexposed (n = 28)	P Value
Digit span	45.59 ± 9.57	53.78 ± 8.45	<0.01
Digit symbol	42.59 ± 5.97	56.35 ± 8.35	<0.01
Aiming II	47.37 ± 7.76	52.26 ± 11.40	>0.05
Block design	43.15 ± 8.65	55.87 ± 7.18	<0.01
Mental arithmetic	46.81 ± 10.93	52.73 ± 8.61	<0.01
Simple reaction time	44.83 ± 12.34	54.43 ± 4.34	<0.01
Choice reaction time	44.04 ± 9.29	55.11 ± 7.77	<0.01
Flicker fusion frequency	46.58 ± 2.20	52.93 ± 13.01	<0.01
Mood states	44.99 ± 8.80	56.10 ± 6.75	<0.01
Total scores	403.96 ± 35.02	489.46 ± 14.14	<0.01

SOURCE: Sheng et al. (1987).

exposed and nonexposed groups (Table 3) (Sheng et al., 1987). In considering the confounding effects of sex, age, educational background, and working age, a stepwise regression analysis was used in the statistical analysis. The results showed that the most significant factor contributing to the reduction in test scores is likely to be lead exposure. The multiple regression can be expressed as Y (total score) = $3.3138 - 1.2127x_2$ (lead exposure) + $0.1568x_4$ (schooling age). Similar research on lead has been conducted at Nanjing Medical College (Wang et al., 1985). A series of neurobehavioral tests were administered to 43 lead-exposed workers, and the same number of nonexposed workers from a machinery factory were used as controls. The majority of tests were related to memory, perception, and vigilance functions. Scores derived from the touch memory, numeral repetition, block design, digit span, and letter cancellation tests were significantly lower in exposed than in the control group. The scores of touch memory and memory quotient (M.Q.) were negatively correlated to the cumulative level of lead exposure, with correlation coefficients of $r = -0.679$ ($p < 0.05$) and $r = -0.854$ ($p < 0.05$), respectively. The study suggests that memory tasks and psychomotor tests might be used as screening indicators for early signs of central nervous system dysfunction due to a low level of lead exposure.

The impact of low-level lead exposure on the development of children's intelligence has attracted great interest among behavioral researchers. A follow-up study of the neuropsychological effects of such exposure has been conducted by the Department of Child and Adolescent

Health, Shanghai Medical University (Wang, 1987) using a test battery of WISC-R and additional testing of reaction time: 157 school-age children, including 89 males and 68 females, were divided into four groups based on their PbB levels—group I with a PbB level of less than 10 µg/dL; group II, 10–20 µg/dL; group III, 20–30 µg/dL; and group IV, more than 30 µg/dL. Subjects were tested over a period of two years. Results showed a close relationship between lead exposure level and tested variables. Children who had elevated PbB performed less well on WISC-R than their counterparts with lower PbB levels. The overall IQ values for groups I, II, III, and IV were 109, 97, 78, and 72, respectively. Confounding factors (e.g., age, sex, socioeconomic status, genetic background, parent health, and habits) were taken into account by using a stepwise regression technique. It is clearly evident that exposures to low lead levels significantly disturbed the development of intelligence as measured by all the tests administered. Reductions in perceptual organization, verbal understanding, inference, and eye-hand coordination were the most affected (Table 4).

Air pollution from industrial emissions and lead contaminants from parent's work clothes are the major sources of children's exposure. Table 5 summarizes the results of a group of nursery children subjected to different levels of exposure (Shen, 1988). Preventive measures for control of industrial pollutants and worker health education to improve personal hygienic behavior are needed to minimize the risks to children from lead exposure. Results of neurobehavioral toxicity testing could provide warnings and the basic evidence for action to remove these risk factors that threaten the life quality of human beings at an early age.

Simulated Video Display Terminal (VDT) Operation

Video display terminals are used in a variety of occupations. It is estimated that there are approximately 8 million VDT operators worldwide using more than 10 million VDT units. In China, VDT use is being expanded in industry, agriculture, commerce, management, public service, and scientific research. Professionals in occupational health have expressed great concern over possible deleterious effects resulting from work with VDT. One of these effects involving psychological factors was the major subject of a research project first reported in 1985 (Liang et al., 1985). Owing to the difficulty of choosing a sample of subjects from the relatively homogeneous work situation at that time, a simulated television screen inspection situation was used in our pilot study designed to discover possible links between work with screen inspection and psychological effects. It is hoped that this

TABLE 4 Comparison of Scores from WISC-R12 Test of Children with Different PbB Levels

PbB (μg/dL)	Information	Similarities	Arithmetic	Vocabulary	Comprehension	Digit Span
<10	11.69 ± 4.27	12.69 ± 2.18	9.50 ± 3.43	10.69 ± 2.47	11.50 ± 2.99	10.31 ± 2.77
10–20	9.87 ± 4.22	10.72 ± 2.86	8.26 ± 3.42	9.57 ± 2.75	9.13 ± 2.69	8.43 ± 3.19
20–30	7.65 ± 3.92	7.40 ± 3.40	7.54 ± 3.30	7.00 ± 2.69	7.15 ± 2.58	7.38 ± 2.96
>30	6.33 ± 4.19	7.04 ± 3.10	6.38 ± 3.72	5.42 ± 3.31	6.17 ± 2.48	7.04 ± 2.79
F value	8.34*	22.70**	3.21*	20.55**	15.06**	5.04**

PbB (μg/dL)	Picture Completion	Picture Arrangement	Block Design	Object Assembly	Digit Symbol	Maze
<10	11.75 ± 3.47	12.44 ± 2.50	9.63 ± 3.91	10.31 ± 2.80	0.81 ± 2.17	11.50 ± 3.46
10–20	11.17 ± 3.35	10.55 ± 2.60	8.71 ± 3.56	9.64 ± 2.88	9.16 ± 3.03	10.97 ± 3.22
20–30	8.58 ± 3.09	8.56 ± 2.77	6.69 ± 3.35	8.00 ± 3.00	7.27 ± 3.16	9.75 ± 3.53
>30	8.13 ± 3.73	7.33 ± 2.90	6.54 ± 4.00	8.00 ± 3.57	7.46 ± 2.72	8.42 ± 3.03
F value	9.60**	16.88**	5.33**	4.64**	5.90**	4.58**

NOTE: * $p < 0.05$; ** $p < 0.01$.

TABLE 5 Biological Effects Among Children Exposed to Different Levels of Lead in a Nursery Near a Battery Plant

Exposure Rank	N	Age (yr)	PbB (μg/dL)	FEP (μg/dL)	ZPP (μg/dL)
0	20	7.46 ± 0.18	9.21 ± 3.18	54.32 ± 23.46	59.62 ± 20.44
1	9	4.57 ± 0.19	16.19 ± 7.78	44.27 ± 24.16	55.86 ± 14.75
2	24	5.60 ± 0.26	23.90 ± 10.65[a]	47.88 ± 28.68	61.74 ± 18.28
3	6	5.20 ± 0.46	35.70 ± 18.05[a]	161.10 ± 67.51[b]	162.68 ± 83.42[a]

[a] $p < 0.01$.
[b] $p < 0.05$.

SOURCE: Wang (1987).

TABLE 6 Psychobehavioral Tests Used in Video Display Terminal Study

Function	Test
Memory	Digit span
Intelligence	Mental arithmetic
Perceptual ability	
Attention	Digit symbol
Cognition	Figure cancelation
Psychomotor performance	
Speed	Simple and choice reaction time
Dexterity	Finger tapping
Visual spatial ability	Block design
Mood	Feeling tone checklist

SOURCE: Liang et al. (1984).

pilot study will serve as a basis for a systematic study of VDT operation in the actual work environment.

Two hundred fifteen employees working in a television screen assembly factory and a television parts manufacturer were selected and categorized into four groups according to their job characteristics, i.e., screen inspection (full-time work with display screen), assembly work, screen inspection combined with assembly work (part-time work with display screen), and auxiliary jobs (as controls). The test battery consisted of items measuring a variety of psychological functions outlined in Table 6. The results may be summarized as follows:

- Findings from the feeling tone checklist clearly indicated that television screen inspectors were subjected to significantly greater mental stress, mainly evident as minor fatigue (comparison with controls, $u = 3.97$, $p < 0.01$)
- Total scores derived from the tests seemed to be affected by three major variables which, ranked in order of the magnitude of these effects, were operation time, screen inspection, and working age. The effects can be expressed as a multiple regression equation: Total score = $137.7562 - 4.0302 X_1$ (screen inspection) $- 6.0151 X_6$ (operation time, hours) $- 0.8954 X_7$ (working age, years).
- Results of several pairs of tests were highly correlated (e.g., digit symbol and digit span, reaction time and finger tapping). This implies that the two correlated tests measure similar types of psychological functions and provides a clue for selecting an appropriate combination test battery for further study.
- Psychological test batteries can prove useful adjuncts to other criteria (e.g., ergonomic and hygienic indicators) for evaluation of harmful effects related to work with VDTs.

Static Magnetic Fields (SMF)

Various subjective symptoms and functional disturbances have been reported in workers involved in the manufacture of permanent magnets and those working near industrial equipment using high currents, including irritability, fatigue, headache, loss of memory, bradycardia, tachycardia, and decreased blood pressure. To study such effects in greater detail, 63 workers exposed for 1–2 years to such environments (electrolysis of sodium choloride) were investigated by using a neurobehavioral test battery similar to those described above and an examination of the functional status of the autonomic nervous system. Results indicated that workers exposed to static magnetic fields had poorer memory than the controls. Of the 63 workers exposed to a higher level of flux density of magnetic field (median, 242 gauss; range, 103–1,975 gauss), 37 had lower scores on the digit span and digit symbol tests than the remaining 26 workers with a lesser exposure (median, 78 gauss, range 26–200 gauss) ($p < 0.05$). These effects occured concurrently with the changes in autonomic nervous system function (Liang et al., 1984).

With advances in technology, the growing number of devices generating magnetic fields are creating new occupational health challenges. In China, no regulatory standards have yet been established. Based on the results of neurobehavioral testing and other findings from medical examinations, we would tentatively recommend that occupational exposure to static magnetic fields at workplaces should not exceed 100 gauss (0.01 T) for 8-hour whole-body exposure. This recommendation corresponds to the standard that was adopted in the USSR in 1978 (World Health Organization, 1987).

NATIONAL AND INTERNATIONAL COOPERATION

A five-year program toward the development of testing methods and criteria for early detection of occupational neurotoxic illness, sponsored by the Chinese Academy of Preventive Medicine, has been in operation since 1986. Our department has been involved in implementation of the neurobehavioral parts of the program. The development and validation of neurobehavioral toxicity tests are two of our major tasks. Emphasis has been placed on two activities:

1. establishing a test battery consisting of relatively simple measuring instruments mainly using paper and pencil, oral presentation, and performance as modes of response; and

2. creating a computer-administered neurobehavioral evaluation system for occupational epidemiology.

The material that follows discusses a proposed computer-administered system consisting of the 17 test items described briefly in Table 7. Development of the battery has involved collaborative efforts at an international level.

A variety of neurobehavioral and neurophysiological tests are available for evaluating effects on the working population of exposure to chemical and physical agents. However, the standardization of procedures for test administration and scoring has often been lacking. As a result, the reliability and reproducibility of some results have been less than desirable. To improve the situation, in 1983 the World Health Organization (WHO) proposed a Neurobehavioral Core Test Battery (NCTB) (Johnson et al., 1987). The goal was to develop a relatively inexpensive, simple, and appropriate battery to study neurobehavioral effects of industrial chemicals on working populations. The U.S. National Institute for Occupational Safety and Health (NIOSH) has been designated one of three oversight centers. In 1987 the Department of Occupational Health, Shanghai Medical University, was the first applicant in China to receive approval from WHO and NIOSH to participate in the NCTB project. Phase I of an evaluation program has been conducted since then in Shanghai. Its objective is to study the applicability of the NCTB to working populations generally (i.e., those not exposed to harmful agents) to establish its relatively low cultural bias and

TABLE 7 Task Performed on the Computer-Administered System

Function	Test (Source)	Task
Intelligence	Mental Arithmetic (Gon, 1981)	Subject is asked to add "3" to the original figure starting from "1" and indicate the answer by pressing the key, say, "4." Then add another "3" to the "4" and indicate the second answer by pressing key "7." Fifteen consecutive calculations are required.
	Serial Add/Subtract (Baker et al., 1985b)	Two randomly selected digits and signals either "−" or "+" are displayed in the same location on the screen. Subject is asked to perform the indicated addition or subtraction and enter the results. Digits and signals are present for 250 ms, separated by 200 ms, with the next trial beginning immediately after the key is pressed. Fifteen performances are required.
Memory	Visual Retention (Baker et al., 1985a)	The screen presents a Benton figure for 10 s, followed by four similar figures from which subject must indicate the figure previously seen and enter the number of the correct figure within 10 s.
	Pattern Memory (Baker et al., 1985a)	A blocklike figure is present on the screen for 10 s, followed by three similar blocks. One of the later blocks is identical to the former one. Subject is asked to identify the figure previously seen and press key for right number within 15 s. Ten performances are required.
	Paired Associate Learning (Xi et al., 1984)	Twelve paired-associate words are displayed on the screen at a rate of one pair per 5 s. Then 20 pairs of words which contain the 12 pairs previously seen are presented randomly. Subject is asked to press key "1" when anyone of the 12 presents.
	Digit Span (Baker et al. 1985b)	The subject is asked to press key recalling a forward or backward series of digits which are given through an earphone at a rate of 2 digits/s.
	Memory Scanning (Maizlish et al., 1985)	A short list of one-digit figures is displayed on the screen for 1 s, and a figure varying from 0 through 9 is then singly presented for the same length. Subject is asked to indicate whether the single figure is included in the short list or not by pressing key "Y" or "N."

Function	Test (Source)	Task
	Continuous Recognition Memory (Maizlish et al., 1985)	A series of three-digit figures are displayed separately at a rate of once per second. Two-thirds of the series are repeatedly present. Subject is asked to indicate whether the series repeatedly present has been previously displayed by pressing key "1" or "0."
	Memory Span (Maizlish et al., 1985)	Forty-two progressively prolonged series of three- to nine-digit figures are included. Subject is asked to recall the forward series of digits, attaining 50% of the given series without mistake.
Visual Perception	Two-Letter Search (Thorne et al., 1985)	Two target letters are present at the top of the screen with a string of 20 letters in the middle of the screen. Subject is asked to determine whether both letters are present in the string or not by pressing key "1" or "0." Fifteen trials are required.
	Symbol-Digit Substitution (Baker et al., 1985a)	Nine symbols and the correspondent digits are present at the top of the screen which never disappear. Subject is asked to press the digit keys corresponding to the symbols in a test list displayed on the screen.
	Length Discrimination (Baker et al., 1985a)	Two different-length lines of 4.5 and 5.0 cm are presented separately at an interval varying from 0.3 to 0.7 s and lasting for 0.7 s. Subject is asked to press key when the longer line appears. The target line is randomly present 25 times during 10-min trials.
Psychomotor Performance	Simple Visual Reaction Time (Baker et al., 1985a)	Subject is asked to press key when a 2×2 cm^2 red square appears on the screen. The interval of interstimulus varies from 3 to 11 s. Sixty-four trials within 6 min are required.
	Simple Auditory Reaction Time (Fang and Liang, 1987)	Hetrotones are randomly added to the regular white noise. Subject is asked to press key when the dissenting tone appears through an earphone.
	Four-Choice Series Reaction Time (Thorne et al., 1985)	Subject is given a box having four light-emitting diodes in a square array. A single light of the four is lit, varying from 2 to 7 s. Subject is asked to press the corresponding key, thereby initiating the next trial.

Table 7 continues

TABLE 7 Continued

Function	Test (Source)	Task
Psychomotor Performance (continued)	Finger Tapping (Anger, 1985)	Subject rests the index finger and middle finger on a red button then taps the button as fast as possible. The number of taps is recorded during 10-s tests for five trials.
	Cursor Tracing (Baker et al., 1985a)	Subject is asked to use a joystick to trace a large sine wave pattern on the screen. A cursor is moving horizontally at a constant rate, while the subject controls the vertical motion of the cursor with the joystick.

TABLE 8 WHO Neurobehavioral Core Test Battery

Test	Functional Domain Tested
Profile of mood states	Affect
Simple reaction time	Attention/response speed
Digit span	Auditory memory
Santa Ana dexterity	Manual dexterity
Digit symbol	Perceptual-motor speed
Benton visual retention (recognition form)	Visual perception/memory
Aiming (Pursuit Aiming II)	Motor steadiness

SOURCE: World Health Organization (1986).

determine the feasibility of its use by personnel with limited training. By following the WHO protocol for evaluating the NCTB, 282 working adults, including 135 males and 147 females, not exposed to neurotoxic chemicals were examined with the individual items outlined in Table 8. Tests were carried out by using the Chinese version of the relevant instructions and forms, but strictly guided by the original concepts and principles of the operating guide for administering the NCTB. Staff involved in the project were faculty members with medical careers and medical students from the School of Public Health, Shanghai Medical University (Chen et al., 1989). Results from the phase I evaluation are summarized below.

Mood States

Results presented in Table 9 show that scores obtained for the item "vigor" decreased progressively with increasing age in both male

TABLE 9 Scores Derived from Profile of Mood States Questionnaire

Mood States	Age Groups (yr)							
	16–25		26–35		36–45		46–55	
	Male	Female	Male	Female	Male	Female	Male	Female
Tension-anxiety	7.4 ± 4.5	6.4 ± 4.0	7.2 ± 3.7	6.9 ± 3.4	8.3 ± 4.5	8.4 ± 4.6	8.5 ± 4.7	11.2 ± 5.9
Depression	9.5 ± 7.3	7.2 ± 8.2	9.3 ± 6.6	7.3 ± 6.3	9.2 ± 6.2	9.4 ± 8.3	10.6 ± 8.3	13.3 ± 8.5
Anger	9.6 ± 5.8	8.0 ± 6.1	10.2 ± 6.4	8.8 ± 6.0	9.8 ± 5.9	10.2 ± 7.2	9.9 ± 5.5	12.6 ± 7.3
Vigor	19.6 ± 4.9	19.8 ± 4.6	16.7 ± 5.3	16.3 ± 5.1	15.3 ± 4.2	14.1 ± 4.5	14.7 ± 4.2	14.0 ± 3.6
Fatigue	5.3 ± 4.5	6.1 ± 3.4	6.1 ± 4.5	6.9 ± 4.1	5.9 ± 2.5	7.8 ± 5.0	7.4 ± 4.5	9.7 ± 4.5
Confusion	5.7 ± 3.6	4.3 ± 2.7	5.1 ± 3.2	5.5 ± 3.3	5.8 ± 2.5	6.9 ± 4.6	6.8 ± 3.8	8.9 ± 3.5

and female workers. If such statistics are to be used, they should be specified more completely: e.g., what is the F value, d.f. is in a 1-way ANOVA across "age"? There were no other significant age differences in the remaining five mood states in male workers. Inspection of Table 8 shows, however, that scores for these same items increased with increasing age in female workers.

Performance Tests

Results of performance tests (specified in Table 8) are summarized in Table 10. Both age and sex seemed to be important factors affecting the visual reaction speed and eye-hand coordination. All parameters of reaction time decreased with increasing age in both male and female groups. There was a tendency for performance speed to be faster in males than in females, particularly in the younger age groups. Strong evidence existed to show that age played a more important role than sex in affecting the test scores of Digit Span, Santa Ana Manual Dexterity, Digit Symbol, and Benton Visual Recognition tests, a tendency for test scores to increase with increasing age again being apparent in both male and female groups. By comparison, sex differences did not show a major impact. Effects of aging on the Pursuit Aiming II test were consistent with those in the other performance tests; female subjects tended to outperform males.

Results of the phase I evaluation of the NCTB may be summarized as follows:

- The NCTB was shown to be an applicable tool for use across cultures in various countries. Scores derived from tests in subjects with similar, but not equal, educational background did not show significant differences. In some cases, the Profile of Mood States (POMS) seemed to be too complicated, some subjects being puzzled by equivocal adjectives as they appeared in translated version. The problem could be attributed to the inevitable modification rather than simply a translation of the original meaning.
- Age is supposed to be one of the factors affecting neurobehavioral performances during the life span. Our results indicate that test performance peaked within the age range 16–25 years, remained relatively constant at 26–35 and 36–45 years, and dropped precipitously at 46–55 years. This pattern is very similar to those previously reported (Johnson et al., 1987).
- Differences attributed to sex were not as marked as those related to age, although some tests did show significant differences between male and female groups. For instance, Pursuit Aiming II seemed to be performed better in the younger group of female workers

TABLE 10 Scores Derived from Neurobehavioral Core Test Battery

	Age Groups (yr)							
	16–25		26–35		36–45		46–55	
Items Tested	Male	Female	Male	Female	Male	Female	Male	Female
Simple Reaction Time (RT)								
Mean RT (ms)	238.8 ± 19.0	260.8 ± 21.6	247.5 ± 23.3	267.4 ± 34.2	250.9 ± 22.7	261.9 ± 25.1	254.5 ± 30.9	270.4 ± 34.5
Fastest RT (ms)	171.7 ± 24.0	189.4 ± 23.9	178.0 ± 27.6	192.7 ± 26.3	185.6 ± 20.1	192.7 ± 23.5	188.6 ± 18.4	188.6 ± 34.5
Slowest RT (ms)	411.4 ± 94.3	465.4 ± 99.2	461.4 ± 185.8	443.9 ± 91.6	435.9 ± 83.6	448.2 ± 128.8	442.0 ± 129.6	493.9 ± 109.2
Digit Span	19.3 ± 2.7	18.6 ± 3.4	17.7 ± 2.9	17.5 ± 2.3	17.8 ± 3.0	17.9 ± 2.9	17.7 ± 2.8	16.6 ± 2.8
Santa Ana Dexterity								
Preferred hand	42.5 ± 5.4	40.4 ± 3.7	40.5 ± 4.9	40.5 ± 4.1	39.3 ± 4.6	37.5 ± 5.6	36.7 ± 4.5	35.6 ± 4.4
Nonpreferred hand	37.1 ± 4.7	35.4 ± 4.1	37.9 ± 5.3	36.6 ± 4.2	34.5 ± 3.9	33.3 ± 3.7	33.7 ± 4.8	32.5 ± 4.2
Digit Symbol	65.1 ± 9.6	66.4 ± 7.1	55.6 ± 9.4	59.8 ± 10.0	52.7 ± 9.3	57.2 ± 11.2	53.1 ± 11.5	49.1 ± 12.6
Benton Visual Retention	8.9 ± 0.7	9.3 ± 0.8	9.0 ± 1.0	8.8 ± 1.1	8.8 ± 1.1	8.7 ± 1.1	8.8 ± 0.9	8.4 ± 1.1
Pursuit Aiming II								
Sum of correct dots	234.8 ± 28.0	237.4 ± 27.7	234.7 ± 31.2	255.5 ± 33.8	209.7 ± 18.7	228.4 ± 38.9	202.0 ± 33.0	210.7 ± 26.1
Sum or incorrect dots	7.5 ± 7.9	7.8 ± 7.0	9.9 ± 8.7	14.4 ± 15.9	15.0 ± 16.7	10.1 ± 8.7	11.0 ± 8.8	18.0 ± 13.9
Total number of dots	442.2 ± 32.2	245.2 ± 30.5	244.5 ± 33.8	270.1 ± 42.2	224.7 ± 28.2	238.4 ± 40.4	213.3 ± 36.7	228.6 ± 28.8

NOTE: Results are all expressed as scores, with the exception of simple reaction times which are expressed in miliseconds.

compared with the same age groups of male workers. However, other measures (e.g., Reaction Time) were found to be performed more adequately by most age groups of male workers compared to females.

Phase II evaluation, focusing on the assessment and demonstration of the ability of the NCTB to identify neurotoxicity in the working population exposed to neurotoxic chemicals such as CS_2 and toluene, is now being conducted.

EXPECTED DIRECTIONS: BARRIERS AND RECOMMENDATIONS

Developments in the Near Future

Neurobehavioral tests have been used in occupational and environmental health sciences in China since the early 1980s. The immediate needs for the futher development of neurobehavioral tests in this country are expected to be those described below.

Developing and Validating Test Methods

There exist a variety of test methods derived from WAIS, WMS, WISC, and others in the repertoire of traditional psychology. Problems related to these diverse test methods include insufficient standardization of techniques, in both test administration and scoring, careless study design, and improper data analysis. The results are that data obtained by different investigators are not adequately reproducible and comparable. Therefore, it is imperative to develop more test batteries based on critical standardization, unification, and validation requirements. Although the WHO-NCTB has partially filled this gap, consideration must be given to validating other conventional methods beyond those included in the NCTB.

Generating New Measuring Instruments

There is a need to generate more individual devices that can be used as quantitative and objective instruments for assessing subtle changes in memory, psychomotor behavior, perception, vigilance, etc. There are very real questions about the value of self-reported symptoms and even performance scores in groups of workers who are alert to the possible effects of suspected neurotoxins and other occupational hazards. Therefore, it is reasonable to develop devices to improve test objectivity and reliability in order to minimize such contamination. To achieve this approach, interdisciplinary cooperation should to be encouraged. For example, we collaborated with the Shanghai Development Center of Science and Technology in developing a portable electronic-based device that has the capacity to test flicker fusion frequency (FFF), tone discrimination, and reaction time. A pilot study in workers exposed to lead or organotin and workers with higher mental demands during work has shown the device's remarkable advantages in terms of steadiness, accuracy, and facilitating field study (Fang and Liang, 1987). Integration of subjective measuring instruments with relatively objective assessments might provide a better neurobehavioral test battery for use by most of the Chinese researchers in the near future.

Building a Computerized Evaluation System

At present there is an increasing interest in the use of computerized test methods, primarily because of their accuracy and objectivity. One example of a computer-based neurobehavioral evaluation system designed for use in occupational and environmental epidemiology has been described by Baker and his associates (Baker et al., 1985a). We are now involved in a joint project aimed at programming the 17 test items listed in Table 7. Because the computerized approach is new, it is understandable that questions are being raised about the role it can properly play as a substitute for more conventional methods (Hanninen, 1985; Iregren, this volume). Critical validation of the computerized system is urgently required, and traditional tests should not be totally discarded until the superiority of the new ones has been satisfactorily established. Keeping abreast of developments in conventional and in computerized methods should go hand in hand.

Limitations and Barriers

Despite increased international contact and domestic practices, neurobehavioral toxicology is still a developing discipline. This is particularly true in the following aspects:

Variation of Test Sensitivity from Study to Study

Although significant differences in test scores between exposed and control groups have been reported in most investigations, the sensitivity of any one test appears to vary from study to study. For example, FFF is often taken as a behavioral test to evaluate eyestrain and mental fatigue; the test finds differences between exposed and control groups up to 4–6 Hz in some studies reported by European investigators (Betta et al., 1983). We have not found differences in fusion frequency of more than 2 Hz between groups we have studied. When sensitivities of measuring instruments vary during their use by different investigators, can national or international norms be established for designating impairment?

Bias of Learning Effect

Effects of practice (learning) have been the most annoying barriers, frustrating the repeated use of tests. Such effects are especially apparent in follow-up studies within a short time or in comparison studies between shifts. Two ways of overcoming this difficulty are (1) to create more test batteries that use different procedures but reflect similar behavioral functions, and (2) to partial out the learning effects by comparing their realtive magnitudes in exposed and control groups when two similar, but different, groups are used. Clearly, test batteries would have to be validated.

Controlling Confounding Factors

It is well known that there are more possible confounders in neurobehavioral studies than in other types of studies, and they vary from one study to another. In addition to the recognized confounders (e.g., age, educational background, and socioeconomic status), insidious confounders (e.g., motivation, mental status, and attitude of the examinee toward the test) must be considered and are more difficult to control. One approach to minimize such bias involves pretest interviews and preeducation. This involves the collaboration of investigators with physicians from the workers' clinic, with representatives of labor unions, or with foremen in the workshop. The purposes of such education should be to develop proper attitudes, to clarify possible misunderstandings of the real objective of testing, and to emphasize the importance of the evaluation in preventing occupational risks. Even with such precautions, questions arise about possible biases introduced by exclusion from a sample of those not willing to participate.

Interpreting Results Obtained from Neurobehavioral Tests

Neurobehavioral tests are used mainly as adjuncts to the conventional tools of medical surveillance in working populations. The purposes of the surveillance are the early detection of adverse functional effects, the evaluation of a work environment, and the monitoring of existing safety regulations. The following are some challenges to these purposes: Can the results obtained only from neurobehavioral tests without subjectively supporting evidence be convincing? Can adverse neurobehavioral effects be considered as a warning signal or forerunner of irreversible damage? Can regulation of occupational exposures be based on neurobehavioral findings alone? These critical issues continue to be debated and, thus, constitute barriers to the wide use of neurobehavioral tests.

Comments and Recommendations

Training and Education

As mentioned above, neurobehavioral toxicology is still in its infant stage. A number of important questions remain unanswered. For example, in general, it is not clearly known how to choose the test battery that is likely to be most specific and sensitive in revealing subtle effects induced by a given chemical at various exposure levels. It is also not entirely clear as to how test procedures may be standardized to ensure proper application in this field. Finally, questions continue to be raised about how much basic science of psychology is required for a researcher in neurobehavioral toxicology. There is an urgent need for international efforts toward offering short-term courses on "Neurobehavioral Toxicology and its Field Practice" for researchers involved in this specific area.

Special priority should be given to providing opportunities for researchers working in developing countries to participate in international courses or meetings. At least five international meetings on behavioral toxicology have been held since 1975. Unfortunately, few participants from developing countries attended those meetings. This situation has undoubtedly blocked intercommunication between researchers from developed and developing countries. It is my hope that channels of communication can be created and maintained with support from both national and international organizations.

Exploring the Possibility of Combining Quantitative Biochemical Markers and Neurobehavioral Testing

Too little is presently known about the modes and sites of action of toxic substances and how these relate to neurobehavioral effects. Few studies on human behavioral toxicology are supported by objective biochemical indicators, in particular, the neurotransmitter indicators. It has been reported that hormonal responses are specific in their behavioral effects: e.g., adrenaline relates to arousal, noradrenaline to irritation, cortisol to distress (Singer, 1983). This suggestion implies that urinary hormonal indices may be more sensitive in reflecting mood states than questionnaires. In a more specific example, six children with increased lead absorption—among whom two were hyperirritable, one was hyperactive, and the rest were clinically asymptomatic—were studied by Silbergeld and Chisolm (1976). Their results showed that the urinary catecholamine metabolites, homovanillic acid (HVA) and vanillylmandelic acid (VHA), were increased fivefold in the daily output. These two reports suggest that behavior markers may provide objective indicators supporting the findings of neurobehavioral tests. If so, I would like to encourage international collaboration to elucidate possible links between these two classes of variables.

Development, Validation, and Characterization of Methods

Neurotoxicology is a complicated field of research, and the neurobehavioral toxicity test, as one of the research tools of neurotoxicology, is even more complicated. The development, validation, and characterization of a variety of neurobehavioral methods for the detecton of early changes of the central nervous system still remain in an exploratory phase. Therefore, methodological studies and a more systematic approach to the testing are urgently needed. As a new explorer, I would be very appreciative if international efforts could emphasize the "basic construction" of methods of neurobehavioral toxicity tests. For example, in China, solutions pertinent to the three following matters will soon be sought: (1) standardization of neurobehavioral tests to ensure full comparability between studies; (2) systematic validation of neurobehavioral tests that have been widely used; (3) publication of an operation guide in addition to the guide for the WHO/NIOSH Neurobehavioral Core Test Battery.

SUMMARY

In summary, this discussion of regional factors in the development of neurobehavioral toxicity tests has focused on three broad issues.

Status of Research and Development of Neurobehavioral Toxicity Tests in China

Emphasis is placed on the need for neurobehavioral approaches avialable to professionals facing new challenges in the fields of occupational and environmental health. Results of several case studies involving exposures to CS_2, lead, VDTs, and static magnetic field only summarize the findings of our preliminary studies but show the enthusiasm of researchers interested in applying neurobehavioral testing methods to the exploration of new fields of potential risks of toxic exposures.

Expected Directions in the Development of Behavioral Tests of Neurotoxicity

A two-tract system involving both manual tests and computer-administered procedures is expected to be the top choice and future direction in the development of neurobehavioral toxicity tests in our region. In comparing the Finland Test Battery, the WHO/NIOSH NCTB, and the proposed computerized system, we believe that there are advantages and disadvantages intrinsic to both manual and computerized systems, as well as to the two manual test batteries we have used (Table 11).

Problems and Recommendations

In considering the problems we face in our region, recommendations are made in two broad areas, one focusing on the basic methodological issues of standardization, validation, and characterization of testing methods, and the second on the search for more objective supporting indicators. In addition, strengthening and improving international collaboration, aimed at benefiting researchers in developing countries, is urged.

TABLE 11 Comparison of Neurobehavioral Tests Used

Evaluation Variables	Finnish Battery	NCTB	Computerized System
Applicability	2.0	3.0	1.0
Expenditure	2.5	2.5	1.0
Time-consuming	1.5	1.5	3.0
Acceptability	2.0	2.0	1.0
Accuracy	1.5	1.5	3.0
Culture effectiveness	1.5	3.0	1.5
Integrity	3.0	2.0	1.0
Total	14.0	15.5	11.5

NOTE: The higher the score, the beter is the evaluation of the battery in terms of the variables used.

In conclusion, this chapter is an attempt to convey the voice of researchers in this region and express their willingness to accept the roles that behavioral toxicologists can play in protecting life quality from the deteriorating effects of behavioral hazards.

REFERENCES

Anger, W. K. 1985. Neurobehavorial tests used in NIOSH-supported work-site studies, 1973–1983. Neurobehav. Toxicol. Teratol. 7:359–368.

Baker, E. L., L. R. Letz, and A. Fidler. 1985a. A computer administered neurobehavioral evaluation system for occupational and environmental epidemiology. J. Occup. Med. 27(3):206–212.

Baker, E. L., R. L. Letz, A. T. Fidler, S. Shalat, D. Plantamura, and M. Lyndon. 1985b. A computer-based neurobehavioral evaluation system for occupational and environmental epidemiology: Methodology and validation studies. Neurobehav. Toxicol. Teratol. 7:369–377.

Betta, A., A. D. Santa, C. Savonitto, and F. D. Andrea. 1983. Flicker fusion test and occupational toxicology: Performance evaluation in workers exposed to Pb and solvents. Human Toxicology 2:83–90.

Chen, Z. Q., Z. H. Yu, and Z. H. Cao. 1989. Study on the norm of WHO neurobehavioral core test battery in working population. J. Ind. Hyg. Occup. Med. 15(1):4–8.

Fang, Y. F., and Y. X. Liang. 1987. A portable electronic-based system for behavioral field testing of audiovisual function. J. Ind. Hyg. Occup. Med. 13(6):344–347.

Geyer, H., I. Scheunert, and F. Rorte. 1986. Bioconcentration potential of organic environmental chemicals in humans. Regul. Toxicol. Pharmacol. 6:313–347.

Gon, Y. X. 1981. P. 469 in Basic Science of Mental Medicine. Hunan Province: Publishing Company of Science and Technology.

Gu, X. Q. 1985. The role of early detection in the prevention of occupational disease: A review of the work in the People's Republic of China. Scand. J. Work Environ. Health 11(4):7–9.

Hanninen, H. 1985. Twenty-five years of behavioral toxicology within occupational medicine: A personal account. Am. J. Ind. Med. 7:19–30.

Hanninen, H., and K. Lindstrom. 1979. Behavioral test battery for toxicopsychological studies used at the Institute of Occupational Health in Helsinki. Institute of Occupational Health Reviews 1.

He, F. S. 1987. Occupational neurotoxicology—Current problems and trends. In The XXII International Congress on Occupational Health, 27 Sept.–2 Oct.

Johnson, B. L., E. L. Baker, M. El-Batawi, R. Gilioli, H. Hanninen, A. M. Seppalainen, and C. Xitarax. 1987. Recommended neurobehavioral test methods. Pp. 217–241 in Prevention of Neurotoxic Illness in Working Populations. New York: John Wiley & Sons.

Liang, Y. X., X. P. Jing, G. G. Shen, Y. Li, and R. Z. Li. 1983. Health effects of low level CS_2 exposure in viscose rayon workers—An epidemiological study on psychological effects. Acta Academiae Medicinae Primae Shanghai 10(1):15–20.

Liang, Y. X., X. P. Jing, Z. Q. Chen, Y. P. Lu, and G. G. Zhen. 1984a. A survey on the function of nervous and cardiovascular systems among workers exposed to low concentration of CS_2. Acta Academiae Medicinae Primae Shanghai 11(4):251–255.

Liang, Y. X., X. P. Jing, and G. G. Shen. 1984b. The behavioral pattern of workers exposed to static magnetic field. J. Ind. Hyg. Occup. Med. 10(5):274–277.

Liang, Y. X., Y. F. Fang, and F. L. Wang. 1985. A study on psychobehavioral evalua-

tion for inspectors of TV screens. P. 195 in Proceedings of the International Scientific Conference: Work with Video Display Units, Part II.
Maizlish, N. A., G. D. Laugold, L. W. Whitehead, L. J. Fine, J. W. Albers, J. Goldberg, and P. Smith. 1985. Behavioral evaluation of workers exposed to mixtures of organic solvents. Brit. J. Ind. Med. 42:579–590.
Shen, Y. Z. 1988. Risk assessment of Pb exposure. Ph.D. thesis, Department of Occupational Health, Shanghai Medical University.
Sheng, Y. Z., Y. L. Wang, and Y. L. Fang. 1987. Relationship between occupational exposure to lead and its effects. J. Ind. Hyg. Occup. Med. 13(2):83–86.
Silbergeld, E. K., and J. J. Chisolm. 1976. Pb poisoning: Altered urinary catecholamine metabolites as indicators of intoxication in mice and children. Science 192:153–155.
Singer, G. 1983. Stress and compensation. Proceedings of seminars held at La Trobe University and Macquaris University.
Thorne, D. R., S. G. Genser, H. C. Sing, and F. W. Hegge. 1985. The Walter Reed performance assessment battery. Neurobehav. Toxicol. Teratol. 7:415–418.
Wang, L. 1987. Effects of Pb exposure on the development of children's intelligence. Ph.D. thesis, Department of Child and Adolescent Health, Shanghai Medical University.
Wang, Q. L., Y. S. Jiang, X. Q. Xu, and Y. Z. Jiang. 1985. The application of performance test in evaluation of Pb exposure. Chinese J. Preventive Medicine 19(1):34–36.
Weiss, B. 1983. Behavioral toxicology and environmental health science—Opportunity and challenge for psychology. Am. Psychologist 1174–1187.
World Health Organization. 1986. Field Evaluation of WHO Neurobehavioral Core Test Battery—A Solicitation Proposal. Geneva: Office of Occupational Health.
World Health Organization. 1987. Standards and their rationales. Magnetic Fields EHC:26–127.
Xi, S. N., Z. X. Wu, and C. H. Sun, eds. 1984. Pp. 31–32 in Clinical Handbook of Memory Scale. Beijing: Institute of Psychology, Chinese Academy of Sciences.
Yu, D. W., and X. Q. Gu. 1987. General introduction to occupational health service for small-scale industry in rural area in China. In Report of Regional Seminar on Occupational Health Service in Small Scale Industries, Singapore, 18–23 August 1986. Manila: WHO Regional Office for the Western Pacific.

Regional Issues in the Development of the Neurobehavioral Core Test Battery

Renato Gilioli and Maria G. Cassitto

For many decades, occupational diseases have been the focus of research in occupational health, with the underlying understanding that improvement in production technologies would eventually result in the total control of diseases as a natural consequence. Although this concept still retains a strong validity, over the past years there has been mounting evidence to show that along with a decrease in traditional occupational diseases, an increase has occurred in environmental as well as work-related diseases with new characteristics. These clinical entities are certainly less severe but still sufficiently incapacitating so as to need early detection and prompt intervention.

Although these considerations mainly refer to industrialized countries, developing countries are now facing problems very similar, or perhaps even worse, than those encountered in the earlier stages of industrialization by now industrially advanced countries. These considerations are common to all occupational health domains and even more so to neurobehavioral toxicology. Thus, environmental and work-related disease—the problem of neurotoxic disorders—presents a number of distinctive features.

First, although it is certain that the central nervous system (CNS) and the peripheral nervous system are the target of many industrial and environmental chemicals, no clear-cut evidence exists of a direct quantitative dose-effect or dose-response relationship between the level of exposure or degree of absorption and a predictable neurobehavioral outcome. Second, most of the combined effects (i.e., additive, synergistic, and antagonistic) of neurotoxic agents are still unknown or, at best, uncertain. Many other interfering or confounding factors are known, but their role is yet to be fully understood. The World Health Organization (WHO) Monograph on Neurotoxic Illness (Johnson et al., 1987) explores these factors in detail and shows how much work still has to be carried out in this direction. Third, another central feature is individual response not only to the neurotoxic agents but also to the behavioral instruments used. This response can be modified by age, level of education, and cultural background. This third issue is far from being clarified. Those factors that are relevant within the same cultural group require much greater attention when dealing with groups belonging to different cultures. Over the last five years, considerable attention has been paid to this aspect and relevant material has already been obtained. In fact, after many years of research designs characterized by heterogeneous collection of similar or identical but variously assembled tests, yielding heterogeneous results, the need for unifying these methodologies has been widely recognized. This willingness to give up at least part of one's own research instruments for the benefit of greater uniformity and the means of realizing it have been the object of much international effort.

INTERNATIONAL INITIATIVES

The ad hoc international initiatives discussed in this chapter were prepared by the first International Symposium on Neurobehavioral Methods in Occupational Health held in Como and Milan, Italy in 1982. This type of approach has proved fruitful and resulted in a triennial series of symposia, the second held in Copenhagen (1985) and the third in Washington, D.C. (1988). A fourth symposium will be held in Tokyo in 1991. These symposia are aimed at acquiring the latest developments in neurobehavioral methods as well as disseminating basic pertinent knowledge to developing countries in different regions of the world.

In Europe, the first official meeting held by an international agency was organized in 1983 by the Health and Safety Directorate of the European Economic Community when experts from the then 10 member states were convened to deal with this specific topic.

In the same period, substantial results were obtained by the WHO Office of Occupational Health in Geneva. Along with the U.S. National Institute for Occupational Safety and Health (NIOSH), WHO convened an international workshop in Cincinnati in 1983 to evaluate existing information on the neurotoxicity of workplace chemicals with a double purpose: (1) to transfer, after sound critique, the already existing knowledge to health personnel in industrialized and developing countries, and (2) to analyze existing methodologies and foster the adoption of common instruments. In fact, an agreement was reached on a minimum core battery, the WHO Neurobehavioral Core Test Battery (NCTB), to be used and validated in all countries.

To help this dissemination, WHO developed an operational guide that describes this instrument in detail, its characteristics and administration procedures. Although the monograph (Johnson et al., 1987) has recently been published, the guide is, at the moment, distributed by WHO to interested groups. According to WHO intentions, the battery is given as a package to be used as a whole; it is short, easy to handle, and meant to measure those functions that are more easily impaired by exposure to neurotoxic agents—namely, mood, attention, visuomotor skills, and memory. A factor analysis performed on the data from 950 subjects resulted in the extraction of five factors: namely, emotional status, motor response speed, visuomotor coordination, sustained attention, and active information processing. The commonalities are very high and reflect important intercorrelations. The reliability and sensitivity of this battery are not optimal, as recognized by the panel of experts who fostered its adoption. However, the tests chosen have the advantage of being simple and well known and, above all, have a minimal loading in cultural factors, which are the most important confounders in interpreting results. The WHO panel of experts also recommended integration of this battery with additional and more specific tests whenever possible.

Thus, two results may be obtained: other more specific measures can (1) increase the sensitivity of the instrument and (2) allow for the use of tests reflecting more specifically the intellectual abilities and emotional traits that might have been determined by the cultural history of the subjects. This approach would still permit the performance of highly needed multicentered studies.

VALIDATION STUDIES

To implement this program on prevention of neurotoxic disorders, WHO in 1986 presented a program of international validation of the WHO-NCTB aimed at regional validation and use of the instruments and development of an international reference data bank of normal values, if feasible, with special reference to developing countries. This validation program is under the operational responsibility of three oversight centers. In the United States, Kent Anger of NIOSH coordinates domestic validation as well as that of South America and China. In

Europe, Helena Hanninen of the Helsinki Institute of Occupational Health and Maria Cassitto of the Milan Institute of Occupational Health are responsible for the other countries willing to participate in the validation program. General coordination of the entire program has recently been assigned by WHO to the Milan center. Within this general frame, some considerations arising out of our experience with this program can help define its potentialities and limits in terms of regional issues and uniformity of interventions. The first recommendation to interested groups is to spend a practice period in one of the three oversight centers. The underlying rationale is that the same, even simple, tests can substantially change when different instructions are given. Moreover, according to the testers' background, different aspects of the tests or of the subjects' behavior will be noticed, thus creating a totally different testing situation. Because the tests in this battery have long been used by most groups involved in neurotoxicity studies, questions may arise as to whether there is a real need to participate in a practice exercise.

An example should clarify this issue: One of the tests in the battery gives measures of manual ability in terms of both speed and accuracy. Instructions stressing either or both of these aspects will result in different performances. A satisfactory number of correct answers can be expected from instructions stressing accuracy, whereas a high number of answers with many errors are obtained with instructions stressing speed. An intermediate level of correct answers and errors will result from the third type of instructions.

In developing countries, qualified personnel skilled in psychological testing are generally lacking. As a result, either nurses or physicians themselves will have to administer the battery. In both cases, training is essential. Difficulties in this case are mainly financial: travel and living expenses cannot always be covered either by the oversight center or by the trainee. Despite these difficulties, a number of short training courses have already been organized by the three oversight centers to provide both theoretical and practical backgrounds aimed at diagnosing impairments of the higher-order nervous functions found in numerous neurotoxic disorders. Theoretical training is essential to provide the operators with the framework within which the use of the battery acquires meaning and usefulness.

If one is already skilled in psychological measures, the difficult step is to move from the traditional use of testing the abilities in view of either job fitness or cerebral dysfunction in severe neurologic diseases to a more subtle analysis of the slightest functional impairments. In fact, as is well known, minor neurobehavioral changes may be often seen in exposed populations which can proceed to further serious deterioration. Similarly, unexperienced personnel must also become so sensitive to this issue as to take into proper account even the slightest dysfunction that can be revealed only by accurate testing. However, it must be stressed that one has to avoid the risk of reading the psychometric outcome as though it were equivalent to a common analytical determination, such as blood sugar. In our courses, it has been realized that these objectives can be attained when training not only is theoretical but also is integrated by testing actual cases and protocols in which participants are asked to record the subjects' verbal and operational behavior in all its aspects. Optimal training would imply a real-life situation, but simulated cases can often be equally effective. To help demonstrations, a videotape was produced in our laboratory showing the entire neurobehavioral examination of a patient at our clinic hospitalized for possible carbon disulfide intoxication. In this way, the subject's behavior can be discussed step by step, thus facilitating the above-mentioned change in the tester's attitude.

So far, local characteristics have been somewhat overlooked for the advantage of uniformity and international standardization. We believe that, once the attitude toward the instruments tends to be the same and the interpretation of the results has common reference guidelines, greater attention to cultural dimensions will be profitable and will result in substantial improvements in the already existing tests or in the development of more advanced ones. According to our experience, we are not likely to face important cultural barriers to the implementation of this program in our regions. Rather, the first data coming from the validation show that, in six European countries, there are significant differences in the attitude of the subjects toward psychological testing in general. These differences in attitude result in differences in the effort taken and, consequently, in the scores. This is an important issue that must receive due consideration in all international efforts to unify methodologies.

The only difficulty for all groups involved in the validation effort has been with POMS (a mood measure), which is a test consisting of a long list of adjectives that do not always have a precise counterpart in all languages. This is a test developed in the United States, and it is interesting to realize that in other English-speaking countries, the same adjective may have different shades of meaning. In Italian, we had to adapt some words by interpreting them in the light of the original meaning. In Polish, some words had to be eliminated because the precise equivalent was lacking. Further, it is worth stressing the fact that other adaptation problems are easily predictable in countries where the very perceptive structures are differently developed. As mentioned earlier, this battery has a minor loading on strictly cultural factors as transmitted by current educational programs. The battery is basically an instrument that stimulates psychomotor and auditory memory functions. However, at present it is not known whether these functions can be measured with these same instruments in countries where even the written language patterns are so perceptibly different. Local validation of the battery will thus become an invaluable tool to clarify these points.

As stated earlier, the WHO-NCTB was developed with the primary aim of screening early signs of impairment of higher nervous function. Accordingly, when the WHO-EURO Office of Environmental Health Service planned multicentered studies of CNS impairments due to chemical exposure, the WHO-NCTB became part of the epidemiological protocol. The first investigation was devoted to the evaluation of CNS effects in a certain number of developing countries, Poland, Czechoslovakia, Hungary, Bulgaria, Yugoslavia, Greece, and Turkey; Israel and Italy were also studied. In spite of unavoidable difficulties in adaptation and implementation, the performance of the study groups showed a good level of achievement in harmonization of the methods. The results will be released by WHO in 1990.

Another study under advanced planning involves the use of WHO-NCTB for the evaluation of exposure to methylmercury from seafood consumption in countries adjacent to the Mediterranean. In this case, the battery will probably be applied only to a specific group of subjects (i.e., fishermen) after careful analysis of methylmercury accumulation in their hair has shown significant levels.

As a general comment, these studies have proved to be important steps forward in the solution of one of the major problems of neurobehavioral studies, i.e., the standardization and more rational use of international and national resources.

An extra effort, however, needs to be made to overcome the natural individual resistance to giving up one's research tools for the benefit of international collaboration. This resistance often stems from a justified consideration of one's own experience which may be just as valid as others'. However, the inherent value of a group decision lies in its being demonstrated by many and devoted to a common aim.

AUTOMATED TESTING

Over the last decade, several studies have also been carried out by our group, mainly on large populations, with different aims, including validation of methodologies; search for subtle effects on higher nervous functions possibly arising out of long-term, low-dose exposure to industrial chemicals whose neurotoxicity is still undetermined; and medical surveillance of populations currently at work and exposed to known neurotoxic chemicals. All the studies mentioned share the need for sensitive and specific instruments to be administered to as numerous a group as possible. To this end, the increasing diffusion of relatively inexpensive, portable microcomputers has offered an extremely useful means to acquire homogeneous data especially when multicentered studies are needed. A number of independent groups have faced the problem and developed ad hoc automated programs. At the 1988 symposium in Washington, D.C., about 15 automated systems were presented, the most widespread of which is the Neurobehavioral Evaluation System (NES) developed at Harvard by Baker and Letz.

A few years ago, a computerized version of six out of the seven tests of the WHO-NCTB was set up in our institute and termed MANS (Milan Automated Neurobehavioral System).

For practical reasons, such as the impossibility of developing new tests in reasonable time, and the need for training testers in regular paper and pencil form, we have chosen to adopt all but one of the tests suggested by WHO-NIOSH because of their long and worldwide use; further, the validation of MANS is simplified by comparison with the paper and pencil form. In fact, the correlations found between the paper and pencil and the MANS tests range from 0.73 to 0.79. The use of computerized techniques will facilitate the development of new and more adequate tests by exploiting the graphic potentialities of computers. This will probably reduce the impact of cultural differences because words can be replaced by images having a wider universal meaning. For the time being, MANS is available in two languages, English and Italian, while versions in German and Greek are in progress.

A first international application of MANS has been activated by the International Commission on Occupational Health through its Scientific Committee on Neurotoxicology and Psychophysiology to evaluate the effects on the central nervous system of organic solvent exposure of paint manufacturers and users. The countries participating in the study are, at present, the United States, the United Kingdom, West Germany, and Italy.

In general, diagnosis of the impairment of higher nervous functions from environmental and industrial exposure needs a thorough evaluation by means of clinical, neurobehavioral, and neurophysiologic investigations. A team of professionals with qualifications in neurotoxicologic assessment is necessary. In practice, very few teams conforming to the proposed model exist in Europe or, to our knowledge, in other countries as well. In Europe, this type of expertise has been developed particularly in both western (Sweden, Finland, Denmark, Germany, and Italy) and eastern (Poland, USSR, Hungary, and Czechoslovakia) countries. An organizational model that has proved effective in occupational health is the self-sufficient interdisciplinary type of institute; this model is based on availability at the institute of all the medical, biological, bioengineering, and ergonomics specialties.

For instance, the Institute of Occupational Health of the University of Milan has, among its different specialties, a Center of Neurotoxicology which is composed of one neurologist, two psychologists, one physician specializing in occupational medicine, and technicians for neurophysiological and behavioral evaluations. The activities of the center include (1) assessment of neurological and behavioral deficits in inpatients of the Clinic of Occupational Medicine, hospitalized for both diagnostic-therapeutic and compensation purposes; (2) medical surveillance of high-risk groups of workers exposed to neurotoxicants; (3) screening early effects on the CNS in working populations exposed to chemicals whose neurotoxicity is still under scrutiny; (4) search for effects on the CNS in case of exposure in the general environment; and (5) training medical students and postgraduate students in occupational medicine.

NOTE

Those interested in participating in the validation program should request the "Solicitation Proposal Document" by writing: WHO Collaborating Centre, University of Milan, Institute of Occupational Health, Via San Barbara 8, Milan, Italy 20122.

APPENDIX

COMPUTERIZED VERSION (MANS) OF SIX OUT OF THE SEVEN TESTS OF THE WHO-NCTB

The POMS contains adjectives that describe different feelings about Tension, Anger, Depression, Vigor, Fatigue, Confusion, and Sociability. The subject indicates the intensity of his mood state, during the week preceding the test session, by pressing for every adjective the spacebar connected to a five-point scale ranging from "not at all" to "extremely."

The Simple Reaction Time is performed by pressing the spacebar as promptly as possible when a red light is shown on the screen. Stimuli are administered at intervals varying between 1 and 5 seconds for 6 minutes. A practical trial is given.

The Digit Span is presented in its traditional form. The digits appear on the screen one at a time; a sound indicates the end of the series after which the subject is required to repeat the digits on the keyboard.

The Digit Serial has been added to evaluate the ability to memorize and recall eight or nine digits. The same series is displayed for a maximum of 12 times until the subject repeats it correctly. The presentation of the digits on the screen is the same as in the Digit Span.

The Symbol Digit based on the stimulus material used in the classic WAIS subtest, requires, in its computerized version, pairing digits to symbols for 1.5 minutes.

The Benton Visual Recognition requires the subjects to recognize a geometrical pattern immediately, after 5 seconds of observation, among three other similar drawings presented together. The 10 critical patterns are the same used in the Visual Retention test.

The Aiming Pursuit is performed by means of a graphic tablet. A sheet of circles is fixed on the tablet and the program is set up in position. The subjects have to make quick and accurate movements to place the pen inside each circle following the pattern given on the test sheet. The circles are larger than those in the original version of Fleischman in order to be proportional to the stylus point.

REFERENCES

Baker, E. L., and R. F. White. 1985. The Use of Neuropsychological Testing in the Evaluation of Neurotoxic Effects of Organic Solvents. Working Group on Chronic Effects of Organic Solvents on CNS and Diagnostic Criteria. Copenhagen: WHO Regional Office–Nordic Council of Ministers.

Camerino, D., M. G. Cassitto, E. Desideri, and G. Angotzi. 1981. Behaviour of some psychological parameters in a population of a Hg extraction plant. Clinical Toxicology 18(11):1299–1309.

Gilioli, R., M. G. Cassitto, and V. Foa, eds. 1983. Neurobehavioral Methods in Occupational Health. Proceedings of the International Symposium on Neurobehavioral Methods in Occupational Health, Como and Milan, 1982. Pergamon Press.

Hanninen, H. 1983. Psychological test batteries—New trends and developments. In Neurobehavioural Methods in Occupational Health, R. Gilioli, M. G. Cassitto, V. Foa, eds. London: Pergamon Press.

Hartman, D. E. 1988. Neuropsychological Toxicology. London: Pergamon Press.

Hooisma, J. 1987. Abstracts of the First Meeting of the International Neurotoxicology Association, Lunteren, The Netherlands.

Johnson, B., ed. 1987. Prevention of Neurotoxic Illness in Working Populations. New York: John Wiley and Sons.

Juntunen, J., ed. 1982. Occupational neurology—First International Course on Occupational Neurology, Finland. Acta Neurologica Scandinava 66: Suppl. 92.

Letz, R., and E. L. Baker. 1987. Computer-administered neurobehavioral testing in occupational health. In Seminars in Occupational Medicine, Vol. 1, No. 3. New York: Thieme Medical Publishers.

Lezak, M. D. 1983. Neuropsychological Assessment. New York: Oxford University Press.
Lindstrom, K. 1980. Changes in psychological performances of solvent-poisoned and solvent-exposed workers. American Journal of Industrial Medicine 1:69–84.
National Institute for Occupational Safety and Health. 1974. Behavioral Toxicology—Early Detection of Occupational Hazards, C. Xintaras, B. L. Johnson, I. de Groot, eds. Washington, D.C.: U.S. Department of Health, Education, and Welfare.
Putz-Anderson, V., B. A. Albright, S. T. Lee, B. L. Johnson, D. W. Chrislip, B. J. Taylor, W. S. Brightwell, N. Dickerson, M. Culver, D. Zentmeyer, and P. Smith. 1983. A behavioural examination of workers exposed to carbon disulfide Neurotoxicology 4:(1):67–78.
Valciukas, J. A., R. Lilis, J. Eisinger, W. E. Blumberg, A. Fishbein, and I. Selikoff. 1978. Behavioural indicators of lead neurotoxicity: Results of a clinical field survey. Int. Arch. Occup. Environ. Hlth 4:217–236.
World Health Organization. 1983. Prevention of Neurotoxic Illness in Working Populations, B. L. Johnson, ed. New York: J. Wiley & Sons.
World Health Organization. 1984. Protocol for Monitoring and Epidemiological Studies on Health Effects of Methylmercury. Copenhagen: WHO Regional Office for Europe.
World Health Organization, Regional Office for Europe, Commission of the European Communities. 1985. Neurobehavioral Methods in Occupational and Environmental Health. Second International Symposium. Environmental Health Series No.6. Copenhagen.
World Health Organization, Regional Office for Europe, Nordic Council of Ministers. 1985. Chronic Effects of Organic Solvents on the Central Nervous System. Environmental Health Series No. 5. Copenhagen.
World Health Organization. 1986. Epidemiological Study on the Health Effects of Exposure to Organophosphorous Pesticides: Core Protocol. Copenhagen.
World Health Organization. 1986. Principles and Methods for the Assessment of Neurotoxicity Associated with Exposure to Chemicals. Environmental Health Criteria 60. Geneva.
World Health Organization, Regional Office for Europe. 1987. Organophosphorous Pesticides: An Epidemiological Study. Environmental Health Series No. 22. Copenhagen.
World Health Organization/National Institute for Occupational Safety and Health. 1983. Prevention of Neurotoxic Illness in Working Populations. Proceedings of a WHO/NIOSH Workshop, Cincinnati.

Issues in the Development of Neurobehavioral Toxicity Tests in India

Vinod Behari Saxena

Developing countries are a vast entity and each country has its own specific problems, geographical, economic, social, legal/policy, as well as technical, so far as occupational health is concerned. Each country has to explore the resources it can make available to support the development of neurobehavioral tests of toxicity for early detection of occupational health and hazards. Part IV of this volume, dealing with regional issues in the development of neurobehavioral toxicity tests, is concerned with future development of behavioral tests of neurotoxicity in occupational health. To understand the topic fully there are several issues to be discussed:

1. The status of research and development
2. Expected directions in the development of behavioral tests of neurotoxicity
3. Barriers to the development of behavioral tests of neurotoxicity
4. Recommendations for the development of behavioral tests of neurotoxicity

These will be considered within the general context of the Indian subcontinent.

STATUS OF ENVIRONMENTAL RESEARCH AND DEVELOPMENT IN INDIA

Recognition of a Need

India is well recognized as being among the developing countries of Asia. It is comparable to developed countries only in the field of space research. Basically it is an agricultural country that is now advancing very rapidly in the development of its industrial sector. Only in one state Uttar Pradesh has the number of industries (both private and public) approximately doubled within a span of six years (from 5,334 in 1981 to 9,529 in 1987). Cottage industries also play an important role. One can very easily consider that an explosion of industrial development in India has occurred during the last decade. With industrial development, the number of workers in industries has also approximately doubled. Statistics for 1983 (Labor Bureau, 1985) show that the labor population was 33.4 percent of the total population in the country as a whole, whereas it was 29.2 percent of the population in Uttar Pradesh. With the advancement of scientific methodology in agriculture and industries, new technology is being adopted everywhere in the country to increase production and to save precious time and labor. A large number of chemicals are being introduced, but unlike the developed countries, the harmful side effects of this industrialization are not being taken into consideration. Workers engaged in, and affected by, toxic substances are not given protection by their employers, be they government or private organizations. Due to lack of standards in health, hygiene, and safety practices, occupational health is very low on the list of national priorities. Workers are exposed to relatively high doses of neurotoxic substances without adequate protection. Because of ignorance and, possibly, of no way to cope with the problems, very little or no attention is paid to early neurobehavioral changes.

The first awareness of pollution and the need for environmental protection started with the great jaundice epidemic of 1956 in Delhi, resulting in the establishment of the Central Public Health Engineering Research Institute (CPHERI) at Nagpur in December 1958–January 1959, and later as the National Environment Engineering Research Institute (NEERI) under the Council of Scientific and Industrial Research (CSIR). As an institution, NEERI has been a forerunner in the country to focus attention on air, water, and land pollution.

The second step in the environmental movement in India had, for all practical purposes, its beginning in 1972, the year of the Stockholm conference. During the last decade, the environmental movement has also received legislative recognition in the form of laws for the prevention of water and air pollution. Promulgations of the Environment (Protection) Act of 1980 and the Factories Amendment Act of 1987 are particularly important developments because they seek to make the agencies responsible for the monitoring and control of pollution more effective by conferring on them greater powers than they have hitherto enjoyed.

The third phase started in 1984 when the methyl isocyanide (MIC) accident of the Bhopal gas tragedy on December 2–3 forced the government and other agencies to review occupational health and safety programs and created awareness in the general population about the effects of toxic substances. Epidemiological studies have been carried out since the MIC accident, but because of ignorance of behavioral toxicology, not much attention has been paid to neurobehavioral changes, which are not being assessed by any well-developed neurobehavioral test battery.

Centers for Occupational Health and Safety

In the field of occupational health and safety, the five following institutes/centers have been operating in India during the last few decades, but none has a separate department/unit dealing with neurobehavioral changes resulting from exposure to toxic substances and their assessment with the help of a neurobehavioral test battery:

1. National Institute of Occupational Health (NIOH); Indian Council of Medical Research, Ahmadabad
2. All India Institute of Hygiene and Public Health, Calcutta
3. Industrial Toxicology Research Centre (ITRC), Lucknow
4. Central Labour Institute (CLI), Bombay
5. Occupational Health Services (OHS), Bharat Heavy Electricals, Ltd. Centre, Tiruchirappalli

These institutes are working mainly in the field of epidemiological studies, animal studies, neurological studies, and general occupational health and hygiene problems. Only the OHS center of BHEL and one OHS unit of ITRC have been carrying out some work in neurobehavioral toxicology since 1980 but without developing their own test battery or even with any total battery from the developed countries. Not until December 1987 and January 1988 did NIOH, Ahmadabad, establish a Department of Behavioral Toxicology, which will begin studies in the near future.

The health problems of Indian workers are especially serious. Workers are more prone to accidents, etc., because many are illiterate or, even when literate, are not aware that the environment in which they are working may, in any way, be harmful to them. They continue working without any awareness of the toxicity of their work environment. The situation is further complicated by the fact that the majority of owners of industries are themselves not aware of the danger to which they are exposing their employees. The role of psychology in the industrial setup is restricted to placing the right worker on the right job and to matters affecting productivity. Psychologists have little to do with detection of toxic effects caused by exposure of workers to toxic substances in industries and in agriculture.

When such are the state of awareness and concerns about occupational health—where even conventional precautions and care are not being observed—the very limited status of the developing field of behavioral toxicology can easily be understood. The field has just reached the country, and the majority of scientists, policy makers, and industrialists, as well as the general public, do not even know that such a field exists, let alone applying it. Moreover, when people are told about them, studies in neurobehavioral toxicology are generally considered an extravagance. In brief, it can be stated that research work in this field is in an early budding stage.

DEVELOPMENT OF A PROGRAM AT KANPUR UNIVERSITY

It was not until 1980 that the Department of Psychology, Pandit Prithi Nath (PPN) College, Kanpur University, started work in this discipline, stimulated by the research work of Dr. Helena Hanninen. At present about six research investigators are working in this field. Most of the research is ex post facto field studies planned to assess effects of various toxic substances on different behavioral parameters with the help of neurobehavioral tests. Up to the present time, the effects of the following toxins have been explored: organic solvents, larvicides, pesticides, and insecticides; lead, chromium oxide, and carbon monoxide. Findings of five studies are summarized here in brief as examples of research that can be helpful in initiating a systematic program during its early stages of development.

All the studies were conducted not with the objective of treatment referral or withdrawal from exposure, but to detect whether toxic substances in the workplace had caused neurobehavioral changes in a population of exposed workers. The exposures had been to chronic low doses over long terms. No environmental monitoring or biomonitoring was done for want of facilities. All tests used in the studies were selected in view of the objectives of studies and guidelines proposed for the selection of tests by the World Health Organization (WHO):

- Tests known to measure functions that are affected by several neurotoxic agents
- Tests that have yielded positive results in earlier, corresponding studies
- Consideration of costs in terms of time and money
- Tests that are reasonably motivating and relatively independent of the subjects' cultural and educational backgrounds.

Out of the ten proposed functional domains, only four to five have been studied for want of test facilities and practical requirements necessary for the study.

One of the studies was conducted by Kumar and Saxena (1985). It was designed to look at the effect of long-term exposure to organic solvents on the performance of two groups of workers (100 workers in each group) from gasoline filling stations and the paint industry. Performance was measured by using a psychological test battery containing Benton Visual Retention test, Bourdon-Wiersma Vigilance test, Santa Ana Dexterity test, and a screening questionnaire. Results were compared with those of comparable groups of unexposed workers matched on mean age, education, and salary, with exposure duration of (1) up to five years and (2) more than five years. The results of the study revealed that exposure to gasoline is more harmful for memory (immediate, as well as delayed) than exposure to paint solvents when exposures were as long as five years. Both chemicals became equally toxic when the duration of exposure increased, although gasoline caused no further impairment whereas paint solvent led to continuing deterioration. In perceptional accuracy/vigilance measures the same phenomenon was found. In this case, however, gasoline caused impairment but not as much as paint solvents. As the duration of exposure increased, gasoline remained the more toxic of the two. This trend was not found in cases of manual dexterity, where both solvents were equally toxic irrespective of the duration of exposure and both maintained their toxicity even after five years. It may be concluded that exposure to both solvents causes impairment of behavioral functions in the exposed worker, with gasoline being the more harmful of the two at least in case of immediate memory and perceptual accuracy.

Another of the studies (Singh and Saxena, 1987a) was designed to determine the effect of long-term, low-dose exposure to a mixture of organic solvents (used as thinner for paint) on memory and subjective symptoms. The scores of 24 spray painters from an automobile workshop were recorded on Postgraduate Institute (PGI) memory questionnaire and a questionnaire on subjective symptoms. These results were compared with the scores of a group of workers from the upholstery section of the same workshop. Significant differences were found in four (memory total, mental balance, concentration, and recall) of ten aspects of memory, with a better performance by the unexposed workers. On comparing the extent of impairment of different aspects of memory, attention and concentration (digit span) were found to be most affected. Subjective symptoms did not clearly differentiate between the two samples, significant differences were found in only 3 of 40 symptoms. In total, however, the results indicate some deterioration in psychological functioning of workers as a result of exposure to organic solvents.

Exposure to Larvicides, Pesticides, and Insecticides

Singh and Saxena (1987b) have reported results of yet another study planned to measure possible effects of long-term exposure to larvicide (organophosphate), fenvalerate (synthetic pyrethroid), and Gammexane (organochlorine). The performances of three groups of workers (100, 30, and 22, respectively) from related departments were measured on a psychological test battery comprising PGI memory test, Santa Ana Dexterity test, Bourdon-Wiersma Vigilance test, a questionnaire on subjective symptoms, and a screening questionnaire. The results were compared with those of a comparable group of unexposed workers matched on mean age, education, and salary. Results were as follows: (1) The group exposed to larvicide was found to have significantly poorer performance on memory, vigilance, and dexterity tests than the control group. The exposed workers also reported certain subjective symptoms indicating that exposure to larvicide impaired psychological functioning. (2) The group exposed to fenvalerate was found to be affected in measures of memory, manual dexterity, and perceptual accuracy. (3) The third group exposed to Gammexane was found to have impaired memory, manual dexterity, and vigilance. They also reported the presence of such symptoms as hostility and destructiveness.

Comparisons among the three exposed groups were made with the help of multiple regression analysis in which the effect of potentially confounding variables such as age and education were partialled out. All three groups of compounds were found capable of causing impairment in the psychological functioning of workers, organophosphate compounds being the most toxic followed by organochlorine and synthetic pyrethroid compounds, respectively.

Exposures in the Leather Industry

Siddiqi and Saxena (1988) studied on 300 workers, 250 of whom were occupationally exposed to chemicals used in the leather industry (50 workers for each stage of tanning: liming, pretanning, tanning, dying, and finishing) and 50 workers as a control group matched on mean age, education, salary, and length of service. Exposure duration was defined in terms of length of service (i.e., less than five years and more than five years). The following measures were used a screening questionnaire, the Santa Ana Dexterity test, a subjective symptoms questionnaire, and EPI. Results showed that exposure to chemicals in all five stages of tanning causes impairments in behavioral functions of the workers. Exposure to chromium oxide during the tanning process had its most adverse effects on perceptual accuracy and manual dexterity. Remote memory, recent memory, and delayed recall were not affected. When results were interpreted in terms of exposure duration, it was observed that exposure up to five years caused impairment in the performance of workers on all the functional domains under study except remote memory and recent memory. Mental balance, digit span, delayed recall, immediate recall, retention of similar pairs, retention of dissimilar pairs, visual retention, and overall memory were affected by exposures even after five years. In brief, it may be stated that deterioration increases with increasing exposure in behaviors involving memory, manual dexterity, perceptual accuracy, subjective symptoms, and neuroticism.

Effects of Exposure to Lead

A study by Saxena and Saxena (1988) focused on effects of longterm, low-dose exposure to lead on the performance of 150 workers from a battery factory. Performance was measured by a psychological test battery comprising the Benton Visual Retention test, digit span, Bourdon-Wiersma Vigilance test, Santa Ana Dexterity test, EPI, and NSQ. These measures were compared with those of a control group of unexposed workers matched on mean age, education, and salary. Duration of exposure was (1) less than 5 years, (2) between 5 and 10 years, and (3) more than 10 years. The results established that exposure to lead was detrimental to measures of memory (both visual and auditory), perceptual accuracy/vigilance, and manual dexterity. The longer the duration of exposure, the greater was the impairment in these functional domains. Significantly lower levels of extroversion, neuroticism, tendermindedness, depression, and anxiety were observed in the unexposed group of workers than in exposed workers. Duration of exposure differentially affected some but not all behaviors. For example, subjective symptoms (factitious behavior, suspiciousness, impatience, self-absorption, and attacking others) were significantly higher in the exposed group of workers than in the unexposed group when the exposure duration was more than 10 years.

EXPECTED DIRECTIONS IN THE DEVELOPMENT OF BEHAVIORAL TESTS OF NEUROTOXICITY

Most of the tests in the studies described above were adopted from the test battery being used at the Institute of Occupational Health, Helsinki, and from that suggested by WHO. As far as the development of test batteries in India is concerned, the work done has been negligible. The only instrument developed so far is a questionnaire on subjective symptoms, which still needs standardization. It covers subjective symptoms related to affective behavior, ingestive behavior, social behavior, attention, arousal, and consciousness. We have studied the potential usefulness a long list of available psychological tests: intelligence, interest, adjustment, aptitude, job involvement and satisfaction, anxiety, and personality. We have examined questionnaires, projective tests, performance tests for motor ability (speed and coordination), and such standardized tests as WAIS, DAT, 16PF, EPI, TAT, and CAT. Our conclusion is that the majority of the tests are not applicable to the total population of the country because they are not culture free in their current forms and, more specifically, because they present language problems.

Similarly, from the literature of behavioral toxicology it has been noticed that all agencies in developed countries involved in studying the effects of neurobehavioral toxicity are using their own behavioral test batteries for different toxic substances. In the recent past, WHO has suggested a test battery that can be used and developed by all nations and services. This may be proper and applicable in nations where the working conditions and the control of occupational hazards are of high level. However, for certain regions/nations where little, if any, attention has been paid to such conditions, the working conditions are poor; regulations concerning exposure and control of occupational hazards are minimum; and occupational health care is not among the top priorities of government. In these developing regions, occupational health questions need deep and exhaustive exploration in view of the geographic, economic, social, psychological, and legal perspective of occupational health. Only after deciding about acceptable Threshold Limit Values (TLVs) for different toxic substances, can the WHO test battery be used, with suitable modifications and adaptations, for the identification of workers adversely affected. It is evident that in so far as development and adaptation of neurobehavioral tests are concerned, developing countries have contributed nothing or very little. This is attributable not to a lack of willingness or the capability of researchers but rather to the unfavorable circumstances in developing countries that do not recognize the important contributions neurobehavioral toxicology can make to the prevention of tragic occupational hazards. Thus it seems that a giant task lies ahead. If we really are to be on a par with developed countries, we will have to work hard.

It is useful to consider some of the directions in which actions are now needed. Priority should be given to the development of a test battery for preliminary screening comprised of elementary tests covering all the functional domains described by WHO. It should then be applied to find out the domains that are being affected by exposure to various toxic substances. This battery should be a general test battery to be used in all epidemiological studies. It should be easy to administer, of about half an hour's duration, and portable. The battery will be helpful in narrowing down the areas of psychological functions to be studied in depth.

The construction, development, and adaptation of tools of a more sensitive nature, covering all the functional areas found during preliminary screening to be affected by various toxicants, are necessary. Guidelines prescribed by WHO should be taken into consideration. These tests should be sensitive enough to detect even the slightest behavioral disorder, and they should be applied periodically to screen out susceptible workers. It would be desirable if these tests could be used in laboratory studies and experimental research, with volunteers performing under various concentrations of the toxic substances being studied. In this way, maximum allowable concentrations can be determined.

Consideration should be given to the development of a test battery whose scores can be compared with tests being used in other countries. This is essential in order to encourage the exchange of technical knowledge, possibly saving time and money in needlessly replicating studies. Comparing our results with those of other countries can provide checks on the validity and reliability of our tests and can make possible the comparison of work conditions.

The preparation of culture-free tests that can be used by different countries is another important consideration. Such instruments require preliminary trials to find out if they are culture free. These types of tests are essential because India is such a vast country having various subcultures, each with its own language, that if we want to enact uniform rules regarding occupational health services, a uniform screening battery is necessary.

In brief, it may be concluded that in the development of behavioral tests of neurotoxicity all tests must possess the following characteristics: (1) the tests should be simple to administer; (2) they must be unambiguous in interpretation; (3) measures derived from them must be indicators of nervous system dysfunction; (4) tests should be comprehensive; and (5) as far as possible, they should be culture free.

BARRIERS TO TEST DEVELOPMENT

The expected direction of development of neurobehavioral tests of neurotoxicity depends on overcoming a number of obstacles. Until—and unless—a suitable test development environment is created, with a interdisciplinary approach and coordinated research programs, the scientific and systematic development of tests will not be possible. There are several obstacles that need to be addressed before carrying out such a research program. These were discussed in depth by participants in the First Asian Conference on Behavioral Toxicology and Clinical Psychology arranged by the Department of Psychology, Kanpur, in January 1987 (Saxena, 1988). They are summarized below.

Public Awareness

The first and most important constraint is the lack of awareness of neurobehavioral toxicology and its usefulness among persons who should be concerned. A favorable attitude toward the field and a policy of support by governments are a must. Neurobehavioral toxicologists can contribute significantly through research and publication in scientific journals. However, such works are not likely to reach beyond those involved in the field itself. Implementation of scientific findings is the responsibility of industry, labor, and human resource departments of the state and central governments. To create awareness among the population as a whole is primarily a function of government. Without this support, effective measures to assess risks and control exposures to toxic conditions are not possible.

Interdisciplinary Coordination

The modern scientific era is based on interdisciplinary approaches to the development of systematic bodies of knowledge. India, too, is using this approach in many fields, but the field of occupational health and safety is an exception. Engineering departments design safe machines to minimize the risks of accident. Medical departments provide medical care to the worker on the basis of manifest symptoms.

Social welfare departments engage in general social welfare activities. There is no recognized role for psychologists in India except for the selection of workers.

The government provides protection in cases of fatal and serious accidents only through compensation and general medical care under employment insurance schemes, facilities available only to registered workers and not to daily wage earners or unregistered workers of many small and cottage industries.

The importance of interdisciplinary coordination is especially apparent in relations between neurobehavioral toxicologists and medical specialists. General medical personnel—even those who are working as industrial medical officers—are not ready to accept that behavioral tests of toxicity can detect neurological damage or that these tests can be used for the early detection of damage. Because, at present, all epidemiological studies are carried out by medical personnel, fuller understanding and collaboration between the two disciplines are essential if the development of behavioral tests of neurotoxicity is to be possible.

Lack of Appropriate Standards

The present baseline Maximum Allowable Concentrations (MACs) and Threshold Limit Values (TLV's) are based on the international standard indices. None have been established by Indian researchers for Indian workers. One reason is that there is no single agency for the nation as a whole to oversee the development of suitable standards. The only alternative is to apply standards established in other regions of the world which differ in many ways from conditions in India.

We need to plan the development of suitable standards basically in provinces and industries where the risk of toxic exposures may be great. Implementation of research plans is totally different from planning them, of course. As noted earlier, a whole network of awareness, government policy, and interdisciplinary cooperation needs to be in place if satisfactory MACs and TLVs are to be established so that standards of occupational health and safety may be raised.

Illiteracy

Most of the population working in industry, specifically in mining, tobacco, and those industries where toxicants are used, is illiterate. Workers are often hesitant in responding to test batteries or questionnaires. A hidden fear is that, if they fail on the tests, they will be removed from their jobs. Thus, both the employee and the employer are hesitant to cooperate in carrying out research. Moreover, there are certain taboos prevailing as far as mental health is concerned, and workers are especially sensitive to questions related to mood, affect, etc.

Drug Abuse and Stress

As in other developing countries, Indian workers suffer from malnutrition, drug abuse, and the like. So it is difficult to ascertain whether the particular dysfunctioning observed in a person occurs because of exposure to a toxicant, because the worker is a drug addict, or because the worker's general health is extremely poor. Similarly, it is difficult to isolate the effects of heat, noise, and other stresses from the effects of exposures to toxic substances. This makes the development and standardization of tests and the comparison of test results difficult.

Control of Exposure

The lack of awareness of neurobehavioral toxicology in India has meant that volunteers are not readily available to assist in the development of appropriate tests. The development of sensitive tests and the assessment of reactions to varying concentrations of potentially toxic substances would be facilitated if experimental studies could be conducted.

Field studies are also difficult to conduct with full accuracy and effectiveness because (1) concentration levels of toxic substances in industry are variable over both short and long durations; (2) it cannot be said with certainty that a worker is exposed to only one toxic substance even in the workplace, and he may also reside in a highly polluted area; and (3) it is difficult to find comparable control groups of unexposed workers from the same industry.

Financial Support

Neurobehavioral toxicologists do not get financial assistance from any government or social agencies. Without support, the development, standardization, and adaptation of behavioral tests of neurotoxicity are not possible. Without public or private financial support the development of neurobehavioral toxicology in India will be slow and will never come up to par with the state of the discipline in developed countries.

RECOMMENDATIONS FOR DEVELOPING COUNTRIES

In view of the state of development of behavioral tests of neurotoxicity and the facilities available in developing countries the following recommendations are made.

Training and Education

The most important factors indicating the need for training and education in neurobehavioral toxicology are the lack of awareness about neurotoxic chemicals and the shortage of trained personnel in developing countries. As with occupational health in general, three levels of education and training are needed: public awareness, training for the specific procedures, and professional specialization. It is well recognized that developing countries differ widely in their ability to fulfill these needs. To create public awareness of neurobehavioral toxicology is a matter of general education; the objective is to provide an understanding of the principles and contributions this field may make in fostering correct attitudes in the workplace, where exposure to neurotoxic materials may occur, and to permit proper participation in any preventive action taken. The target groups include legislators, policy makers, managers, and workers. Specific task training may include administrators, managers, workers, designers, medical and paramedical personnel, chemists and engineers, and occupational health and safety technicians. Specialization involves the education of fully qualified professionals in the field of neurobehavioral toxicology.

Training programs in neurobehavioral toxicology should be directed to existing problems and to the current level of practice of occupational health in general and neurobehavioral toxicology in particular. When training programs are established, the development of behavioral tests of neurotoxicity and their application will be greatly facilitated. An important step can be taken when agencies of developed countries and international agencies such as WHO hold workshops, seminars, and conferences to support the objectives discussed above and also to train persons for the further development of neurobehavioral tests.

Inclusion of Neurobehavioral Toxicology as a Full Partner in Occupational Health Programs

Occupational health programs differ among developing countries and range from very early to relatively advanced stages. Neurobehavioral toxicology needs to be included as a part of occupational health practice regardless of its stage of development in a country. This principle can only be followed if the practical aspects of neurobehavioral toxicology are explained and implemented through appropriate actions at national levels. The World Health Organization and the International Labor Organization should use their influence until this goal is achieved.

Applied Research

Applied research is needed in developing countries to clarify the actual magnitude of the problems surrounding the development of behavioral tests of neurotoxicity. Research should focus on applied rather than basic questions. Developing countries have to learn from both the positive and the negative experiences of the developed nations. They should not repeat the mistakes made by the latter but should attempt to take shorter routes to reach a stage of modern occupational health care appropriate for their special needs.

Communication Networks

Development of test batteries is essential both for preliminary screening and for more extensive investigation. Comparison with findings obtained in other countries can facilitate the process. For this purpose it would seem advisable to encourage the creation of communication networks both internally and internationally. A center should be established in each developing country by appropriate international organizations, with the responsibility of fostering the development of neurobehavioral tests. Such centers should exchange information, passing it along to researchers working on the development of neurobehavioral tests. Each center should also get feedback from field workers and the results obtained from the development of such tests. Thus, the exchange of knowledge within a country and among countries will facilitate the development of usable and acceptable behavioral tests of neurotoxicity internationally.

REFERENCES

Kumar, P., and V. B. Saxena. 1985. Disturbances in Psychological Functions of Workers Occupationally Exposed to Organic Solvent. Unpublished doctoral thesis, Kanpur University, India.

Labour Bureau. 1985. Pocket Book of Labour Statistics. Chandigarh, India: Ministry of Labour and Rehabilitation, Government of India.

Saxena, A., and V. B. Saxena. 1988. Behavioral Effects of Occupational Exposure to Lead. Unpublished doctoral thesis, Kanpur University, India.

Saxena, V. B., ed. 1988. Proceedings of the First Asian Conference on Behavioral Toxicology and Clinical Psychology. Kanpur, India: Alka Printers.

Siddiqi, J., and V. B. Saxena. 1988. Behavioral Disorders in Workers Exposed to Chemicals (Chromium Oxide and Dyes) Used in Leather Industry. Unpublished doctoral thesis, Kanpur University, India.

Singh, J., and V. B. Saxena. 1987a. Memory and subjective symptoms in automobile painters. Indian Psychological Review 32(2):5–12.

Singh, J., and V. B. Saxena. 1987b. Psychological Performance and Subjective Symptoms of Workers Occupationally Exposed to Past Control Substances. Unpublished doctoral thesis, Kanpur University, India.

Regional Issues in Neurobehavioral Testing:
An Overview

Ann M. Williamson

It is clear in reviewing the other three chapters in this section on "regional issues" that there are great differences in the status and application of neurobehavioral testing among regions. Of the regions represented, most activity is occurring in Europe, mainly in research on the neurobehavioral effects of particular toxic substances and in the development of tests that can be used across regions. Comparatively less activity is occurring in the less developed regions of China and India. The activity that is occurring there is directed at researching neurotoxicological problems that are common in each particular region. For example, in India, which has a predominantly agricultural base, a considerable proportion of research has been on the effects of pesticides and larvicides. Virtually no research has been directed at developing tests that are specifically designed for the local population. All tests used in studies in both regions were developed in Europe, Scandinavia, or the United States.

This difference in degree and focus of activity is largely due to the comparative age of the area as a field of inquiry in each region. The chapters by Liang and Saxena demonstrate that, despite the lack of activity in test development in their regions, there is an obvious awareness of the issues involved. The arguments put forward in both chapters highlight regional differences as important factors to be considered in the development of new tests and test batteries.

EFFECTS OF REGIONAL DIFFERENCES

There are a number of primary issues that must be considered in a discussion of the effect of regional differences on test design and development.

Exposure Characteristics of the Region

The main types of neurotoxic hazards are likely to differ from region to region. Carbon disulfide, for example, one of the earliest occupational neurotoxins to be studied by neurobehavioral methods (Hanninen, 1971), is used widely in Europe and the United States but is used to a much smaller extent in Australia. Consequently, the need to develop tests for this particular hazard in Australia is limited and will be overshadowed by more commonly encountered hazards such as lead, organic solvents, and organophosphorus pesticides. Similarly, the need to study the neurobehavioral effects of abalone diving (Williamson, et al., 1987), although not a common occupational problem in Australia, would not even occur in many other countries because professional diving is uncommon for practical or commercial reasons.

Regions also differ in the degree of exposure to particular neurotoxins. In developing countries where occupational hygiene standards have not reached the same levels achieved in some of the developed countries, poisonings are not uncommon. Thus, the perceived need for neurobehavioral testing is less because frank clinical symptoms provide clear evidence that the substance is neurotoxic.

Combinations of hazards are also likely to be unique to particular regions. The effects of thermal stress, for example, may well interact with the neurotoxic action of a substance and, as a result, confound the measurement of neurotoxic effect. The extent of the interaction, of course, depends on the degree of thermal stress, but in regions such as the tropics where extreme temperatures are common, neurobehavioral test development must take this into account.

Ethnic Composition of the Region

Regions differ in the relative homogeneity of ethnic backgrounds and languages. Saxena pointed out the difficulties of testing in situations such as in India, where there are many different language groups, and called for the development of culture-free tests that could be used to compare results across the country.

The need for culture-free tests becomes even more important in regions such as Australia where it is not uncommon to have 20 or more language groups in a single work force of a few hundred. Testing can become very difficult in ethnically heterogeneous regions due to the problems of ensuring (1) that test instructions are comprehended by workers of each language group and (2), where different forms of the tests must be used for different groups, that they are in fact identical. The difficulty experienced in attempting to adapt measures of mood for different language groups, discussed by Gilioli and Cassitto, where there may be no direct equivalents in the new language to certain words used in the original version is a perfect example of the latter point. Even where the same words exist across language groups it is quite likely that the meanings are different.

The problem of language differences across cultures is clearly important because one of the primary aims of the World Health Organization (WHO) Neurobehavioral Core Test Battery initiative is to obtain cross-cultural baseline measures on a group of tests. If this can be achieved, we will have a set of tests that can be applied in many countries. The development of tests that have multiple language equivalents should be extended however and, where possible, the language problem should be avoided by the use of performance tests that are not language based. The exception to this, of course, is where testing for verbal abilities is the main object of a test.

Educational and Socioeconomic Characteristics

The educational and socioeconomic characteristics of workers differ among regions. This factor presents a number of problems. The degree of illiteracy differs between regions such that it is relatively uncommon in the developed countries but still common in the developing countries. Illiteracy presents much the same problem for testing as does language difference except that nonlanguage-based tests are essential not just advantageous. In both cases there is the additional problem of the approach to neurobehavioral testing of workers who are not only unused to being tested but may also not understand properly why they are being tested or what they are being asked to do.

The worker's approach to testing and the test situation has become increasingly important with the greater use of computer-assisted testing. In this case, both literacy and computer literacy need to be taken into account because unfamiliarity with computers is likely to have the same effect on testing as illiteracy does. Socioeconomic status is also an issue in this area because it is well established that many of the factors associated with low socioeconomic status, in particular, independently affect behavior (Cravioto and Delicardie,1973). Malnutrition, for example, has a marked effect on most aspects of human performance (Brozek, 1978) and would undoubtedly confound any attempts to measure the neurobehavioral effects of a specific toxic substance.

Attitudes to Testing in the Region

Regional differences may also be evident in workers' attitudes to tests and their motivation to participate in them. In areas where health information is limited and mental illness is regarded with fear or suspicion it may be difficult to convince many workers to participate in a testing program. Where employers have little knowledge about the potentially toxic effects of substances in their workplaces or take little responsibility in informing workers by such means as workplace training courses or material safety data sheets (MSDS), or where employees have little awareness of their "right to know" about the degree of safety of substances in the workplace, it is likely that reluctance to participate in testing programs will be encountered. It is important, therefore, that test batteries be developed for those conditions that are acceptable both to the workers who are to be tested and to the employers who will permit testing to be carried out in their workplaces. Until mass education programs have widespread impact in these areas, it is important to take such factors into consideration when designing test batteries or attempting to apply existing ones in new regions.

Test motivation, and in some cases even performance, will also be influenced by the real or perceived implications of finding neurobehavioral impairment. Where workers differ in political reactivity, motivation to participate and to perform in toxicity testing will also differ. In regions where job security is uncertain, workers may be unwilling to participate or motivated to deny symptoms for fear of losing their jobs. On the other hand, in regions where workers compensation systems are liberal, there is a risk that workers may fake poor performance and "malinger" in order to obtain the benefits of the compensation system.

The latter problem, where it occurs, presents a significant impediment to accurate testing for neurobehavioral impairment, particularly of individual workers. It is virtually impossible to unequivocally diagnose faking; however, the American Psychiatric Association's (1980) DSM III lists a number of features associated with faking that can provide a guide. In addition, Lezak (1983) describes a number of tests that can be added to a test battery where faking may be a problem, which will provide an estimate of the degree of faking by a subject.

Any attempt to make international comparisons or to translate tests or even test results between countries needs to take into account such issues as those discussed above. Failure to acknowledge differences between regions on exposure, ethnic, socioeconomic, and educational dimensions, as well as in attitudes to neurobehavioral testing, will certainly limit the quality and applicability to the local population of the results of studies when they are performed in a new country or region.

STATUS OF WORK IN THE RESEARCH AREA

The amount of activity in the area of neurobehavioral toxicity testing in different countries is very variable and limited to distinct regions of the world. Clearly a great deal of work is going on in Europe and North America relative to the rest of the world. As a consequence it is in these regions that the most impact has been made on Threshold Limit Values (TLVs) and neurobehavioral test development has been a high priority. On the face of it, the limiting factor to activity in this area would appear to be whether the region is developed or not. Certainly the chapters in this section describing the position of neurobehavioral testing in China and India support this concept. It is likely, though, that the extent of development in a region is not the only factor governing the amount of activity in this field.

In Australia, for example, which would be classified as a developed country, there is relatively little activity in neurobehavioral toxicity testing and, partly as a consequence, there has been little impact on permissible health standards. In addition, until very recently, there has been limited awareness by the working population about neurotoxicity as an occupational problem. Although this situation is changing, it has been due largely to the action of political and industrial relations forces that have traditionally focused on issues other than occupational health. One major factor governing the degree of emphasis and responsibility that governments will take on the issue of neurobehavioral testing is, of course, how important it is considered to be relative to other issues. In countries where basic public health is still an enormous issue and where people are regarded as lucky to even have a job, it is not surprising that neurotoxicity problems are regarded as a fairly low priority until a major incident occurs such as in Bhopal. Other health and safety factors have assumed precedence in some countries where, even though most basic health problems have been brought under control, neurotoxicity testing still lags behind other health and occupational health issues. This, coupled with the general belief that neurotoxicity testing is looking at comparatively subtle changes, results in very low priorities being assigned to the area.

Nevertheless, attitudes to assessment of chemical hazards are changing around the world and some neurotoxicity testing is being conducted using animal models for certain categories of chemicals (notably new pesticides). Also, as animal welfare concerns increase, toxicology evaluators are demanding more information from conventional toxicity studies. One such source of information is the results of neurotoxicological investigations where ongoing study of sensitive indicators may provide a unique and timely measure of potential hazard.

Education and Training

A higher profile can be achieved for neurotoxicity testing through education of workers and the public. This, however, requires experts who are themselves aware of the need for neurotoxicity testing and the issues involved. It also requires financial resources to implement large-scale education programs. In countries where, for whatever reason, the area is largely overlooked, there are often very few experts to take on this role and very few training opportunities to increase their number. With few experts to lobby governments, the allocation of financial resources to improve this situation is unlikely to occur.

This is clearly a circular problem and one whose solution probably requires the exchange and promulgation of research findings and information about activities in other countries. Opportunities for this to occur have certainly increased in recent years. There have been a number of international meetings of experts involved in neurobehavioral toxicity testing since 1980, mostly conducted under the auspices of the WHO. These have provided the conditions for invaluable discourse between workers in the field and have had an obvious impact on the quality of research being conducted. Unfortunately, most of these meetings until now have been held in the Northern Hemisphere, thus severely limiting the involvement of workers from the Southern Hemisphere due to the costs of travel to the meetings. The cost of attending meetings is also likely to be a limiting factor for scientists from developing countries even in the Northern Hemisphere. In fact, scientists from any country in which neurobehavioral toxicology is not a priority are likely to encounter difficulties in obtaining any kind of support, either financial or in time off from work, to attend such meetings.

Apart from these scientific meetings the only other organized mechanism for the exchange of neurotoxicity research findings and test methods is through the scientific literature. Additional mechanisms include the institution of visiting fellowships at some of the more advanced research centers, as well as scholarships for postgraduate and postdoctoral students. Specialized training courses such as that conducted at the University of Milan under the sponsorship of the

WHO are a very efficient, if brief, mode of providing information on at least some aspects of the subject. Training courses are an especially important way of increasing exchange between workers in the field because the information they provide can form the basis of training programs in the trainee's home country. For this reason, more comprehensive training courses should be encouraged in one or two centers around the world.

Is the Concept of a Global Test Battery Feasible?

With the development of the WHO Neurobehavioral Core Test Battery, the concept of a global test battery that could be used for multiple toxins in virtually any country becomes more of a reality. Obviously there are a number of problems such as those listed in the beginning of this chapter that will have to be worked through before this actually occurs. It is unlikely that problems such as cultural differences will ever be entirely overcome. Although it is probable that actual tests can be designed to be culture free (in fact, a number already exist for some neurobehavioral functions), the problem of differences in attitudes to testing will probably never be overcome.

Nevertheless the position is definitely better now than at the end of the 1970s because the amount of international exchange and collaboration has increased markedly. The field of neurobehavioral toxicology is currently in the position of being able to at least plan in advance and guide the development of neurotoxicity testing in countries where it has not yet made much impact. This will take time, but the opportunity to do so should be seized now.

REFERENCES

American Psychiatric Association. 1980. Diagnostic and Statistical Manual of Mental Disorders (DSM III), third edition. Washington, D.C.

Brozek, J. 1978. Nutrition, malnutrition and behaviour. Annual Review of Psychology 29:157–177.

Cravioto, J., and E. R. Delicardie. 1973. Nutrition, behaviour and learning. World Reviews in Nutrition and Dietetics 16:80–96.

Hancock, P. A, and J. O. Pierce. 1985. Combined effects of heat and noise on human performance: A review. American Industrial Hygiene Association Journal 46:555–566.

Hanninen, H. 1971. Psychological picture of manifest and latent carbon disulphide poisoning. British Journal of Industrial Medicine 28:374–385.

Lezak, M. D. 1983. Neuropsychological Assessment. New York: Oxford University Press.

Williamson, A. M., B. Clarke, and C. Edmonds. 1987. The neurobehavioural effects of professional abalone diving. British Journal of Industrial Medicine 44:459–466.

PART V
Recommendations for Further Research and Testing

Environmental Modulation of Neurobehavioral Toxicity

Robert C. MacPhail

INTRODUCTION

There is an inevitable tension in neurobehavioral toxicology research between focusing experimental attention inward and outward. This tension is largely due to a clash between those who view behavior as a by-product of nervous system activity and those who view it as a by-product of dynamic forces operating in the environment. My opinion is that although both views may be justifiable, the development of a comprehensive conceptual and empirical framework in neurobehavioral toxicology must promote and capitalize upon both approaches. I would like to provide evidence of the relative neglect of environmental variables in neuroscience, the importance of environmental variables in modifying the effects of toxicants, and a framework for systematically determining the importance of environmental variables in modifying the effects of toxic chemicals on acquired behavior. Finally, I would like to speculate on how that framework can be used to address some of the important research issues in the future.

Some introductory remarks are in order concerning a framework for describing behavior in relation to the environment. Consider, for example, a continuum such as that shown in Figure 1 in which behavior is related to stimulus events. At one extreme, behavior is represented principally as a reaction to stimuli. Such behavior is elicited, and the procedures by which elicited behavior is acquired and maintained involve respondent conditioning. At the other extreme,

FIGURE 1 Relationship between behavioral response classes (R) and controlling stimulus events (S) in the environment.

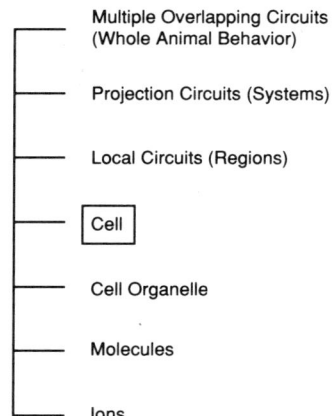

FIGURE 2 Levels of organization in the nervous system.

SOURCE: From Shepherd (1989).

behavior is principally represented as the result of consequent (rather than antecedent) stimuli. This type of behavior is emitted, and the procedures by which emitted behavior is acquired and maintained involve operant conditioning. Intermediate between these extremes, the relationship between controlling stimulus events and behavior is unclear. Under these circumstances, behavior is said to be spontaneous, although it should be recognized that to label a behavior spontaneous is only to affirm our ignorance regarding its controlling variables.

An implicit assumption in neuroscience is that behavior is a byproduct of nervous system activity. Although the steadily increasing incorporation of behavior analysis into neuroscience research is salutary, the lack of appreciation of the importance of the environment is not. For example, in his textbook on neurobiology, Shepherd (1989) presented the hierarchical scheme for nervous system organization shown in Figure 2. Behavior, according to this scheme, represents the integrated output of multiple neuronal processes and pathways. Such a scheme implies that a neurological lesion can produce multiple effects

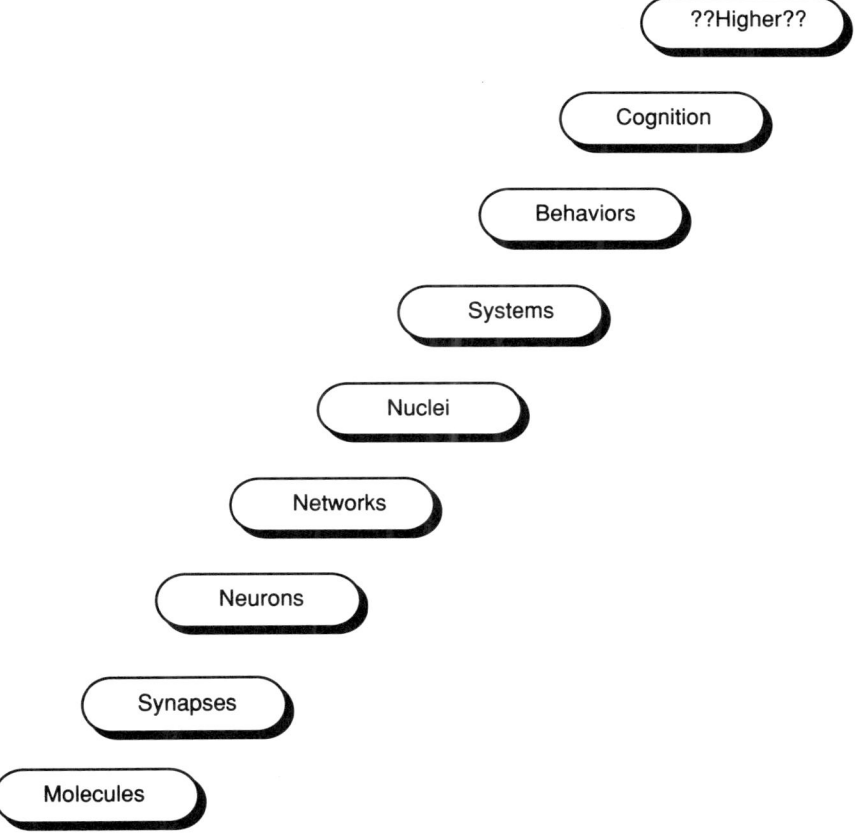

FIGURE 3 Relationship between behavior and the nervous system: levels of progressive complexity.

SOURCE: From Bloom (1988).

on behavior and, conversely, that a change in behavior can come about because of changes in the activity of different neuronal pathways. Although such a scheme may have substantial heuristic value, it fails to recognize the important role the environment plays in determining behavior. Similarly, Bloom (1988) recently presented a more elaborate organizational scheme of nervous system function. That scheme, shown in Figure 3, also completely fails to recognize the fact that behavior and its environmental context are for all intents and purposes inseparable.

Presenting these schemes should in no way be construed as an effort to disparage attempts to define the role of behavior in neurobiology. It should be clear, however, that environmental considerations have been largely neglected in neurobiology research. A comprehensive approach in neurobehavioral toxicology can come about only by jointly defining the intrinsic and extrinsic determinants of behavior. The reasons environmental considerations are important in toxicology can be made clear with a few examples taken from the literature.

MODULATION OF LETHALITY

The dramatic influence of environmental conditions on the lethal effect of amphetamine-like drugs has been well established. Events such as crowding, loud noise, high ambient temperature, and electric shock can substantially increase the lethal potency of sympathomimetic amines. In a relatively recent demonstration, for example, Landauer and Balster (1982) showed that the LD_{50} of d-amphetamine was substantially lower when mice were housed in groups of 12 than in isolation (2.8 versus 82 mg/kg, respectively).

Siegel et al. (1982) showed that environmental events could also modify heroin lethality. In this experiment, rats received a series of escalating dosages of heroin so that they could survive an otherwise lethal dosage. Heroin was always administered on alternate days in a distinct environment, whereas the heroin vehicle was administered in an equally distinct environment on intervening days. The rats were then challenged with a much larger dosage of heroin. For half the rats this dosage was administered in the heroin-associated environment, whereas for the others it was administered in the vehicle-associated environment. The same dosage of heroin was also given to naive rats that had no history of heroin exposure. The results of this experiment are shown in Table 1. Heroin killed almost all of the drug-naive rats (96.4 percent). Although prior exposure to heroin produced fewer mortalities, the degree of protection was much greater for rats receiving heroin in the heroin-associated environment (ST) than for those receving it in the vehicle-associated environment (DT). This experiment showed, therefore, that environmental and chemical variables played an equally important role in modifying heroin-induced lethality.

Poulos and Hinson (1982) showed that environmental circumstances could also profoundly modify the ability of a drug to alter a nonlethal form of behavior. Rats repeatedly received haloperidol in one distinctive environment and vehicle in another. The regimen was designed to induce tolerance to the cataleptic effect of haloperidol. The rats were then treated with haloperidol in either the haloperidol-associated

TABLE 1 Rat Mortality After Injection of Heroin at 15 mg/kg

Group	Number of Rats	Mortality (%)
ST[a]	37	32.4
DT[b]	42	64.3
Control	28	96.4

[a]Heroin-association environment.
[b]Vehicle-associated environment.

SOURCE: Siegel et al. (1982).

TABLE 2 Cataleptic Response of Each Group to Haloperidol (1.5 mg/kg) at Each Assessment Interval (Values Are Means ± Standard Errors)

Time After Haloperidol Injection	Duration of Cataleptic Response (s)		
	Control Rats	Rats Tested in Drug-Associated Environment	Rats Tested in Saline-Associated Environment
25	109.8 ± 16.8	22.8 ± 5.4	108.7 ± 16.5
50	136.8 ± 13.5	45.8 ± 11.3	125.8 ± 13.6
75	169.6 ± 5.7	68.3 ± 15.8	157.6 ± 10.9

SOURCE: Poulos and Hinson (1982).

or the vehicle-associated environment, and the duration of catalepsy was measured. Haloperidol was also administered to drug-naive rats. The results are shown in Table 2. When compared to drug-naive rats, the other rats were either almost completely tolerant or completely intolerant to haloperidol depending on whether they received it in the haloperidol-associated or the vehicle-associated environment. Clearly, then, the environment can have profound effects on the behavioral consequences of chemical exposure.

MODULATION OF OPERANT BEHAVIOR

Lethality (or survivability) and catalepsy are rather gross and global aspects of behavior. What about more subtle aspects? To what extent do environmental variables modulate toxicant effects on more subtle forms of behavior? This question can best be addressed through research on schedule-controlled operant behavior. Conditioned behavior should offer advantages over unconditioned behavior because of the degree to which environment-behavior relationships can be specified. As Dews (1962) said, "To express a preference for working with conditioned behavior is thus merely to express a preference for working with well-controlled situations rather than vague ones." Schedule-controlled operant behavior is particularly appropriate because of the degree to which the relevant environmental controlling variables can be identified, specified, and manipulated. Such a degree of specification promotes reproducibility and sensitivity, and fosters mechanistic approaches. The basic paradigm and the controlling variables are shown in Figure 4. A wealth of data exists on the importance of each of these variables in the control over operant behavior.

Schedule-controlled operant behavior has been used extensively and profitably in evaluating the effects of a wide range of drugs on conditioned behavior. A fundamental concept in this research involves drug-behavior interactions (Sidman, 1956), or the joint dependence of drug effects on the drug and the variables maintaining behavior. More than three decades of research have shown that drug-behavior

1. Schedule-Controlled Behavior: Operant Paradigm
$$S^D * R \rightarrow S^R$$
2. Controlling Variables:
    A. S^D: Discrimintative Stimuli
        1. Qualitative difference
        2. Quantitative difference
    B. R: Response Effects
        1. Topography
        2. Ongoing rate
    C. S^R: Reinforcer Effects
        1. Qualitative differences
        2. Quantitative differences
    D. R → S^R: Schedule Effects
        1. Differing schedules
        2. Differing parameters of the same schedule
    E. S^D: S^R Effects
    F. Historical Variables
        1. Long term
        2. Short term (context)

FIGURE 4 Operant paradigm and the variables determining schedule-controlled behavor. Behavior (R) occurring in a stimulus environment ($S^D * R$) produces changes (→) in that environment. Changes that maintain or strengthen behavior are called reinforcers (S^R).

interactions prevail widely in behavioral pharmacology. In fact, research is available to show that each and every variable represented in Figure 4 has been able to modify the effects of drugs on schedule-controlled behavior. A fundamental question to be addressed in neurobehavioral toxicology is the extent to which the effects of exposure to environmental chemicals may also be influenced by these variables (MacPhail, 1985).

PESTICIDES

Some of our recent work showing how schedule variables modify the effects of pesticides is described here. Pesticides represent an extremely diverse array of substances, many of which (e.g., insecticides) are specifically designed to adversely affect the nervous system. Although there are, in general, considerable data on organophosphate, carbamate, and organochlorine pesticides, many newer pesticides—for example, formamidine and pyrethroid insecticides and triazole fungicides—have not undergone thorough evaluation.

In one experiment, the efects of the formamidine insecticide chlordimeform were determened on the schedule-controlled behavior of pigeons (Leander and MacPhail, 1980). Pigeons performed under a multiple schedule in which fixed-interval reinforcement alternated with fixed-ratio reinforcement. Figure 5 shows that chlordimeform generally decreased overall fixed-interval responding in a dosage-dependent manner. The decreases in overall response rate were also accompanied by a disruption of the fixed-interval pattern of responding. Intermediate dosages of chlordimeform produced either no change or increases in fixed-ratio responding, while larger dosages uniformly decreased responding. Estimates of the dosage producing a 50 percent reduction in responding were, depending on the pigeon, between 30 percent and 300 percent greater for fixed-ratio than for fixed-interval performance. These data show therefore that the particular schedule maintaining responding can influence the magnitude of the effect produced by chlordimeform.

In another experiment, the effects of three formamidine insecticides on schedule-controlled behavior were compared (Moser and MacPhail, 1986). The three formamidines included chlordimeform, amitraz, and formetanate. Formetanate was unique in that it also contains a carbamate moiety. Rats performed under a multiple schedule in which presentation of two fixed intervals of different length alternated throughout the session. The results of this experiment are shown in Figure 6. The three pesticides could be differentiated on the basis of their multiplicity of effects. Chlordimeform substantially

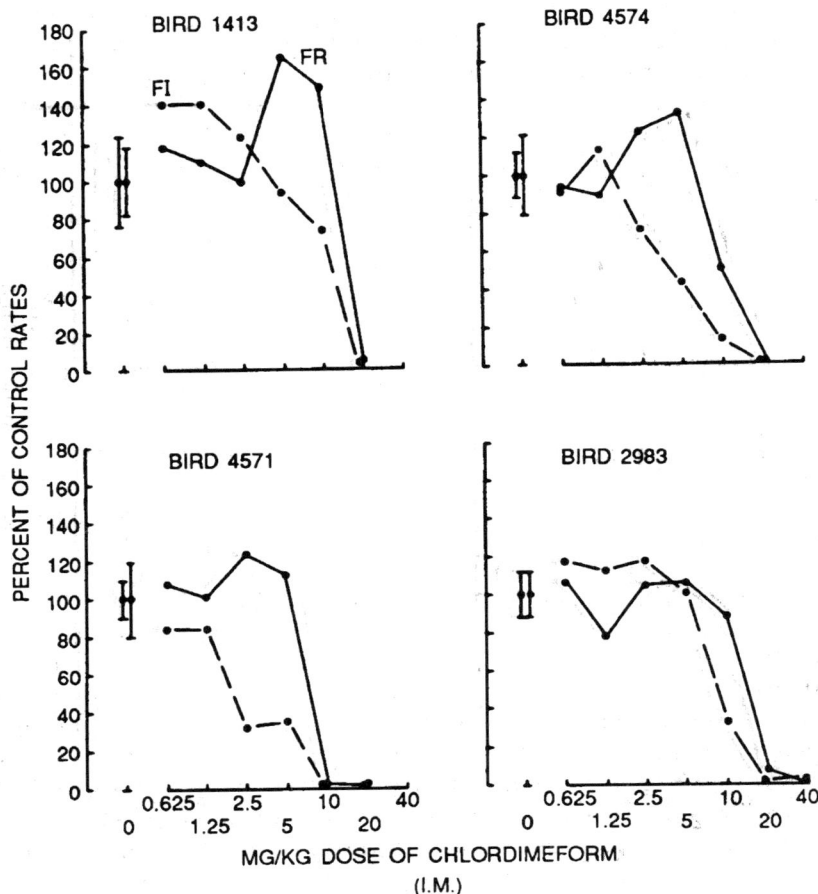

FIGURE 5 Schedule-dependence of the effects of chlordimeform on operant behavior. Dosage-response functions were determined for the effects of chlordimeform in pigeons responding maintained with fixed-interval (FI) or fixed-ratio (FR) reinforcement. Data are plotted as percent of vehicle-injected control (0 mg/kg) response rates. In absolute terms, FR response rates were much higher (2.5–4.8×, depending on the pigeon) than FI response rates. Brackets at 0 mg/kg represent ±1 SD around the control mean. Filled circles represent effects on FR responding and unfilled circles represent effects on FI responding.

SOURCE: From Leander and MacPhail (1980).

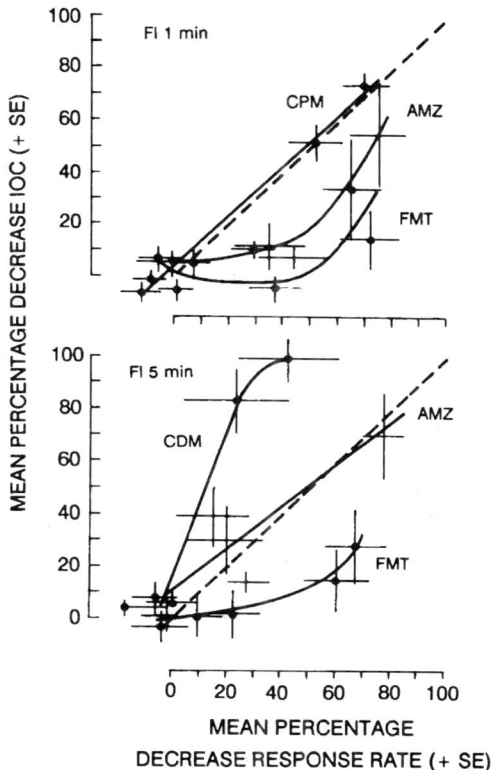

FIGURE 6 Differentiation of the effects of formamidine insecticides on schedule-controlled behavior. Dose-response functions were determined for chlordimeform (CDM), amitraz (AMZ), and formetanate (FMT) on the performance of rats maintained under a multiple fixed-interval/fixed-interval schedule of reinforcement (FI 1 min, FI 5 min). Disruptions in response patterning (decreases in index of curvature, IOC) are plotted as a function of corresponding decreases in overall effects. Dashed lines represent hypothetical equivalent effects.

SOURCE: From Moser and MacPhail (1986).

disrupted the temporal pattern of responding maintained under the long fixed interval while producing a much smaller change in overall rate of response. Formetanate, on the other hand, selectively decreased overall response rate while producing little disruption in the temporal pattern of responding. Amitraz produced intermediate effects. Although the upper panel shows that differences in effect were also obtained under the short fixed interval, it is clear from this figure how the schedule parameter can magnify differences in the effects of chlordimeform and, to a lesser extent, amitraz. The effect of formetanate, on the other hand, did not appear to depend on the schedule parameter. This finding, along with the fact that overall rate but not pattern was mainly affected, suggested that formetanate acted more like a carbamate than a formamidine. This prediction was confirmed by pharmacological blocking experiments showing that muscarinic receptors were involved in the action of formetanate (Moser and MacPhail, 1987).

An analysis of toxicant effects on fixed-interval patterns may play an important role in neurobehavioral toxicology. A particularly dramatic example of the relativity of toxicant effects on fixed-interval behavior involves the triazole fungicide triadimefon. Triadimefon was found to increase levels of motor activity in rats (Crofton et al., 1988; Moser and MacPhail, 1989), and to induce stereotyped behavior in rats following large dosages (Moser and MacPhail, 1989). These effects suggested that triadimefon acted in a way similar to that of the psychomotor stimulants. Because psychomotor stimulants have been shown to disrupt temporal patterns of responding, triadimefon's effects were determined on performance maintained under the multiple fixed-interval schedule described above (Allen and MacPhail, 1988). Triadimefon produced a dosage-related disruption in the temporal pattern of responding under the long fixed interval. Response patterning was almost completely eliminated following the largest dosage. Patterning under the short fixed interval, however, was only slightly affected by triadimefon. This selectivity of effect is remarkable if one considers that there was no difference in the extent of temporal patterning maintained by the two schedules under baseline conditions. The dosage disrupting response patterning by 50 percent was estimated to be five times smaller for performance under the long fixed interval. The effect of triadimefon was also remarkable in that overall rates under the long fixed interval changed very little despite the dosage-related disruption of response patterning.

Findings such as these suggest that relatively subtle environmental variables may have substantial effects on the behavioral consequences of some pesticide exposures. It is tempting to speculate that many of the other variables shown in Figure 4 could also substantially influence the effect of formamidine and triazole pesticides on schedule-controlled behavior, whereas the importance of these variables in modifying the effect of carbamates may be relatively negligible (see also MacPhail, 1985). Questions remain concerning the relative importance of these variables in determining the effects of a wide range of other environmental chemicals.

RESEARCH DESIGN

Research designed to evaluate environmental contributions to toxicant effects will likely have both a practical and a fundamental impact in neurobehavioral toxicology. In a practical sense, results such as the above serve to define more completely optimal conditions of testing to ensure that a toxicant's effect will be detected if indeed it exists. Hazard identification studies could be improved greatly if sufficient detail were given to describing and controlling the environmental context in which behavior is evaluated. Fundamentally, however, this type of approach is critically important not only in helping to define the full range of effects associated with chemical exposures, but also in determining the behavioral and neurological mechanisms by which these effects are produced. An integrated conceptual framework such as that shown in Figure 7 can then be used to focus research

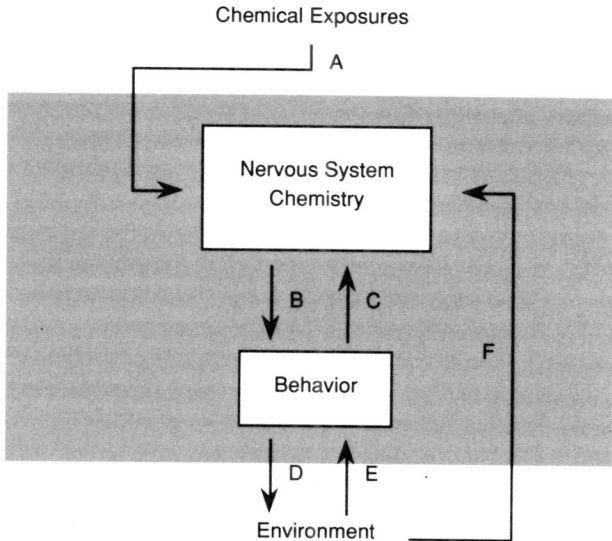

FIGURE 7 Conceptual framework for relating behavior and chemical exposures to the environment and the nervous system. Chemical exposures produce changes in nervous system chemistry (A) that can then lead to behavioral changes (B). Performance of certain behavior can also affect nervous system chemistry (C). Mutual interrelationships exist between behavior and the environment in which it occurs (D,E). Environmental variables can also affect nervous system chemistry without clear behavioral concomitants (F). One important feature not readily apparent in this scheme is that behavior operating on the environment may also alter the intensity or frequency of exposure to environmental chemicals.

effort as well as to assimilate new information derived from diverse research efforts. Such information can then serve as the basis for making informed decisions in estimating risks due to chemical exposure and the steps needed to effectively and efficiently regulate exposures.

REFERENCES

Allen, A. R., and R. C. MacPhail. 1988. Effects of the triazole fungicide triadimefon on schedule-controlled behavior: Comparison with methylphenidate. Paper presented at the annual meeting of the Association for Behavior Analysis.

Bloom, F. E. 1988. Neurotransmitters: Past, present and future. FASEB Journal 2:32–42.

Crofton, K. M., V. M. Boncek, and L. W. Reiter. 1988. Hyperactivity induced by triadimefon, a triazole fungicide. Fund. Appl. Toxicol. 10:459–465.

Dews, P. B. 1962. Monoamines and conditioned behavior. Pp. 143–151 in Symposium sur les Monoamines et le Systeme Nerveux Central de Ajuriaguerra. Geneva:George et Cie.

Landauer, M. R., and R. L. Balster. 1982. The effect of aggregation on the lethality of phencyclidine in mice. Tox. Lett. 12:171–176.

Leander, J. D., and R. C. MacPhail. 1980. Effect of chlordimeform (a formamidine pesticide) on schedule-controlled responding of pigeons. Neurobehav. Toxicol. 2:315–321.

MacPhail, R. C. 1985. Effects of pesticides on schedule-controlled behavior. Pp. 519–535 in Behavioral Pharmacology: The Current Status, L. S. Seiden and R. L. Balster, eds. New York: Alan R. Liss.

Moser, V. C., and R. C. MacPhail. 1986. Differential effects of formamidine pesticides on fixed-interval behavior in rats. Toxicol. Appl. Pharmacol. 84:315–324

Moser, V. C., and R. C. MacPhail. 1987. Cholinergic involvement in the action of formetanate on operant behavior in rats. Pharmacol. Biochem. Behav. 26:119–121.

Moser, V. C., and R. C. MacPhail. 1989. Neurobehavioral effects of triadimefon, a triazole fungicide, in male and female rats. Neurotoxicol. Teratol. 11:285–293.

Poulos, C. X., and R. Hinson. 1982. Pavlovian conditioned tolerance to haloperidol catalepsy: Evidence of dynamic adaptation in the dopaminergic system. Science 218:491–492.

Shepherd, G. M. 1989. Neurobiology. New York: Oxford University Press.

Sidman, M. 1956. Drug-behavior interaction. Ann. N.Y. Acad. Sci. 65:282–302.

Siegel, S., R. E. Hinson, M. D. Krank, and J. McCully. 1982. Heroin "overdose" death: Contribution of drug-associated environmental cues. Science 216:436–437.

Computerized Performance Testing in Neurotoxicology: Why, What, How, and Whereto?

Francesco Gamberale, Anders Iregren, and Anders Kjellberg

WHY MEASURE PERFORMANCE?

Behavioral performance tests have been developed primarily to assess the psychophysiological efficiency of the individual. Most frequently, the tests have been used for diagnostic purposes in clinical contexts or for the purpose of personnel selection. During the last 15 to 20 years, however, performance tests have been applied with increasing frequency to assess functional changes in the central nervous system (CNS) induced by exposure to unfavorable work environmental conditions.

Ever since the early 1970s, when the use of psychometric techniques made it possible to link together deterioration in human performance and the inhalation of solvent vapor (Astrand and Gamberale, 1978), psychometric tests have been widely and successfully used in many countries in the study of solvent toxicity (Anshelm Olson, 1985; Gamberale, 1985; Iregren, 1986b) as well as in the study of the toxicity of numerous other chemical compounds including anesthetic gases (Biersner, 1972), agricultural chemicals (Rodnitzky et al., 1975), and metals (Roels et al., 1987).

The growing interest in the measurement of performance is most probably due to the sensitivity shown by these methods in unveiling changes in the human organism that otherwise would not be detected. By now, the evidence that these changes are some of the earliest indicators of the occurrence of health effects has become unequivocal. As a consequence, the measurement of performance has come to be regarded by many as a device of major importance for monitoring hazards to health and safety in the work environment. This development appears to agree with the ideas promulgated by the World Health Organization (WHO) that health does not mean only absence of disease but also optimum physical, mental, and social well-being and, moreover, that health means not only freedom from pain and disease but also freedom to maintain and develop one's functional capabilities.

At our institute, the measurement of performance has undoubtedly constituted the main method of studying the effects on the CNS of low dose exposure to the chemical substances that are frequently found in the work environment. Thus, we have applied performance tests of various kinds in experimental laboratory studies as well as in field studies and in cross-sectional epidemiological investigations. Through the years, we have made use of performance tests in experimental inhalation studies of industrial solvents such as toluene (Gamberale and Hultengren, 1972; Iregren, 1986a), methylchloroform (Gamberale and Hultengren, 1973), styrene (Gamberale and Hultengren 1974), white spirit (Gamberale et al., 1975a), methylene chloride (Gamberale et al., 1975b), trichloroethylene (Gamberale et al., 1976), xylene and ethylbenzene (Gamberale et al., 1978), methyl isobutyl ketone (Wigaeus-Hjelm et al., 1990), and toluene in combination with p-xylene (Anshelm Olson et al., 1985) or with ethanol (Iregren et al., 1986). Psychometric tests have been applied directly at the worksite in two investigations of workers in the plastic boat industry exposed to styrene (Gamberale et al., 1976b; Kjellberg et al., 1979), in studies of steelworkers (Anshelm Olson et al., 1981) and of workers in the paint industry (Anshelm Olson, 1982) exposed to solvent mixtures, and in two investigations of nurse anesthetists exposed to anesthetic gases (Gamberale and Svensson, 1974; Kjellberg and Strandberg, 1979). Furthermore, we have used comprehensive batteries of behavioral tests to investigate possible long-term effects of chronic exposure to organic solvents among car and industrial spray painters (Elofsson et al., 1980), workers in a jet motor factory (Knave et al., 1978), and rotogravure printers (Iregren, 1982).

Behavioral tests are also used to an ever-increasing extent to study the effects of work environmental conditions other than exposure to neurotoxic substances. Thus, unfavorable effects on performance of environmental factors—such as noise, vibration, cold, heat, electric and magnetic fields, and physical work load—have been demonstrated in laboratory experiments as well as in field studies.

Our experience with the use of psychometric tests to study the effects of nonchemical agents in the physical work environment is relatively modest. However, behavioral tests have been successfully applied at our laboratory in experiments on the effect on performance of exposure to noise (Kjellberg and Wide, 1988) and to moderate cold (Enander, 1987). A series of experimental studies on the effects of different climatic conditions on performance is now in progress (Gamberale et al., 1988b). Finally, we have used psychometric tests to investigate the possible effects on workers of acute (Gamberale et al., 1988a) and chronic exposure (Knave et al., 1979) to electric and magnetic fields.

HOW TO MEASURE PERFORMANCE?

A working group within the WHO has recommended (WHO, 1987) a battery of tests to use in the search for neurotoxic effects in working populations. The main criterion applied by the WHO in selecting the tests to be included in the battery was that the tests should have proven their sensitivity in empirical investigations.

To facilitate a widespread application of the methods, no tests that required complicated technical equipment for their administration were included in the battery. A further requirement was that the tests should be selected among those commonly used by the clinical practitioner for diagnostic purposes. In practice, these requirements limited the choice to the tests in the Wechsler Adult Intelligence Scale (WAIS battery). Thus, no attention was paid to tests developed especially for use in laboratory experiments and quasi-experimental field studies of the effects of exposure to neurotoxic substances. In our opinion the tribute paid to the clinical practitioner by selecting among traditional manual or paper-and-pencil tests has had a negative effect on the sensitivity of the WHO battery to detect neurotoxic effects.

Against this background it is difficult to understand why some groups working with the development of computerized tests feel the need to refer to this WHO list of tests as a rationale for their test implementations (Cassito, 1985; Letz and Baker, 1986). It is obvious that such a strategy leads to inadequate utilization of the possibilities offered by computerized testing.

Another problem should be considered when implementing existing traditional tests on computers, namely, that the correlations between the results obtained with the two versions of the tests (i.e., the computerized and the paper-and-pencil tests) often are quite low. This low correspondence may be due to several inevitable differences between the resulting test protocols with regard to stimulus presentation as well as response input. To mention one example, Beaumont

(1985) investigated the effects of various response modes on the results in a computerized Digit Span test. He found substantial differences between responses entered via the ordinary keyboard, an external keypad, and a touch sensitive screen.

Several of the testing systems applied within the area of neurotoxicology use fairly simple psychomotor tasks, and reaction time or response latencies are generally used as the outcome variables. It has been argued, however, that performance on more complex cognitive tasks should be more sensitive to disruption by exposure to toxicants. Still, in the experience at our laboratory, a test of Simple Reaction Time (SRT) has proved to be generally the most sensitive test.

The greater sensitivity demonstrated by the tests of relatively simple mental functions does not necessarily imply that these tests tap the CNS functions most vulnerable to neurotoxic substances. Instead this greater sensitivity may be due primarily to the higher reliability of these tests compared to tests measuring complex cognitive functions. A substantial contribution to the reliability of the tests of simple mental function stems from the fact that performance parameters in these tests are usually based on a large number of items. These circumstances concerning the sensitivity and reliability of different types of tests should be taken into consideration especially when analyzing results in terms of possible differential deficits (e.g., Chapman and Chapman, 1978).

Several groups have recently developed computer-based performance evaluation systems for use in neurotoxicology (e.g., Baker et al., 1985; Braconnier, 1985; Cassito, 1985; Eckerman et al., 1985; Iregren et al., 1985; Laursen and Jorgensen, 1985), and some laboratories use similar systems in related fields (e.g., Bittner et al., 1985; Irons and Rose, 1985). Furthermore, there are of course several computer-based systems that have been applied in clinical use, two examples of which are those of Acker (1983) and Beaumont and French (1987).

WHAT ELSE TO MEASURE?

The methods of value in early detection of neurotoxic effects or of the effects on the CNS of exposure to other unfavorable physical environmental agents include, besides performance tests, neurophysiological and neurological testing, as well as questionnaires for the assessment of subjective experience. In most investigations, a systematic collection of the subjects' experience when exposed to different experimental or occupational conditions may constitute an invaluable source of information. Most questionnaires for this type of assessment can easily be computerized with maintained reliability and validity (see, e.g., Carr et al., 1981; Lucas, 1977).

With the Swedish Performance Evaluation System (SPES), it is possible to collect three types of self-report data: (1) symptoms of acute as well as long-term exposure, (2) self-rating of mood, and (3) self-rating of performance. The first two types of data aim at the detection and description of possible environmentally induced changes in the subjects' perceptions of their physical and psychological states. The third type of perceptual data are motivational in character and refer to the subjects' motivation, confidence, and effort expended during testing.

PROS AND CONS OF AUTOMATED TESTING

Automated testing in general implies some advantages over traditional paper-and-pencil testing, The automation of tests gives

- an excellent opportunity for strict standardization of test procedures (e.g., instructions, test protocols, and evaluation variables), thus increasing the possibilities for comparisons across studies;
- the ability to perform detailed measurement and analysis of single responses or response components—this type of microanalysis has greatly enhanced the sensitivity of performance tests; and
- increased precision in the measurement procedure; by reducing the influence of the investigator, the reliability and the validity of the results are increased.

Because an automated test may be administered by a technician or a nurse, there is the possibility of reducing the work load of psychologists, who therefore can make better use of their skills. It has also been suggested that automation of the testing procedure would render the testing situation less threatening.

Fully computerized testing procedures provide some additional possibilities as well:

- Computers are flexible, and one system can be used to administer a variety of tests, as well as to perform other routine tasks in the laboratory or clinic.
- Computers facilitate prompt scoring and evaluation of even very complex tests and questionnaires.
- Computers make it possible to adapt the choice of items according to the performance capacity of the individual.
- Computers offer communication possibilities, making data transference for statistical computations or other purposes very convenient.
- Computers are transportable and, in the extreme case, even portable, thus making field testing feasible.

Some of the most frequently mentioned critical comments regarding computerized testing are (1) the supposedly poor rapport established between the subject and the machine; (2) the difficulties in testing large groups; (3) the static form of computerized approaches; (4) the restricted range of stimuli that can be presented; and (5) the restriction in the choice of response media.

With regard to the rapport established, investigations pertaining to this problem generally indicate no difficulties (see, e.g., Carr et al., 1981; French and Beaumont, 1987; Lucas, 1977; or Lukin et al., 1985). The "user friendliness" of a system is not dependent upon whether it is computerized or not, but on the careful design of the system (Hedl et al., 1973). As pointed out by Beaumont (1982), the single most important requirement for successful design is the predictability of the system. Thus, no action on the part of the subject should result in accidental termination of the test.

Due to the still relatively high purchase costs of even a microcomputer system, it is difficult to test large groups of people simultaneously. On the other hand, computerized tests often provide much more information than traditional tests within a specified time period. Thus, the ability to perform simultaneous testing of large groups is not as important with computerized tests.

The objection concerning the static form of computerized tests is by now invalid. There have been successful attempts at making "tailored" (i.e., adaptive) tests, where the test items administered are contingent upon the performance of the subject (see, e.g., Weiss and Vale, 1987). One of the memory tests available with our system, a version of the Digit Span test, functions in an adaptive way.

Until today, most efforts at the implementation of performance tests on computers have used visual stimuli, because the administration of auditory or tactile stimuli, for example, requires the use of rather complicated (probably custom-made) external equipment in addition to the computer. Such additions would naturally make it very difficult to standardize the tests to allow for widespread use.

The few response media available present a similar problem with respect to the development of tests. Because input to a computer is normally made via the keyboard, most testing systems presently use this medium. Attempts to use other means of input usually also imply manual performance, as for example, with joysticks or touch-sensitive screens. At present, the choice of response mode is probably the most limiting factor in the development of computerized tests because the response medium may easily affect test results in unintended ways. A fast and reliable system for processing speech input would in many instances provide the only acceptable solution.

The restrictions regarding stimulus presentation and response input are, of course, steadily diminishing because continuous technical developments make computers increasingly competent. However, standardized tests using new technical possibilities (e.g., speech input) are still several years off because the development of well-standardized tests is a laborious, long-term project.

In spite of the restrictions mentioned, efforts at developing computerized tests and testing systems are steadily increasing, and this type of assessment is currently in use in a wide variety of settings. Computerization has made complicated psychometric techniques available even to persons lacking training as professional psychologists. Therefore, this trend accentuates ethical demands for control of the construction, distribution, and use of these methods (Matarazzo, 1983). The American Psychological Association (APA, 1986) and the British Psychological Association (Bartram et al., 1987) have published guidelines for this purpose.

The experience gained at our laboratory in using computerized tests has generally been positive, although there are of course difficulties (Iregren et al., 1985). One significant problem, which applies equally to traditional tests, relates to the time and effort needed for successful test development. Furthermore, computers are still technically complicated machines, thus requiring special skills of the psychologists and technicians engaged in this development.

Several recent reviews have treated various aspects of computerized testing, and the reader is recommended those by Bartram and Bayliss (1984), McArthur and Choppin (1984), Space (1981), and Thompson and Wilson (1982).

DEVELOPMENTS IN SWEDEN

Since 1970, the Division of Psychophysiology, National Institute of Occupational Health (NIOH), Solna, Sweden, has been concerned with the development of psychometric methods suitable for the study of adverse effects of environmental stressors, primarily neurotoxic substances. The first test to be standardized for use in environmental research was an SRT test (Lisper and Kjellberg, 1972). At the start, this test was administered with an electronic apparatus, consisting of a timer, a tape recorder, and a stimulus/response panel. The test was then implemented on other types of testing equipment, and it has been used in most of our investigations. Special equipment, developed in 1973, was used solely in laboratory experiments. This apparatus comprised a paper tape controlled solenoid operated stimulus/response panel and a teletype printer, and it was used for the first time in an investigation of the toxicity of white spirit (Gamberale et al., 1975a). Besides the above mentioned SRT test, tests of Choice Reaction Time (CRT), numerical ability, and memory were implemented on this equipment.

A further step in the development of testing equipment was taken by the introduction of a new stimulus/response panel in 1975. Stimuli were presented on eight rows of 32-LED displays, each capable of showing any alphanumeric character. Responses were entered on a full QWERTY keyboard. This equipment, which made possible the presentation of more complex stimuli as well as the registration of written responses, was used to test different cognitive functions. It was used in several cross-sectional epidemiological investigations on workers exposed to industrial solvents and electromagnetic fields (Elofsson et al., 1980; Iregren, 1982; Knave et al., 1978, 1979). The major disadvantages of this type of equipment were the laborious programming procedure, the fragility of the paper tape, and the time-consuming evaluation of the results. For a review of similar attempts at noncomputerized automated testing, see Denner (1977).

Some of these disadvantages could be overcome when computers became more easily available. However, early computers had other shortcomings with respect to automated testing. They were expensive and difficult to program or handle, and the access via time-sharing terminals made exact timing of response latencies impossible.

With the advent of the microcomputer, new approaches became possible. For the first time, fully automated testing could be performed, with administration of instructions and test items, as well as response registration with precise timing of response latencies and data storing. New demands were made on our performance assessment methods by the acquisition of an exposure chamber, which required a fully automated procedure in the solvent inhalation studies. The equipment used in these experiments consisted of a computer with a black-and-white monitor, a dual disk drive, a modified numerical keyboard, a reaction time panel, and a printer. Three performance tests were used with this equipment. The previously used SRT test and a test of short-term memory were adapted to this computer system, and a new test of CRT was developed.

This system had several shortcomings owing to its technical limitations, e.g., poor picture quality due to low graphic resolution, and poor precision in the timing of response latencies. A small working memory and a slow basic language were also severely limiting factors with respect to test development. For a review of system requirements for computerized automated testing, see Beaumont (1982).

When a new generation of the same computer was introduced, the performance assessment system underwent further development. The new computer was equipped with high-resolution color graphics, a more flexible working memory, and a basic language that facilitated the construction of long sequences of tests. The requirement of a timing accuracy of at least 1 ms was met by using an external clock with program routines in Assembler language. Several of the previously used tests were implemented on this equipment (i.e., SRT, CRT, a memory test, and a test of numerical ability). Furthermore, new tests were developed for use with this computer, e.g., a Complex Reaction Time task using color words as stimuli.

The performance assessment system was further improved in 1984 by using a later version of the computer, which was equipped with greater working memory for graphics and a high-quality color monitor. The number of tests currently available on this system is 14. The SPES has now been transferred to IBM computers to facilitate its use by other research groups.

DESCRIPTION OF THE SWEDISH PERFORMANCE EVALUATION SYSTEM

The SPES consists of a number of semiautomated computerized performance tests and various scales for the subjective evaluation of performance on the tests, of mood, and of different kinds of symptoms. The system is designed to be dynamic and flexible. Thus, it allows the researcher or the practitioner to choose among the tests and the scales, adapting the battery to the specific purpose of the evaluation at hand. The system is also intended to undergo gradual improvement based on analyses of the results of ongoing empirical studies and on future experience with the use of the system.

With few exceptions, the performance tests are nonverbal, i.e., they can be used to assess performance independent of the language of the subjects. Some of the tests, e.g., the Color word vigilance (SPES3:1), the Color Word Stress (SPES3:2), and the Verbal Reasoning test (SPES10) can be easily adapted to other languages and will only require translation of the text files used by the programs. These text files, which are easily edited, contain all the verbal communication with the subject (i.e., instructions on how to perform the test as well as the test items). The only test that requires a completely new construction and standardization if used with non-Swedish speaking subjects is the Vocabulary test (SPES11).

Hardware

Any IBM or IBM-compatible PC, XT, or AT, equipped with an external clock card (SB11 multifunction card, Emulex Corp., 3545 Harbor

Blvd, P.O. Box 6725, Costa Mesa, CA 92626), an overlay to the keyboard, an EGA graphics card, a color monitor (for some of the tests), and an optional printer may be used. A hard disk is not necessary to run the test battery. However, because the full system is too large to be contained within a single diskette, a hard disk is recommended.

Software

The programs, which are available in compiled form on diskette, are written in TURBO Pascal. The system includes a master program referring to a number of different test programs, which can be combined to any preferred sequence. This sequence may in turn be repeated any number of times, according to the design of the investigation at hand. The graphic presentations within this system are developed with the aid of a graphic tools package TURBO PAINT TOOLS © 1986 DATABITEN/P.S. DATAKRAFT.

At present, the system consists of the 14 tests listed in Table 1, in addition to four scales for the measurement of reported mood, symptoms (two scales), and self-rated performance. The tests are Simple, Choice, and Complex Reaction Time (four tests); Search and Memory Test; Symbol Digit; Digit Span; Logical Reasoning; Additions; Finger Tapping (two tests); Vocabulary; Digit Classification; and Digit Addition. A short description of each test may be found in the appendix to this chapter. Anyone interested in further information about the SPES system should contact Anders Iregren, who is responsible for the system development and the distribution of SPES software.

EMPIRICAL BACKGROUND AND APPLICATIONS

Table 2 lists the investigations that have been performed so far by using SPES tests. These include experimental studies in the laboratory, occupational studies of effects from acute or long-term exposure to various agents, two investigations applying SPES tests in clinics of occupational medicine, and two studies directly aimed at the methodological evaluation of the tests.

Standardization Study

A standardization sample of 100 subjects went through SPES1, 2, 3:1, 4, 5, 6, 7, 10, and 12:1 (Kjellberg and Wisung, 1987). A large proportion of the subjects (i.e., 38 persons) were university students, and 62 were employees of NIOH. For 59 of the latter, testing was repeated four to five months later.

TABLE 1 Tests and Scales in the SPES

SPES Code	Performance Tests	No. of Items (+ Practice)	Parameters Extracted	Approximate Time (min)
1	Simple Reaction Time	80 + 16	Mean, SD, decrement	6
2	Choice Reaction Time	112 + 32	Mean, no. of errors	9
3:1	Color Word Vigilance	192 + 16	Mean RT, no. of commissions, no. of omissions	8
3:2	Color Word Stress	192 + 16	Mean RT, no. of commissions, no. of omissions	7
4	Search and Memory	3* (10 + 1)	Mean RT/search level	10
5	Symbol Digit	6 + 4	Mean RT, no. of errors	6
6	Digit Span	Varies	Length of memory span	6
7	Additions	40 (+ 3)	Mean RT, no. of erros	6
8	Digit Classification	240	Mean RT, no. of errors, no. of lags	4
9	Digit Addition	120	Mean RT, no. of errors, no. of lags	7
10	Verbal Reasoning	64	Mean RT, no. of errors	8
11	Vocabulary	45	No. of correct answers	8
12:1	Finger Tapping Speed	2* (3 + 1)	Mean no. of taps/hand	5
12:2	Finger Tapping Endurance	1	Changes of movement time and resting time over test time	3

SPES Code	Self-Rating Scales	No. of Items	Paramters Extracted	Approximate Time (min)
30	Performance	1/test	Percent of maximum performance	1
31	Mood	12	Activity score and stress score	3
32	Acute symptoms	17	No. of symptoms reported	4
33	Long-term symtpoms	38	No. of symptoms reported	6

NOTE: SD = standard deviation; RT = reaction time.

TABLE 2 Studies in Which the SPES Tests and Scales Have Been Used

Authors	Environmental Agent	SPES Code	Number of Measurements	N exposed (+N control)
Experimental Studies				
Lisper and Kjellberg (1972)	Sleep deprivation	(SRT)	2	8
Gamberale and Hultengren (1972)	Organic solvents	(SRT)	4	12
Gamberale and Hultengren (1973)	Organic solvents	(SRT)	4	12
Gamberale and Hultengren (1974)	Organic solvents	(SRT)	4	12
Gamberale et al. (1975)	Organic solvents	(SRT, RTadd)	4	14
Gamberale et al. (1975a, b)	Organic solvents	(SRT, RTadd)	4	14
Gamberale et al. (1976a)	Organic solvents	(SRT, RTadd)	3	15
Gamberale et al. (1978)	Organic solvents	1,(RTadd)	3	15+8
Anshelm Olson et al. (1985)	Organic solvents	1,2	12	16
Iregren et al. (1986)	Organic solvents	1,2,3:1	12	12
Iregren (1986)	Organic solvents	1,2,3:1,32,33	6	26
Enander (1987) (exp 1)	Cold exposure	1,3:1	2/4	12
Emander (1987) (exp 2)	Cold exposure	3:2,8,9	2	12
Kjellberg and Wide (1988)	Noise exposure	10	1	26
Gamberale et al. (1988b)	Heat and cold	1,3:1,3:2,4,8, 10,12,30,31	3	24
Wigaeus-Hjelm et al. (1990)	Organic solvents	1,7,31,32	8	8

Occupation Studies			
Gamberale and Svensson (1974)	Narcose gases	(SRT)	2	20 (+20)
Kjellberg and Strandberg (1979)	Narcose gases	1	2	14 (+14)
Gamberale et al. (1976)	Organic solvents	1	2	106 (+36)
Kjellberg et al. (1979)	Organic solvents	1,7	6	7 (+7)
Anshelm Olson et al. (1981)	Organic solvents	1	6	42
Anshelm Olson (1982)	Organic solvents	1	2	47 (+47)
Gamberale et al. (1988a)	Electric and magnetic fields	1,3:1,5,6	4	24
Knave et al. (1978)	Organic solvents	1,7	1	30 (+30)
Knave et al. (1979)	Electric and magnetic fields	1,7	1	53 (+53)
Elofsson et al. (1980)	Organic solvents	1,7	1	80 (+80)
Iregren (1982)	Organic solvents	1,7	1	34 + 34 (+34)
Clinical Studies			
Hagberg and Iregren (1984)	Organic solvents	1	1	51 (+27)
Iregren et al. (1987)	Organic solvents	1,2,3:1,5,6,10	1	148
Methodological Studies			
Soderman et al. (1982)	Methodological	1	Varies	730
Kjellberg and Wisung (1987)	Methodological	1,2,3:1,4,5,6,7,20,12	2	100

NOTE: Codes in parentheses refer to early, but similar, versions of the tests and scales.

The aim of this study was to assess (1) the reliability of the tests (test-retest as well as homogeneity); (2) the factor analytic structure of the tests; (3) learning and fatigue effects; and (4) sex and age differences, as well as differences among educational levels. Because the educational level of the majority of this group was high, the results could not be treated as norm data, at least not for the tests in which educational level proved to yield significant differences.

Simple Reaction Time Standardization

Data from previous field and epidemiologic studies using the SRT task were reanalyzed for standardization purposes (Soderman et al., 1982). The sample consisted of 730 industrial workers, 306 of whom were exposed to solvents in their work. For 83 of the nonexposed and 149 of the exposed workers, measurements were made twice, before and after a work shift. Numerous different performance indices were assessed with respect to their power to discriminate among exposed and nonexposed workers, age groups, and morning-afternoon performance.

Clinical Validation of the Battery

Six SPES tests (i.e., SPES1, 2, 3:1, 5, 6, and 10) were used as a complement to traditional tests for diagnosing occupational illness due to the chronic effects of long-term exposure to organic solvents (Iregren et al., 1987). A total of 148 cases with suspected solvent-induced illness were tested at four Swedish clinics of occupational medicine over 15 months. The aim of this study was to investigate whether these computerized tests were useful in a clinical situation, from a practical as well as from a psychometric point of view.

Clinical Trial with SPES1

During one year, the SRT test was administered to 51 consecutive patients referred to a clinic of occupational medicine with suspected solvent-induced occupational illness (Hagberg and Iregren, 1984). The performance of these patients was compared to that of a control group.

Other Studies

Many of the tests have been used in several experimental and field studies, as well as in epidemiological investigations, conducted at the Institute of Occupational Health to assess the effects of different environmental factors (see Table 2). Data from these studies have been used to assess the sensitivity and, in some cases, the reliability of the tests. Furthermore, in some cases the validity of the test is supported by research conducted with similar tests at other research institutions.

RESULTS

Standardization Study

Table 3 shows means and standard deviations over repeated measurements of the different performance indices for each of the nine tests included in the standardization study. The mean response times might be viewed as indicators of the degree of difficulty of the tests. Thus, Tapping stands out as the easiest test (a frequency of 66.5/10 s corresponds to a response time of 150 ms), followed by Simple RT, Search and Memory (mean time for each of the 30 letters), Choice RT, Color Test Vigilance, Additions, Symbol Digit, and Reasoning. No response times were recorded in Digit Span.

Reliability Coefficients

Test-retest reliability coefficients and alpha coefficients for the tests are also given in Table 3. The test-retest coefficient shows to what extent the performance level is a stable characteristic of the individual, whereas the alpha coefficient reflects the precision of the individual measurement. The alpha coefficient could thus be viewed as an estimate of the test-retest coefficient, given that the measurement is repeated under identical circumstances. Mean response times in Choice RT, Color Test Vigilance, Additions, and Simple RT all proved to be highly reliable. The test-retest coefficients of Search and Memory, Symbol Digit, Digit Span and Reasoning were all below 0.80. As expected, response times were found to be more reliable than error rates, among which the error rate in Symbol Digit is notable as a very unreliable performance indicator.

Sex Differences

The men had a higher mean educational level, and sex differences therefore were analyzed within the group with academic education. This group contained 31 men and 30 women with a similar age distribution. The women tended to make somewhat more errors in the Choice RT test, to give fewer correct answers in the Reasoning test,

TABLE 3 Means and Standard Deviations for Tests in the Two Test Sessions ($N = 59$)

Test	Session 1 Mean	Session 1 SD	Session 2 Mean	Session 2 SD	t	p	r_{tt}	Alpha
1. Simple RT								
Mean (ms)	257.8	42.0	258.8	37.3	0.34	>0.10	.85	.99
SD	54.4	19.7	54.6	18.9	0.90	>0.10	.63	
Regression	0.22	0.21	0.15	0.19	2.45	0.012	.44	
2. Choice RT								
RT mean (ms)	644.2	109.9	635.6	116.8	1.17	>0.10	.88	.99
Errors	3.2	5.7	3.1	4.5	0.14	>0.10	-.02	.87
3. Color Word Vigilance								
RT hits (ms)	535.2	53.4	539.8	57.6	1.39	>0.10	.90	.96
No. correct	41.1	8.0	41.8	8.6	0.91	>0.10	.74	.91
4. Search and Memory								
RT 1 letter (s)	9.2	2.6	9.0	2.4	1.13	>0.10	.75	.94
RT 2 letters (s)	14.6	3.9	13.3	3.5	3.40	0.001	.71	.95
5. Symbol Digit								
RT 5-10 (s)	25.3	6.5	28.2	8.8	3.74	<0.001	.74	.95
No. errors	0.7	1.2	0.5	1.1	1.15	>0.10	.15	.58
6. Digit Span								
Median span	7.1	1.3	7.4	1.5	1.81	0.076	.70	
7. Reasoning								
RT mean (s)	4.6	1.4	4.6	1.5	0.61	>0.10	.76	.96
No. errors	13.5	8.5	11.0	8.5	2.97	0.004	.70	.91
8. RT additions								
RT mean (s)	2.0	0.9	2.1	0.9	1.66	>0.10	.94	.98
No. errors	3.4	2.6	3.2	2.8	0.64	>0.10	.60	.87
9. Tapping								
Frequency dominant	66.5	8.8	64.9	10.6	2.05	0.045	.82	.91
Frequency Nondominant	58.5	9.0	57.7	10.7	0.90	>0.10	.74	.85

NOTE: Test-retest and alpha coefficients (KR20 for dichotomized variables) ($N = 59$ and 100, respectively). RT = reaction time.
SOURCE: Data from Kjellberg and Wisung (1987).

and to have a lower tapping frequency in the dominant hand. However, only the difference in the Tapping test was significant at the 0.05 level.

Age Differences

To minimize the influence of the effects of sex and education, age effects were also tested in the group with academic education. This group was divided into three ages (29 years or younger, 30–39 years, and 40 years or older). Age differences in response times were tested with one-way analyses of variance, whereas error rates were tested with extended median tests. Significantly longer response times were obtained in the oldest age group for Choice RT and Symbol Digit. A similar tendency ($p < 0.10$) was found for Color Test Vigilance and Tapping (dominant hand). Error rates were very similar in the three groups, indicating that the prolonged response times did not simply reflect a more cautious strategy in the older group.

Educational Level

The composition of the standardization sample did not permit a division into groups with different educational level and similar age and sex distribution. To gain an impression of the importance of educational level, the differences between groups at three educational levels (no academic education, lower academic education, doctoral degree) were analyzed with analyses of covariance using sex and age as covariates. No differences between educational groups were obtained in Simple RT, Color Word Vigilance, Choice RT, or Tapping. The group with the highest educational level performed significantly better than the other two with respect to response times in the Search and Memory test ($p = 0.011$), response times in Reasoning ($p = 0.004$), and length of the Digit Span ($p = 0.014$). The group with the lowest educational level made more errors in the Search and Memory test than the other two groups ($p = 0.0006$). Response times in Symbol Digit and Additions and the number of errors in Reasoning were successively lowered with higher educational level.

Other Individual Differences

No performance differences were obtained between groups who were more or less experienced with work on computers or between subjects who had or had not participated in experiments using computerized tests.

Fatigue and Learning Effects

Training effects and other effects of increased experience with the tests were analyzed both within the tests and between the repeated testings. Within-test changes were analyzed by computing mean response times or errors for successive comparable periods. One-way analyses of variance were performed to test the significance of changes as a result of time on task. The conservative estimate of the level of significance recommended by Greenhouse and Geisser (1959) was used in these tests. The changes between repeated testings were analyzed in the group of 59 subjects who performed the tests twice and were tested with t-tests (Table 3).

In most tests no significant changes occurred after the prescribed training period. A significant training effect was obtained only in Additions, where response times were shortened from the first to the second half of the test ($p = 0.00001$). There was also a tendency for errors to diminish between the two halves of the Verbal Reasoning test ($p = 0.06$). The opposite effect was obtained in Simple RT where response times were gradually prolonged ($p < 0.00001$), being 17 ms longer during the fifth than during the first minute. Similarly, in Color Word Vigilance (CWV), response times were fairly stable until the last minute of the test when they were significantly prolonged ($p = 0.018$).

As shown in Table 3, performance improved between the two sessions in Search and Memory (two letters) and Reasoning (number of errors). The performance decrement in Simple RT was also less in the second session. Surprisingly, response times in Symbol Digit were prolonged in the second session.

Factor Analysis of Response Time

A factor analysis was performed on response latency data from all the tests except Digit Span. The model used was a principal factors analysis with an orthogonal varimax rotation (the resulting factor loading matrix is given in Table 4).

A two-factor solution was chosen, although the second factor was rather weak. Simple RT and Choice RT had the highest loadings in the first factor. Tapping also had a much higher loading in this factor than in the second one. Thus the first factor seemed to represent motor response speed. Search and Memory (SAM) had the highest loading in the second factor, but both Reasoning and RT Additions had higher loadings in this factor than in the first one. Thus, this factor seems to represent decision processes that require more com-

TABLE 4 Factor Analysis of Response Times: Factor Loading Matrix of an Orthogonal Two-Factor Solution

Test	Factor I	Factor II	h^2
Simple RT	.74	.06	.56
Choice RT	.76	.33	.69
CWV	.52	.52	.54
SAM	.21	.83	.74
Symbol Digit	.45	.48	.44
RT additions	.05	.44	.20
Reasoning	.24	.48	.29
Tapping	-.57	-.19	.36
Eigenvalue	3.11	.70	

SOURCE: Data from Kjellberg and Wisung (1987).

plex information processing. However, it should be noted that both RT Additions and Reasoning had very low commonalities. Thus, the response times in the two tests that required the most advanced information processing primarily reflected factors not common with the other tests.

Correlations between the accuracy scores were much lower than between response latencies, primarily as a result of the small variation in accuracy measures in most tests. A factor analysis of these scores therefore did not yield any interpretable factor structure.

Other Data on the Stability of Performance Measures

The stability of performance on the tests can also be evaluated by analyses of data from the experiments by using repeated measurement designs. Table 5 shows the correlation coefficients between successive measurements using Simple RT, Choice RT, and Color Test Vigilance in an experimental study with exposure to toluene in combination with ethanol ingestion (Iregren et al., 1986). A total of 12 subjects were tested three times during each of four sessions, which were separated by intervals of two weeks.

Table 6 shows the consecutive correlations for data from Simple RT and Choice RT obtained in another experiment, evaluating the acute effects of toluene exposure on a sample of 26 spray painters (Iregren, 1986). In this study the subjects were tested in two sessions separated by one week, and the tests were repeated three times within each session.

TABLE 5 Correlation Coefficients for the Relation Between Successive Measurements with Three SPES Tests in an Experimental Study of the Effects from Toluene Exposure and Ethanol Ingestion (decimal points are omitted)

Day	I			II			III			IV	
Testing occasions	1–2	2–3	3–1	1–2	2–3	3–1	1–2	2–3	3–1	1–2	2–3
SRT mean RT	63	97	86	74	97	77	93	91	93	87	91
CRT mean RT	91	90	80	85	83	73	94	73	67	81	75
CWV mean RT	47	79	74	78	92	85	88	91	78	94	91
CWV hits	86	94	80	67	69	74	83	79	93	86	87

SOURCE: Data from Iregren et al. (1986).

TABLE 6 Correlation Coefficients for the Relation Between Successive Measurements with Two SPES Tests in an Experimental Study of Acute Effects from Toluene Exposure in a Group of Spray Painters (decimal points are omitted)

Day	I			II	
Testing occasion	1–2	2–3	3–1	1–2	2–3
SRT mean RT	79	70	70	82	91
CRT mean RT	84	90	91	90	89

SOURCE: Data from Iregren (1986).

Performance data from consecutive sessions in a field experiment on the acute effects of exposure to electric and magnetic fields (Gamberale et al., 1988) are presented in Table 7, from which it can be seen that there are effects of learning as well as of time of day. In this study, Simple RT, Color Word Vigilance, Symbol Digit, and Digit Span were administered in the morning and in the afternoon of two consecutive days to a sample of 24 workers in the electrical industry. Table 8 presents the correlations between consecutive measurements in this study.

In evaluating the data presented in Tables 5, 6, and 8, one should bear in mind the small group sizes ($N = 12$, 26, and 24, respectively) as well as the relative homogeneity of the groups. Both these factors contribute to a restricted range and thereby a reduced variance within the groups, which restricts the possibilities of obtaining high correla-

TABLE 7 Performance on Four Tests in the Field Experiment of Acute Effects of Electric and Magnetic Fields

		Day 1		Day 2	
		a.m.	p.m.	a.m.	p.m.
Simple Reaction Time					
Reaction time (ms)	Mean	253	245	252	242
	SD	29	28	34	29
Variation	Mean	54	52	52	51
	SD	15	15	16	18
Decrement	Mean	16	10	10	12
	SD	21	15	20	14
Color Word Vigilance					
Reaction time (ms)	Mean	540	517	519	501
	SD	45	44	45	39
No. correct	Mean	42.2	45.5	45.3	46.8
Symbol Digit					
Reaction time (ms)	Mean	29.9	25.5	24.9	23.5
	SD	5.8	4.5	4.7	3.5
Digit Span					
No. correct	Mean	7.0	7.4	7.8	8.1
digits	SD	1.2	1.2	1.0	1.4

SOURCE: Data from Gamberale et al. (1988a).

TABLE 8 Pearson Correlation Coefficients for the Correspondence Between Successive Measurements Using Four SPES Tests in a Field Study of Acute Effects of Exposure to Electric and Magnetic Fields (decimal points are omitted)

Day	I		II
Testing occasion	1–2	2–1	1–2
SRT mean RT	89	92	84
CWV mean RT	68	67	50
Symbol Digit RT	86	89	80
Digit Span length	63	61	75

SOURCE: Data from Gamberale et al. (1988a).

tion coefficients. Furthermore, the correlations presented in Table 5 are also lowered by the effects of ethanol in the study. If these facts are considered, the presented data are impressive, and especially for Simple RT, Choice RT, and Symbol Digit they indicate suitability for use with repeated measurements designs.

Simple Reaction Time Standardization

Soderman et al. (1982) give a detailed report of the analyses made of the simple RT data collected in a group of 730 workers.

Reaction Time Distribution

The distribution of RTs is given in Figure 1. The figure is based upon the 67,840 RTs obtained from the 424 workers who were not exposed to industrial solvents. As expected, the distribution is skewed positively, although not extremely so.

Effects of Time on Task

The RT is gradually prolonged in a way similar to that found in the main standardization study. However, the decrement is somewhat smaller, and the mean RT is about 20 ms longer than found in that study.

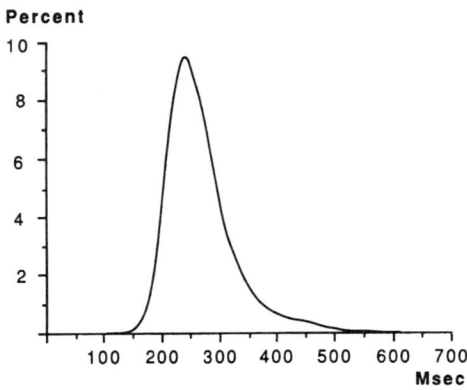

FIGURE 1 Distribution of reaction times in SPES1, Simple Reaction Time (N = 67,840; 424 individuals × 160 RT).

SOURCE: Data from Söderman et al. (1982).

Discriminatory Power of Different Performance Indices

Several statistical indices of reaction time performance were evaluated with respect to their statistical characteristics in a simulated lognormal distribution. Among other things, these analyses showed that all skewness measures gave very unsatisfactory estimates of population values. On the basis of these analyses, seven indices were selected for further evaluation by using the data from the worker group. The indices were the mean (X) and the median (Md) reaction time RT, the standard deviation (s) and the semiinterquartile range (Q), and the regression of the RTs as a function of time on task (Klin). Two measures of the zero point of the RT distribution were also computed but were found to be unsatisfactory. The discriminatory power of the indices was evaluated with respect to age, time of day, and occupational exposure to solvents. In tests of the effects of age and time of day, only data from the unexposed group were used. To obtain a more detailed analysis of the effects on the RT distribution, the RTs of each person were sorted from the slowest to the fastest of the 160 RTs. The discriminatory power of each of these 160 steps in the RT distribution was tested.

Age effects were tested by classification of subjects into three age groups, the oldest group being about 5 years older than in the standardization study. The regression did not show any age effect, whereas the measures of variation and central tendency all discriminated about equally well between the age groups. Figure 2A shows that the differences between the age groups are largest in the upper part of the RT distribution. However, as is evident from Figure 2B, the discriminatory power peaks around the 90th percentile.

Analyses of the data from the 83 subjects who performed the task both in the morning and in the afternoon did not yield a significant effect of time of day in any of the measures. However, the most consistent difference between the two times of testing was found for means and medians ($p = 0.12$) and Klin ($p = 0.11$). In all the measures, afternoon performance tended to be better than morning performance. Figure 3A shows that the difference between morning and afternoon was about the same in the whole RT distribution, with the exception of the lowest and highest percentiles. The discriminatory power (Figure 3B) peaked at the 10th percentile.

Corresponding data for the exposed and unexposed workers are given in Figures 4A and B. The mean, median, and standard deviation were all significantly larger in the exposed group. No consistent effect was found in Klin. Figure 4A shows that the difference be-

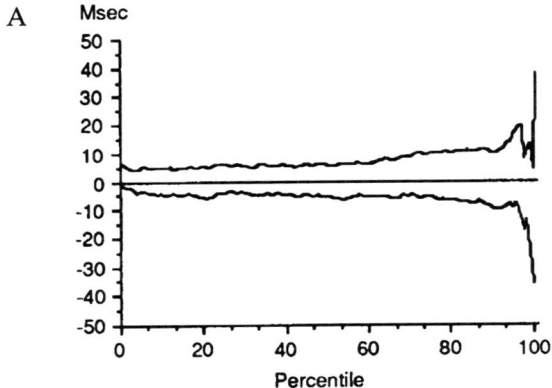

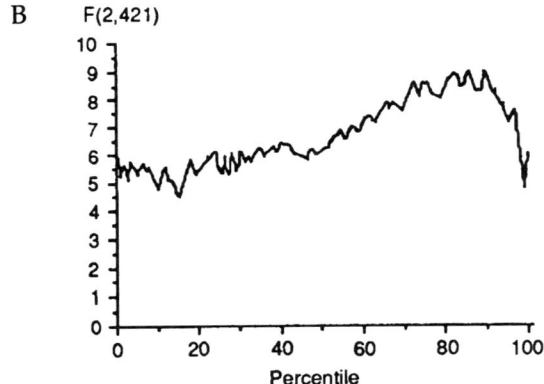

FIGURES 2A and 2B. Mean differences between three age groups in different parts of the reaction time distribution in SPES1 and the significance level for these differences. The lower curve shows the difference between the youngest group (–35 years, N = 129) and the middle age group (36–45 years, N = 97). The upper curve shows the difference between the oldest group (46–65 years, N = 196) and the middle age group. The figure on the right shows the F-values for the differences between three age groups.

SOURCE: Data from Söderman et al. (1982).

tween the two groups was most pronounced in the upper end of the RT distribution. The discriminatory power, however, decreased gradually from the lowest to the highest percentiles. Thus, the parametric measures (mean and standard deviation) generally fared better than their nonparametric counterparts. Given the fact that the maximum discrimination was found in different parts of the RT distribution for

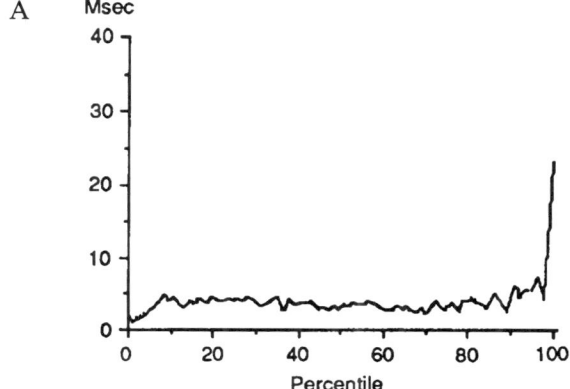

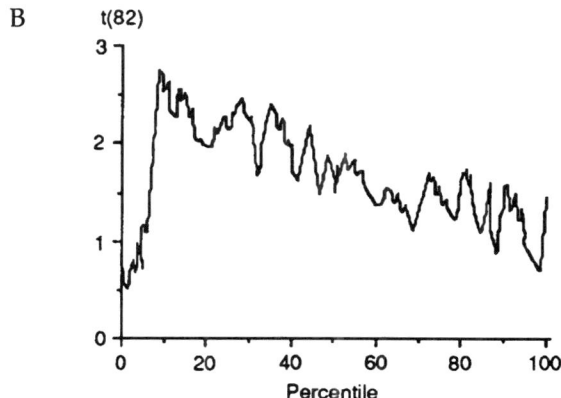

FIGURES 3A and 3B. Mean differences between morning and afternoon measurements in a group of workers not exposed to solvents ($N = 83$) in different parts of the reaction time distribution in SPES1; t-values for the differences are shown in the figure on the right.

SOURCE: Data from Söderman et al. (1982).

the three effects tested, it also seems wise to choose measures that are affected by the whole distribution. In spite of the fact that Klin showed no consistent effect of either age or solvent exposure, it might be worthwhile computing this measure because it was the one that gave the most consistent difference (although insignificant) between morning and afternoon performance.

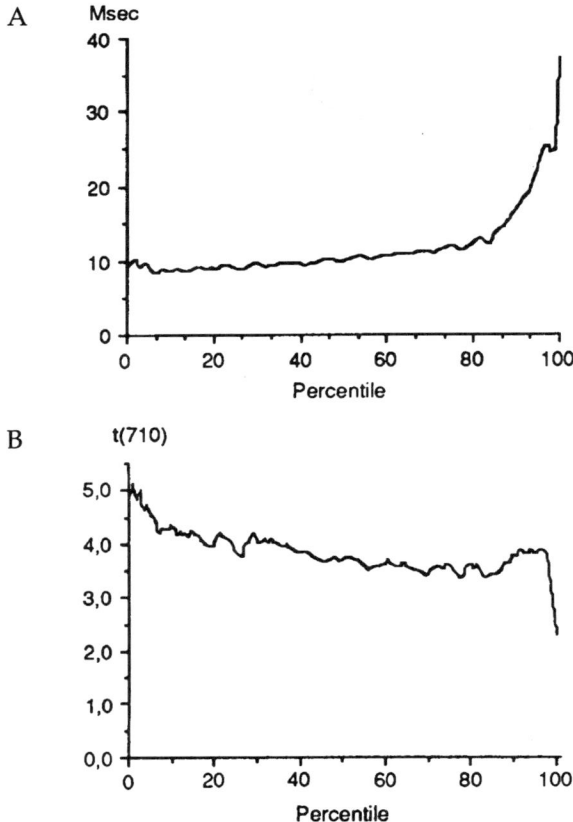

FIGURES 4A and 4B. Mean differences between solvent exposed ($N = 292$) and nonexposed workers ($N = 424$) in different parts of the reaction time distribution in SPES1; t-values for the differences are shown in the figure on the right.

SOURCE: Data from Söderman et al. (1982).

CLINICAL VALIDATION STUDY

General Results

The test battery proved to be simple to use in clinics, and neither psychologists nor patients had any difficulty in utilizing the tests or the equipment. Furthermore, results indicated that computerized tests predicted the diagnosis slightly better than traditional tests.

Descriptive Statistics

Mean values for the different diagnostic groups on the various performance measures are presented in Table 9. The p-values for the group differences from a one-way analysis of variance are also given.

Discriminatory Power

The predictive power of the computerized tests on the diagnosis was tested with a multiple regression analysis. The multiple correlation coefficients ranged from 0.54 to 0.81 for the three psychologists involved. Fairly low though they are, these correlations are still slightly higher than those obtained between traditional tests and diagnosis.

CLINICAL TRIAL WITH SPESL

Table 10 shows performance on the Simple Reaction Time test for the four diagnostic subgroups and the control group. The positive

TABLE 9 Mean Values for Various Performance Measures in the Diagnostic Groups and p Values for Group Differences

Test/Variable	Solvent-Induced Illness			p
	No	Possibly	Yes	
Simple RT				
Mean	333	446	475	0.039
Standard deviation	80	125	139	0.004
Choice RT				
Mean	958	983	1,092	0.003
Color Word Vigilance				
Mean	641	705	710	0.003
No. of misses	6.9	12.5	14.4	0.081
No. of alarms	8.1	7.2	8.6	0.836
Symbol Digit				
Mean	45.7	50.4	55.9	0.142
Estimated RT	37.6	44.1	50.5	0.033
Errors	0.83	0.66	0.79	0.931
Digit Span				
50% level	6.1	5.7	5.1	0.002
Reasoning				
Mean RT	7.9	8.4	7.7	0.723
No. correct	45.7	42.0	41.0	0.510

SOURCE: Data from Iregren et al. (1987).

TABLE 10 Mean and Standard Deviation for Performance on the Simple Reaction Time Test, Group Size and Age for Various Groups in Clinical Try-out of SPES1

Diagnosed Group	Reation Time Mean (SD)	Variability	N	Age Mean (years)	SD
Solvent-induced illness	543 (206)	122	5	48	12
Possibly solvent-induced illness	388 (135)	104	17	51	8
Psychiatric illness	315 (105)	73	10	44	12
Other diagnoses (e.g., low back pain)	268 (49)	59	19	46	14
Control group	242 (25)	46	27	39	7

SOURCE: Data from Hagberg and Iregren (1984).

predicted value for the SRT on this diagnosis is 56 and the negative predicted value is 79, given a cut-off limit from a 99 confidence interval derived from the control group data.

POSSIBLE FUTURE DEVELOPMENT OF SPES

The increasing technical competence of computers will certainly broaden the range of mental abilities that can be tested. However, due to the laborious procedure of test development, well-standardized and validated new tests are still a few years off.

The current tests, which already have provided much useful information about the neurotoxic effects of many substances, will be applied in closer collaboration with representatives from other disciplines. Thus, we will be able to relate performance data to increasingly precise measurements of the exposure to toxic substances, as well as to more sophisticated physiological and neurochemical effect measures. In the long run, this development will increase our understanding of the biological mechanisms behind the functional changes that we observe and will provide still better validation for the performance measures.

However, the immense variety of performance measures in use is probably the single factor that at present has the greatest effect on the rate of growth of knowledge. The possibilities of making comparisons across studies performed at separate laboratories and in different countries are of major importance, and initiatives to facilitate the standardization of computerized tests have been taken within the

European Economic Community. One significant problem in this process is the slightly different primary uses of various tests and test systems, because the intended use of a test naturally affects the way in which it is implemented. However, the development of standardized test protocols is now in progress, and efforts to accomplish this task have been made at our laboratory as well as elsewhere. This volume is a good example of the present strivings.

APPENDIX

Description of Performance Tasks

Simple Reaction Time SPES1 is a sustained attention task measuring response speed to an easily discriminated but temporally uncertain visual signal. The task is to press a key on the keyboard as quickly as possible when a red square is presented on the display. A total of 96 stimuli are administered during 6 min at intervals varying between 2.5 and 5.0 s. The first minute serves as practice, after which performance capacity is assessed for 5 min.

Choice Reaction Time SPES2 is a four-choice RT task similar to SPES1 with the addition of response selection requirements, The stimuli consist of crosses displayed one at a time on the screen. One arm of the cross is always shorter, and the task is to indicate on one of four keys, placed in analogy to the arms of the cross, which arm is the shorter. A total of 144 stimuli are presented at the same intervals as in SRT SPES1, and the first two minutes are excluded as practice trials.

Color Word Vigilance SPES3:1 is a Choice Reaction Time task in which response selection is based on a more complex signal characteristic than in SPES2. It is a task of vigilance type since a response is required only to a minority of the signals. The Swedish word for "red," "yellow," "white," or "blue" (all three-letter words) is presented on the screen. The text can be written in any one of the colors. The task is to press a key as rapidly as possible when there is congruency between the meaning of the word and the color of the text. The interval between consecutive stimuli is 2.2 s, and the 16 possible combinations of words and color are randomly distributed within each sequence of 16 stimuli. Thus, the proportion of critical stimuli is 25 percent. A total of 256 items are presented, and the first 16 are regarded as practice trials.

Color Word Stress SPES3:2 is a version of SPES3 which is constructed to provoke false alarms, and thus primarily measures the ability to inhibit such responses. The stimuli are the same, but the interval between subsequent stimuli is decreased to 1.5 s, and the proportion of critical stimuli has been increased from 25 to 75 percent.

Search and Memory SPES4 measures the speed of comparing stimuli shown on the screen with a set of stimuli retained in memory. One, two, or three letters are presented on the screen for 1, 2, or 3 s, respectively. The task is to reproduce the letters on the keyboard after their disappearance. Following a successful reproduction, a row of 30 letters is presented. The task is to search this row as fast as possible for the occurrence of any of the critical letters, and each appearance is indicated by pressing a key. There may be anything from 0 to 3 critical letters in each row. Altogether there are 33 items, 11 for each number of search letters. The first trial at each level is regarded as practice.

Symbol Digit SPES5 is a revised version of a traditional test of perceptual speed. In one row, a key to this coding task is given by the pairing of symbols with the randomly arranged digits 1 to 9. The task is to key in as fast as possible the digits corresponding to the symbols presented in random order in a second row. Each item consists of nine pairs of randomly arranged symbols and digits, and a total of ten items are presented in all. Performance is evaluated for the last six items of the test.

Digit Span SPES6 is a traditional test of short-term memory capacity. Series of digits are presented on the screen. The digits are presented one at a time with a 1-s presentation time, and the task is to reproduce the series on the keyboard. Depending on the correctness of the answer, the length of the following series is either increased or decreased. The test starts with a series of three digits and is terminated after six changes from a correct to an incorrect answer.

Additions SPES7 measures speed of simple mental arithmetic operations. An addition task comprising three horizontally placed digits is presented on the screen for 1 s. The task is to add the digits as quickly as possible and to indicate the sum on the keyboard. The test includes a total of 43 items.

Digit Classification SPES8 is a continuous CRT task. Digits ranging from 1 to 8 are presented one at a time on the screen. The task is to determine whether the digit presented is odd or even and to respond by pressing one of two appropriately marked keys. As soon as a response is given a new digit appears, and 240 digits are presented in all.

Digit Addition SPES9 is a version of SPES8 requiring more complex processing of the signals. The digits are presented one at a time on the screen for 1.5 s at intervals of 1.8 s. The task is to add the digit currently presented to the previous digit and determine whether this sum is odd or even. The response is given by pressing one of two appropriately marked keys.

Verbal Reasoning SPES10 measures the speed and accuracy of verbal reasoning. Sentences of varying syntactic complexity are presented on the screen. Each sentence describes a relation between the letters A and B, and it is followed by a combination of these letters. The task is to indicate with one of two keys whether the sentence gives a correct description of the relation between the letters A and B. There are 32 different items in a random series which is repeated twice.

Vocabulary SPES11 is a traditional test of verbal understanding. The task is to indicate which of five alternatives is the synonym of a key word. A total of 45 items—15 nouns, 15 verbs, and 15 adjectives—are presented. The words have been selected from a 102-item vocabulary test which was distributed as a paper-and-pencil test to 164 subjects with varying educational background. The selection of words was made with the primary aim of achieving discriminatory power in a low-education group. The words are presented in ascending order of difficulty.

Finger Tapping Speed SPES12:1 measures the maximum rate of repetitive movement. The task is to tap as rapidly as possible on a key at the keyboard with the index finger. The forearm is kept in a fixed position at the table. Eight 10-s trials, with a forced interval of 15 s, are performed while alternating between the preferred and nonpreferred hand. Four trials are given with each hand, and the first trial with each hand is regarded as a practice trial.

Finger Tapping Endurance SPES12:2 is a version of SPES12 in which the change in tapping rate over time is assessed. The task is to tap as rapidly as possible with the index finger on a key. A 1-min trial is performed with the dominant hand, and for each single tap the movement time and the resting time are registered separately. Performance is evaluated with respect to level and to changes over time.

Description of the Self-Rating Scales

Self-Rating of Performance SPES30. Within the system, it is possible to let the subject rate his performance directly after each test. In the standard version, the subject is asked to rate his actual performance in percent of his maximum performance. The question could, however, easily be rephrased.

Self-Rating of Mood SPES31. The scale consists of 12 mood-descriptive adjectives coupled to a six-category response scale. The response categories have verbal labels ranging from "not at all" to "very much." Ratings are given by typing the number of the appropriate response alternative. The questionnaire is based on two more comprehensive Swedish mood adjective check lists (Kjellberg and Bohlin, 1974; Sjoberg et al., 1979) each containing six subscales. Several authors have argued in favor of reducing these six dimensions of mood to two basic dimensions, an Activity or Energy dimension and a Stress or Tension dimension (Kjellborg and Bohlin, 1974; Sjoberg et al., 1979; Thayer, 1978; Watson and Tellegren, 1985). On the basis of previously reported factor analyses, six words were selected for each of the two dimensions. Words in the original questionnaires which have been found to be unfamiliar to, or at least unnatural to use by, nonstudent groups were excluded. A score in each subscale is computed as a mean of the ratings of the six adjectives in the scale.

Acute Symptoms SPES32. This questionnaire contains 17 items regarding symptoms of local irritation as well as symptoms from the CNS. The subject is asked to rate the present intensity of each symptom on a four-point scale.

Long-Term Symptoms SPES33. The questionnaire contains 38 items regarding a wide variety of symptoms, such as vegetative symptoms, concentration deficits, fatigue, tiredness, dizziness, and symptoms of peripheral neuropathy. The subject is asked to rate the frequency of occurrence of each symptom during the last six months on a four-point scale.

REFERENCES

Acker, W. 1983. A computerized approach to psychological screening—The Bexley-Man-Audsley Automated Psychological Screening and the Bexley-Man-Audsley Category Sorting Test. Int. J. Machine Stud. 18:361–369.

American Psychological Association. 1986. Guidelines for computer tests and interpretations. Washington, D.C.

Anshelm Olson, B. 1982. Effects of organic solvents on behavioral performance of workers in the paint industry. Neurobehav. Toxicol. Teratol. 4:703–708.

Anshelm Olson, B. 1985. Early detection of industrial solvent toxicity. The role of human performance assessment. Arbete Halsa National Board Occupational Safety Health 21:1–59.

Anshelm Olson, B., F. Gamberale, and B. Gronqvist. 1981. Reaction time changes among steel workers exposed to solvent vapor. A longitudinal study. Int. Arch. Occup. Environ. Health 48:211–218.

Anshelm Olson, B., F. Gamberale, and A. Iregren. 1985. Coexposure to toluene and p-xylene in man: Central nervous functions. Br. J. Ind. Med. 42:117–122.

Astrand, I., and F. Gamberale. 1978. Effects on humans of solvents in the inspiratory air: A method for estimation of uptake. Environ. Res. 15:1–4.

Baker, E. L., R. Letz, and A. Fidler. 1985. A computer-administered neurobehavioral

evaluation system for occupational and environmental epidemiology. J. Occup. Med. 27:206-212.
Bartram, D., and R. Bayliss. 1984. Automated testing: Past, present and future. J. Occup. Psychol. 57:221-237.
Bartram, D., J. G. Beaumont, P. Cornford, P. L. Dann, and S. Wilson. 1987. Recommendations for the design of software for computer based assessment—Summary statement. Bulletin for the British Psychological Society 40:86-87.
Beaumont, J. G. 1982. System requirements for interactive testing. Int. J. Man-Machine Stud. 17:311-320.
Beaumont, J. G. 1985. The effects of microcomputer presentation and response medium on digit span performance. Int. J. Man-Machine Stud. 22:11-18.
Beaumont, J. G., and C. C. French. 1987. A clinical field study of eight automated psychometric procedures: The Leicester/DHSS project. Int. J. Man-Machine Stud. 26:661-682.
Biersner, R. J. 1972. Selective performance effects of nitrous oxide. Human Factors 43:187-194.
Bittner, A. C., M. G. Smith, R. S. Kennedy, C. F. Staley, and M. M. Harbeson. 1985. Automated portable test (APT system). Overview and prospects. Behav. Res. Methods Instrum. 17:217-221.
Braconnier, R. J. 1985. Dementia in human populations exposed to neuro-toxic agents: A portable microcomputerized dementia screening battery. Neurobehav. Toxicol. Teratol. 7:379-386.
Carr, A. C., R. J. Ancill, A. Ghosh, and A. Margo. 1981. Direct assessment of depression by microcomputer. A feasibility study. Acta Psychiatr. Scand. 64:415-422.
Cassito, M. G. 1985. Review on recent developments and improvements of neuropsychological criteria for human neurotoxicity studies. Pp. 20-24 in Neurobehavioral Methods in Occupational and Environmental Health. Copenhagen: WHO.
Chapman, L. J., and J. P. Chapman. 1978. The measurement of differential deficit. J. Psychiatr. Res. 14: 301-311.
Denner, S. 1977. Automated psychological testing: A review. Br. J. Soc. Clin. Psychol. 16:175-179.
Eckerman, D. A., J. B. Carrol, D. Foree, C. M. Guillon, M. Lansman, E. R. Long, M. B. Waller, and T. S. Wallsten. 1985. An approach to brief field testing for neurotoxicity. Neurotoxicity Toxicol. Teratol. 7:387-393.
Elofsson, S. A., F. Gamberale, T. Hindmarsh, A. Iregren, A. Isaksson, I. Johnsson, B. Knave, E. Lydahl, P. Mindus, H. E. Persson, B. Philipson, M. Steby, G. Struwe, E. B. Soderman, A. Wennberg, and L. Widen. 1980. Exposure to organic solvents: A cross-sectional epidemiologic investigation on occupationally exposed ear and industrial spray painters with special reference to the nervous system. Scand. J. Work Environ. Health 6:239-273.
Enander, A. 1987. Effects of moderate cold on performance of psychomotor and cognitive tasks. Ergonomics 30:1431-1445.
French, C. C., and J. G. Beaumont. 1987. The reaction of psychiatric patients to computerized assessment. Br. J. Clin. Psych. 26:267-278.
Gamberale, F. 1985. The use of behavioral performance tests in the assessment of solvent toxicity. Scand. J. Work Environ. Health (Suppl. 1):65-74.
Gamberale, F., and M. Hultengren. 1972. Toluene exposure. II. Psychophysiological functions. Work Environment and Health 9:131-139.
Gamberale, F., and M. Hultengren. 1973. Methyl-chloroform exposure. II. Psychophysiological functions. Work Environment and Health 10:82-92.

Gamberale, F., and M. Hultengren. 1974. Exposure to styrene. II. Psychological functions. Work Environment and Health 11:86–93.

Gamberale, F., and Kjellberg, A. 1983a. Behavioral performance assessment as a biological control of occupational exposure to neurotoxic substances. Pp. 111–121 in R. Gilioli, M. G. Cassitto, and V. Foa, eds. Neurobehavioral Methods in Occupational Health. Oxford: Pergamon Press.

Gamberale, F., and A. Kjellberg. 1983b. Field studies of the acute effects of exposure to solvents. Pp. 117–129 in The Neuropsychological Effects of Solvent Exposure, N. Cherry and A. Waldron, eds. Hampshire, England: The Colt Foundation.

Gamberale, F., and G. Svensson. 1974. The effect of anaesthetic gases on the psychomotor and perceptual functions of anaesthetic nurses. Work Environ Health 11:108–111.

Gamberale, F., G. Annwall, and M. Hultengren. 1975a. Exposure to white spirit. II. Psychological functions. Scand. J. Work Environ. Health 1:31–39.

Gamberale, F., G. Annwall, and M. Hultengren. 1975b. Exposure to methylene chloride. II. Psychological functions. Scand. J. Work Environ. Health 2:95–103.

Gamberale, F., G. Annwall, and B. Anshelm Olson. 1976a. Exposure to trichloroethylene. III. Psychological functions. Scand. J. Work Environ. Health 4:220–224.

Gamberale, F., G. Annwall, and M. Hultengren. 1978. Exposure to xylene and ethylbenzene. III. Effects on central nervous functions. Scand. J. Work Environ. Health 4:204–211.

Gamberale, F., B. Anshelm Olson, P. Eneroth, T. Lind, and A. Wennberg,. 1988a. Acute effects of ELF electromagnetic fields. A field study on linemen working at 400 kV. Solna, Sweden: National Institute of Occupational Health.

Gamberale, F., H. O. Lisper, and B. Anshelm Olson. 1976b. The effect of styrene vapour on the reaction time of workers in the plastic boat industry. Pp. 135–148 in Adverse Effects of Environmental Chemicals and Psychotropic Drugs, M. Horvath, ed. Amsterdam: Elsevier.

Gamberale, F., A. Kjellberg, and S. Razmjou. 1988b. The Effects of Unfavorable Thermal Conditions on Performance. Solna, Sweden: National Institute of Occupational Health.

Greenhouse, S.W., and S. Geisser. 1959. On methods in the analysis of profile data. Psychometrika 24:95–112.

Hagberg, M., and A. Iregren. 1984. Simple reaction time as a diagnostic aid in psychoorganic syndrome induced by organic solvents. Proceedings from the International Conference on Organic Solvent Toxicity, Stockholm, October.

Hedl, J. J., H. F. O'Neil, and D. N. Hansen. 1973. Affective reactions toward computer based intelligence testing. J. Consult. Clin. Psychol. 40:217–222.

Iregren, A. 1982. Effects on psychological test performance of workers exposed to a single solvent (toluene)—A comparison with effects of exposure to a mixture of organic solvents. Neurobehav. Toxicol. Teratol. 4:695–701.

Iregren, A. 1986a. Subjective and objective signs of organic solvent toxicity among occupationally exposed workers. An experimental evaluation. Scand. J. Work Environ. Health 12:469–475.

Iregren, A. 1986b. Effects of industrial solvent interactions. Studies of behavioral effects in man. Arbete Halsa National Board Occupational Safety Health 11:1–60.

Iregren, A., F. Gamberale, and A. Kjellberg. 1985. A microcomputer based behavioral testing system. Pp. 75–80 in Neurobehavioral Methods in Occupational and Environmental Health. Copenhagen: WHO.

Iregren, A., T. Akerstedt, B. Anshelm Olson, and F. Gamberale. 1986. Experimental exposure to toluene in combination with ethanol intake. Psychophysiological functions. Scand. J. Work Environ. Health 12:128–136.

Iregren, A., O. Almkvist, M. Klevegard, and U. Aslund. 1987. A clinical validation of

six computerized tests for diagnosing solvent caused occupational illness (in Swedish). Arbete Halsa National Board Occupational Safety Health 13:1–37.
Irons, R., and P. Rose. 1985. Naval biodynamics laboratory computerized cognitive testing. Neurotoxicity Toxicol. Teratol. 7:395–397.
Kjellberg, A., and O. Bohlin. 1974. Self-reported arousal: Further development of a multifactorial inventory. Scand. J. Psychol. 15:285–292.
Kjellberg, A., and M. Strandberg. 1979. The effects of anaesthetic gases on reaction time of anaesthetic nurses. Report No. 11. Solna, Sweden: National Board of Occupational Safety and Health.
Kjellberg, A., and P. Wide. 1988. Effects of simulated ventilation noise on performance of a grammatical reasoning task. Proceedings of the 5th International Congress on Noise as a Public Health Problem, Stockholm.
Kjellberg, A., and H. Wisung. 1987. Some metrical properties in a computer administered test battery for use in behavioral toxicology. Report No. 1. Solna, Sweden: National Board of Occupational Safety and Health.
Kjellberg, A., B. Wigaeus, J. Engstrom, I. Astrand, and B. Ljungquist. 1979. Long-term effects of exposure to styrene in a polyester plant. Arbete Halsa National Board Occupational Safety Health 18:1–25.
Knave, B., B. Anshelm Olson, S. Elofsson, F. Gamberale, A. Isaksson, P. Mindus, H. E. Persson, G. Struwe, A. Wennberg, and P. Westerholm. 1978. Long term exposure to jet fuel. A cross sectional epidemiologic investigation on occupationally exposed industrial workers with special reference to the nervous system. Scand. J. Work Environ. Health 4:19–45.
Knave, B., F. Gamberale, S. Bergstrom, E. Birke, A. Iregren, B. Kolmodin Hedman, and A. Wennberg. 1979. Long-term exposure to electric fields. A cross-sectional epidemiologic investigation of occupationally exposed workers in high-voltage substations. Scand. J. Work Environ. Health 2:115–125.
Laursen, P., and T. Jorgensen. 1985. Computerized neuropsychological test system. In Neurobehavioral Methods in Occupational and Environmental Health. Copenhagen: WHO.
Letz, R., and E. Baker. 1986. Computer-administered neurobehavioral testing in occupational health. Sem. Occup. Med. 1:197–203.
Lisper, H.O., and A. Kjellberg. 1972. Effects of 24-hour sleep deprivation on rate of decrement in a 10-minute auditory reaction time task. J. Exp. Psychol. 96:287–290.
Lucas, R.W. 1977. A study of patient attitudes to computer interrogation. Int. J. Man-Machine Stud. 9:69–96.
Lukin, M. E., E. Dowd, B. S. Plake, and R. Kraft. 1985. Comparing computerized versus traditional psychological assessment. Computers in Human Behavior 1:49–58.
Mahoney, E. C., P. A. Moore, E. L. Baker, and R. Letz. 1988. Experimental nitrous oxide exposure as a model system for evaluating neurobehavioral tests. Toxicology 49:449–457.
Matarazzo, J. D. 1983. Computerized psychological testing. Science 221:323.
McArthur, D. L., and B. H. Choppin. 1984. Computerized diagnostic testing. J. Educational Measurement 31:391–397.
Rodnitzky, R. L. , H. S. Levin, and D. L. Mick. 1975. Occupational exposure to organophosphate pesticides. A neurobehavioral study. Archives of Environmental Health 30:98–103.
Roels, H., R. Lauwreys, J. P. Buchet, P. Genet, M. J. Sarhan, I. Hanotiau, M. deFays, and D. Stanescu. 1987. Epidemiological survey among workers exposed to manganese: Effects on lung, central nervous system and some biological indices. Am. J. Ind. Med. 11:307–327.

Sjoberg, L., E. Svensson, and L. O. Persson. 1979. The measurement of mood. Scand. J. Psychol. 20:1–18.

Soderman, E., A. Kjellberg, B. Anshelm Olsen, and A. Iregren. 1982. Standardization of a simple reaction time test for use in behavioral toxicology. Report No. 27. Solna, Sweden: National Board of Occupational Safety and Health.

Space, L.G. 1981. The computer as psychometrician. Behav. Res. Methods Instrum. 13:595–606.

Thayer, R.E. 1978. Toward a psychological theory of multidimensional activation (arousal). Motivation and Emotion 2:1–34.

Thompson, J. A., and S. L. Wilson. 1982. Automated psychological testing. Int. J. Man-Machine Stud. 17:279–289.

Watson, D., and A. Tellegen. 1985. Toward a consensual structure of mood. Psychol. Bull. 98:219–235.

Weiss, D. J., and C. D. Vale. 1987. Adaptive testing. Appl. Psych. 36:249–262.

Wigaeus-Hjelm, E., M. Hagberg, A. Iregren, and A. Lof. 1990. Exposure to methyl isobutyl ketone (MIBK). Toxicokinetics and occurrence of irritative and CNS symptoms in man. International Archives of Occupational and Environmental Health. In press.

World Health Organization. 1987. Prevention of Neurotoxic Illness in Working Populations, B. L. Johnson, ed. New York: John Wiley & Sons.

The Scope and Promise of Behavioral Toxicology

Bernard Weiss

Behavioral toxicology (BT) has almost ceased to be a term that arouses quizzical or bemused expressions in the more orthodox venues of its parent discipline. At the same time, its full scope and potential remain largely unappreciated and unexploited. The range of questions it can be used to ask and the unique perspectives it can provide on certain issues so far exceed what has been demanded of it. This chapter aims to illustrate or identify some of its unused capabilities, and to indicate those that need further development. Its special focus is how behavioral measures have expanded our previous views, based on traditional criteria, of what constitutes toxicity, and the nature of the new issues that this expanded perspective fosters. Foremost among these issues is how behavioral endpoints are to be treated in risk assessment.

Almost the entire risk assessment process is designed around cancer (National Research Council, 1983). What are called systemic toxicants, such as those acting on the nervous system, are evaluated by a wholly different set of principles. The difference stems from a presumed biological dichotomy. The induction of carcinogenesis is assumed to have no threshold. A single molecular event, such as a transcription error in DNA, can generate carcinogenesis and, it is assumed, can arise from the action of a single molecule of a carcinogenic agent. Systemic toxicants, in contrast, are presumed to exhibit thresholds, perhaps at the point at which they overwhelm compensatory mechanisms.

This doctrine of distinct biological modes of action finds expression in the current approaches to risk assessment.

The first step in conventional risk assessment is hazard identification, which can be based on either epidemiological or experimental data. This first step is crucial because of the regulatory apparatus activated when a substance is classified as carcinogenic. The next step, dose-response assessment, almost always is based on high-dose animal or human data. These data are used in extrapolation modeling to compute predictions of cancer probability at low dosages. Because of the biological assumptions, only a zero dose of a carcinogen is assumed to add no increment of risk. Exposure assessment estimates the levels of the agent to which the target population is exposed. Together with the dose-response model chosen for extrapolation, the estimated risk of cancer can then be calculated for that population in a step called risk characterization.

Systemic toxicants, such as those acting on nervous system tissue, are viewed from a totally different perspective. Instead of coupling a risk estimate with an exposure level, some arbitrary threshold is defined, then divided by a safety factor or uncertainty factor to yield an acceptable daily intake. Such thresholds are described as effect levels of various kinds (Klaassen et al., 1986). The no-observed-effect level (NOEL) refers to an exposure level offering no statistically significant increases in either frequency or severity of response in an exposed, compared to a control, sample. A lowest-observed-adverse-effect level (LOAEL) refers to the lowest exposure level producing statistically significant increases in the frequency or severity of adverse responses. Other effect levels are defined by similar standards.

The suitability of effect levels for risk assessment is now being questioned in many quarters and for many reasons. First, how are adverse effects defined? Conventionally, they include any effects that impair function, result in lesions, or inhibit an organism's ability to respond to additional challenges, so that effect levels depend on the specific endpoint chosen as the critical one. Moreover, some critics contend that these do not distinguish between reversible and irreversible effects, between immediate and delayed effects, and between agents that may be rapidly eliminated and those that remain in the body for extended periods. A second objection to the effect threshold concept is statistical. It makes use of only one point on the dose-consequence function rather than the entire function and essentially ignores the size of the experiment. Third, the uncertainty factors by which the NOEL, for example, is divided to provide a safety margin for population exposure, are also arbitrary and fail to make optimal use of experimental data. All of these objections have encouraged speculation that the highly developed cancer model for risk assessment might be adapted for systemic toxicants.

The standard risk assessment protocol derived from carcinogenesis might be modified for neurotoxicants, with extrapolation to the origin (zero dose, zero added risk) replaced by another function, such as a threshold model, if the standard protocol were adequate. Neurotoxicants, however, introduce a complication: the stage of risk characterization, instead of becoming a matter simply of finding the intersection of dose and risk probability, turns into a complex weighing of endpoints and their measures. The complications are especially difficult to resolve when the endpoints are behavioral in content. To grasp this point, it helps to begin with a review of the history of BT and some of its special properties.

ANTECEDENTS OF BEHAVIORAL TOXICOLOGY

Although BT arrived on the scene, at least in the United States (Weiss and Laties, 1975), less than two decades ago, it had a plethora of antecedents that serve to explain its unusual position in toxicology and why it is difficult to mold into the conventional risk assessment process.

BEHAVIORAL PHARMACOLOGY

With the introduction of the minor tranquilizing and antipsychotic drugs in the 1950s, and the demonstration that chemotherapy could be a legitimate option in the treatment of behavior disorders, an intense search began for new agents. It was accompanied by a swelling interest in the behavioral mechanisms underlying the clinical actions of these drugs. These two developments combined to establish a technology and a discipline hospitable to their goals. Behavioral pharmacology grew out of the extensive literature of experimental psychology, particularly that aspect of it called the experimental analysis of behavior and associated with applications of what is known as schedule-controlled or operant behavior (Iversen and Iversen, 1981). Much of the early work in behavioral pharmacology was built on the power of coupling behavior and its consequences in prescribed ways known as schedules of reinforcement. By manipulating correlations between specified behaviors, such as lever presses by rats, and the subsequent delivery of food pellets, it was possible to generate patterns of behavior that proved differentially sensitive to various kinds of drugs and that also provided the basis for analyses of such differential sensitivity.

Translating this technology into one suitable for toxicology proved a fairly easy task because most of the questions posed to behavioral pharmacology were essentially questions in selective toxicity. Toxicology did not grapple, however, with its heritage from the central theme of behavioral pharmacology: toxicant-behavior interactions. It remained centered on whether a particular agent deserves to be labeled neurotoxic. Yet, if any single principle can be identified as the dominant product of behavioral pharmacology, it is that the nature of a behavioral response to a chemical challenge depends on the characteristics of the behavioral situation. At the same time and in the same organism, a drug might elicit one kind of response pattern, such as an increase in rate, under one schedule, whereas it elicited a decrease in rate under another schedule or schedule variant. On the basis of our experience with drugs, the question ought to be how to interpret the modifications produced by exposure. Simply specifying schedule-controlled behavior as one component of a screening battery, while ignoring the interaction, is unlikely to yield significant contributions to BT as a science and could evoke considerable confusion (see MacPhail, this volume) if the resulting data were to be used to calculate some version of a threshold.

Some confusion already exists because of the different aims of BT and behavioral pharmacology (Weiss, 1984). Central nervous system (CNS) drugs are administered therapeutically at doses great enough to influence behavior, so that behavioral pharmacologists study high doses either to try to detect active agents or to differentiate their behavioral effects. A BT study might entail these aims as well, but must always consider the importance of its findings for risk estimates, which implies action at low doses. Wood and Cox (1986), for example, measured the response rates of rats exposed to toluene vapor and performing on a reinforcement schedule that maintained fairly stable rates under control conditions. They chose to study exposure concentrations considerably lower than those used in past investigations with rodents. They observed that toluene exposure, at levels close to those used in experimental studies with humans, and even approximating the threshold limit value, elevated rates above control levels. A typical strategy for dose selection, however, would have begun with rather high levels, would have observed rate decreases, would then have lowered the concentration to a point at which no rate decreases occurred, and would have missed toluene's rate-enhancing properties at low concentrations. We do not have to rely on neurotoxicology alone to document the futility of high doses when the aim is risk estimation. The connection between lead exposure and hypertension emerged from an analysis of the the Second National Health and Nutrition Examination Survey (NHANES II) data, which showed the steepest effects at low doses. Animal data had indicated such a relationship (Victery et al., 1982) but failed to attract attention because the results conflicted with our stereotyped expectations.

What we find currently in most assessments of animal behavior is an arbitrary selection of experimental parameters combined with relatively high exposure levels. We seem to have appropriated a technology without an appreciation of what that transfer of technology requires to make it work.

Workplace Exposure Criteria

The first industrial hygiene legislation on record was prompted by the manifestations of mercury poisoning in miners who worked the famous mines at Idria. Among these manifestations are tremor and a collection of psychological complaints. Behavioral disturbances were also listed in descriptions of many other workplace toxicants by pioneers in occupational hygiene, such as Ramazzini in the eighteenth century and Hamilton in the twentieth century. Formal recognition was embodied in the Threshold Limit Values (TLVs) issued by the American Conference of Governmental Industrial Hygienists (ACGIH). Note its description of the short-term exposure limit (STEL): "the maximum concentration to which workers can be exposed for a period of up to 15 minutes continuously without suffering from... narcosis of sufficient degree to increase accident proneness, impair self-rescue, or materially reduce work efficiency..." (ACGIH, 1974). Such a definition of safety implies quantitative information about performance capacity that would have to be acquired under experimental conditions. Adequate information is sparse. Anger and Johnson (1985) estimate that about 25 percent of the workplace chemicals for which TLVs exist are neurotoxic, but the volume of pertinent toxicity data is far less than warranted by such a role.

The TLVs are supposed to protect against adverse effects during a working lifetime. For substances such as organic solvents, however, they have been based on a combination of impressionistic clinical data, some epidemiology, and observations of acute effects. A chronic syndrome, described extensively in the Scandinavian literature, has also been described. It comprises signs such as a slowing of responses, memory difficulties, and personality disturbances. The validity of such a syndrome has aroused robust debate, but even critics acknowledge its confirmation in workers exposed to carbon disulfide (Grasso et al., 1984).

Several of the chapters in this volume describe this syndrome and the research programs and techniques designed to extend the scientific basis for exposure standards. We must acknowledge the enormous contributions of these programs to shaping our views of how the criteria for workplace safety should be formulated. Yet the entire literature suffers from an inherent conflict between eagerness to apply these views and aptness of the techniques on which these applications depend. Convenience, ease of administration, standardization, and testing time are dominant concerns and, in response, the concerns are met by collecting a series of tests into a battery. Sometimes, however, convenient technology can be misleading.

For example, several investigations have relied on a device called the Optacon to assess somesthetic sensitivity. The device itself, and the psychophysical procedures governing testing, are wholly inadequate for such a purpose, as so cogently discussed by Maurissen (1988). Another example is the various reaction time measures included in many batteries. Again, because of unfamiliarity with the psychophysical literature, authors may fail to specify stimulus values, despite the body of knowledge indicating that stimulus intensity is inversely related to response latency. In fact, reaction time can be used to plot psychophysical functions (Stebbins, 1970), so that one could question whether some of the results with solvents, for example, represent altered "cognitive" function or sensory deficits.

The conflict between convenience and sensitivity obstructs the usefulness of many of these batteries for risk estimation. Is it legitimate to argue that, because they probably yield underestimates of impaired function, safety standards such as TLVs require the application of uncertainty factors based on groups showing deficits attributed to exposure? Or can the argument about sensitivity be used to question the validity of the findings, a tactic used in toxic tort cases? Finally, if an arbitrarily safe exposure standard is the aim of such research, what model is to provide such a standard and what are the associated risks?

Standards in the Soviet Union

Discrepancies between the workplace and community exposure standards accepted in the West, and those prevailing in the Soviet Union, which tend to be much lower, have generated considerable speculation about their sources. One source surely was the image of virtue to be gained by USSR standards that seemed more rigorous than those adopted by capitalist nations. Other sources should be recognized as well, however.

The most important was the doctrine flowing from the history of

Soviet science, and the overwhelming authority of I.P. Pavlov, that measures of CNS function should play a major, even dominant, role in assessing safety and prescribing exposure standards. Tissue damage occupied the corresponding role in the West and still remains dominant. Pavlenko (1975), discussing methods for toxic assessment of the CNS, notes Pavlov's assertion that ". . . the animal organism as a system is able to survive in its natural environment only if a dynamic equilibrium is maintained between this system and its environment. This is achieved, in higher animals, chiefly through the agency of the nervous system and by means of reflexes." Although Western scientists tend to view Soviet data with some skepticism, in part because the standards of scientific publication seem to be looser in the USSR, the Soviet approach still managed to generate enough curiosity in the West to stimulate tests of its validity.

One aspect of Soviet doctrine that still separates it from Western toxicology, however, is the principle that any deviation from baseline functional parameters due to toxic exposure must be interpreted as an adverse effect (Glass, 1975). Western scientists may interpret such effects as evidence of adaptation, much like the change in vital capacity produced by physical training. Simultaneously, however, Western toxicologists must at some time resolve which behavioral endpoints denote toxicity. Some of the more conventional practitioners insist that behavioral changes of a transient nature, unaccompanied by pathology, do not warrant the label of toxicity. Such a narrow definition is probably no longer tenable, but what are its limits?

Ozone occupied such an ambiguous niche not long ago. It was recognized as a potent lower-airway irritant and as a source of pathological changes in the lung at high doses. Questions about its toxicity at low environmental concentrations have been answered satisfactorily only recently. Inhalation toxicologists now can document adverse pulmonary effects, at least as a consequence of chronic exposure, at levels permitted by current regulations. Yet consider the problem of how to interpret findings such as those published by Weiss et al. (1981), Tepper et al. (1982), and Tepper and Weiss (1986). Weiss et al. (1981) trained rats to respond on a fixed-interval schedule of food reinforcement and measured response rates during 6-hour exposures to ozone. They observed a reduction in lever-pressing rate at concentrations of 0.5 ppm and above. To provide a contrast with this kind of sedentary behavior, Tepper et al. (1982) allowed rats access to running wheels during a 12-hour period and exposed them to ozone during the middle 6 hours. A concentration as low as 0.12 ppm, the level deemed by the regulations issued under the Clean Air Act as what might be considered a surrogate for an effect threshold, reduced running. This

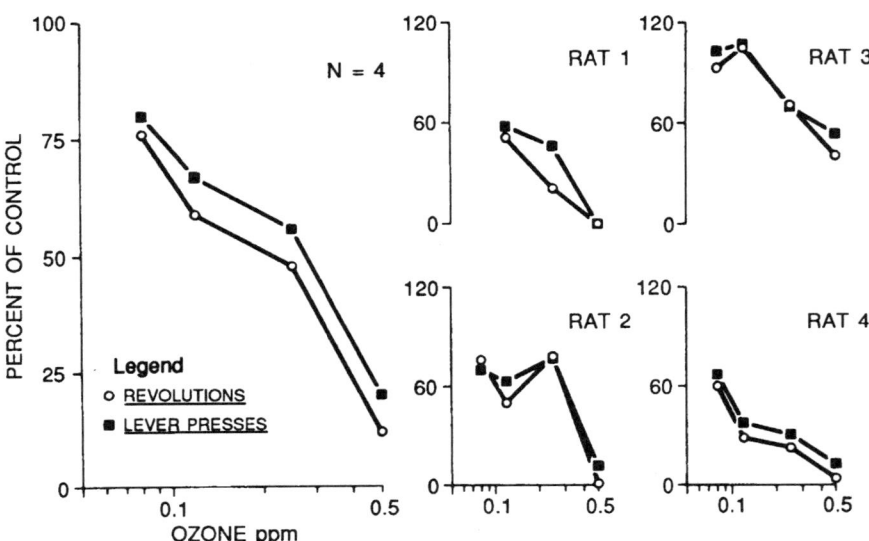

FIGURE 1 Results of an experiment in which rats pressed a lever, attached to the inner wall of a running wheel, to release a brake that locked the wheel. After preliminary training, the rats were required to make five lever presses (a fixed-ratio 5 scheule of reinforcement) to release the brake for a period of 15 seconds. Rat 1 remained on fixed-ratio 1. Access to the running wheel occurred during the last hour of a six-hour period of exposure to ozone. Three of the four rats showed a reduction in both lever presses and wheel revolutions at 0.08 ppm ozone; the Environmental Protection Agency standard is 0.12 ppm.

SOURCE: Tepper and Weiss (1986).

reduction was largely the product of lengthened pauses between bouts of running and can be interpreted, not as toxicity per se, but as behavioral adaptation. That is, reduced motor activity reduces minute volume which, in turn, reduces pulmonary uptake of ozone and, finally, the aversive consequences of running in an ozone-enriched environment.

The later paper (Tepper and Weiss, 1986) described an experiment in which rats pressed a lever, while in the running wheel, to release a brake and so secure an opportunity to run. A reduction in the frequency of this behavior, which can be interpreted even more directly as avoidance of the aversive consequences of exercise, occurred at an ozone concentration of 0.08 ppm (Figure 1). Are these behavioral data to be adopted as evidence of ozone toxicity at these low levels, as they would in the USSR? Do they simply indicate that exercise is aversive at these concentrations? This is one arena in which behavioral observations offer a unique problem for risk assessment and regulation: Can an exposure level that elicits avoidance of the consequences of exposure be considered an adverse functional or toxic effect (Weiss, 1989).

Public Awareness

With the stirrings of the environmentalist movement in the United States, public concern began to shift from the grosser aspects of toxic damage to the more subtle ones, especially those arising as the consequence of low-level, prolonged exposure. Although cancer was featured, it was inevitable that the public would begin to ask questions about the coupling of environmental chemicals and "mental disease," for example. Fifteen years ago, when the Environmental Protection Agency (EPA) was preparing to respond to what later became the Toxic Substances Control Act, legislators in the United States were already drafting requirements that behavioral disturbances be included among the criteria of adverse effects. At present, several legislative initiatives and federal agencies define and regulate chemical exposures and include behavior among the aspects of toxicity to be considered in determining safety. The removal of lead from gasoline can be attributed to the mounting evidence showing an inverse relationship between intelligence test scores in children and indices of lead exposure. The public, however, remains largely uneducated about such issues. Even the media still use terms such as lead poisoning to describe the impact of lead exposure on test scores.

Terms such as lead poisoning imply that risk and the associated calculation of acceptable exposure standards can be defined, like cancer, by number of cases, but the lead issue has been defined by a different metric. The paper by Bellinger et al. (1987), whose results are supported by several other groups, offers a clear example. They compared scores, on the Bayley Scales of Infant Development, of three groups of children 24 months of age. The groups were differentiated by lead concentrations in cord blood: low (a mean of 1.6 µg/dL); medium (a mean of 6.5 µg/dL); and high (a mean of 14.5 µg/dL). As observed in other chapters, a few years ago the high-lead group would have been regarded as a low-lead group. All the groups attained scores above average on the Bayley Scales, but the differences between the high-lead group and the other two groups came to about 8 percent.

Given the above-average scores in the high-lead group, none of the children could be identified as cases, for example, of mental retardation. Also, 8 percent, although statistically significant, is a degree of variation that is often encountered on retesting. The clearest appreciation of this difference is found in its implications for the community.

Figure 2 compares two distributions of intelligence test scores. The

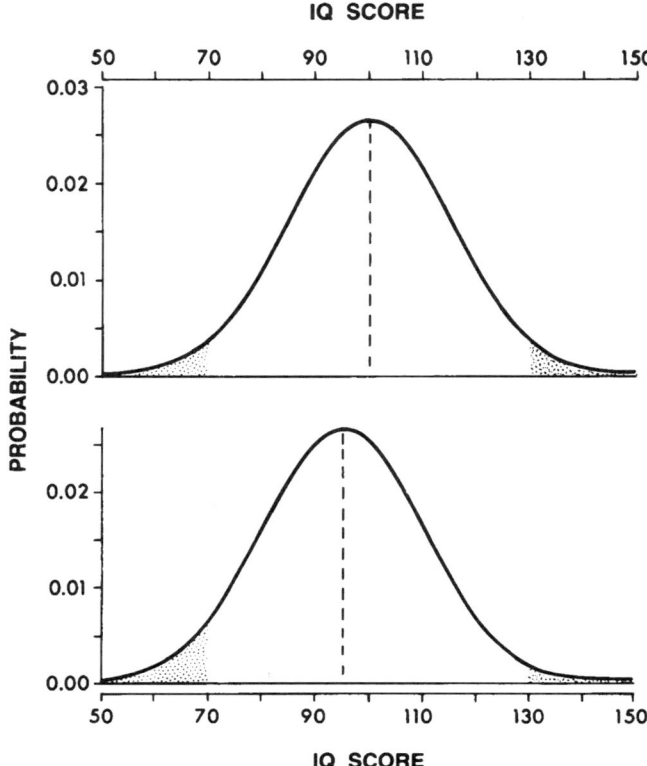

FIGURE 2 Plot describing implications of a 5 percent shift in intelligence test scores. The upper curve depicts the distribution of intelligence test scores for instruments such as the Stanford-Binet and Wechsler Intelligence Scale for Children. Its mean is 100 and its standard deviation is 15. In a population of 100 million, 2.3 million individuals will score above 130, as shown by the stippled area in the upper tail of the distribution. If the distribution is shifted by 5 percent, or one-third of a standard deviation, to a mean of 95, only 990 thousand individuals will score above 130. Bellinger et al. (1987) observed a difference of 8 percent on the Mental Development Index of the Bayley Scales of Infant Development, between children whose cord bloods fell into a group with a mean lead concentration of 14.5 µg/dL and those in groups with lower means.

SOURCE: Weiss (1988).

upper chart depicts a distribution with a mean of 100 (the defined average) and a standard deviation of 15 (as found, say, on the Stanford-Binet). In a population of 100 million, 2.3 million will score above 130. The lower chart depicts a distribution with a mean of 95, or a reduction of 5 percent. In such a population of 100 million, only 990 thousand individuals will score above 130 (Weiss, 1988). The implications for a society are staggering. Yet, they are impossible to convey with the standard model of risk assessment, which counts cases.

Toxic Torts

Litigation is exerting a significant impact on the acceptance of behavioral criteria of toxicity. Courts in the United States are now evaluating suits by workers claiming injury from exposure to organic solvents, metals, pesticides, and other neurotoxic substances. Such claims take the form of impaired intellectual performance, impaired sensory function, subjective complaints, and other indications of nervous system damage. These suits are now prompting segments of industry, which previously had tended to ignore behavioral assays in chemical development and workplace safety, to inaugurate programs responsive to these newer facets of toxicology. Moreover, legislation such as the Gaydos-Metzenbaum bill is prompting further review of the impact of subtle toxicity on worker health. Behavioral toxicology will be forced more and more, as the legal system responds to these issues, to make explicit the sources of its conclusions and to defend them. Legal arguments have a way of challenging vagueness, possibly to the disadvantage of BT and some of its practitioners, because then we will have to offer statements about probability. When workplace exposure is at issue, how convincingly can we argue, for an individual, that a particular collection of signs and symptoms was a likely or unlikely outcome of a work history? What degree or proportion of responsibility can we allocate to the work environment and what proportion to other factors, such as the personal habits of the individual?

The test batteries devised to assay the neurobehavioral consequences of workplace exposure to various substances, which are the substance of the scientific arguments advanced in toxic tort cases, convey a great deal of ambiguity when used to support decisions about individuals. Perhaps such ambiguity is inevitable, given the problem of multiple chemical exposures in many work histories—complicating attempts to extract characteristic test profiles—but an emphasis on differential exposure histories and even rudimentary dose-response analysis would yield more effective instruments in the end.

SPECIAL CHARACTERISTICS OF BEHAVIORAL TOXICOLOGY

One message to be extracted from this list of predecessors is that BT is still in search of an identity. We inherited certain techniques and viewpoints, but still have to synthesize them into a mature discipline.

To move toward such a synthesis requires not just refinements of borrowed technology, but a technology and viewpoint uniquely our own. Viewpoints determine technology, so we will have to examine those that are special to BT and imbue it with some of the properties that make it a unique challenge for risk assessment.

Basic Themes

A review of our history distinguishes two themes, which could be termed validation and amplification, that have emerged with the development of BT. The validation theme is embodied in the process of hazard identification, the first step in the conventional risk assessment process. At this stage, in contrast to cancer, emphasis falls on establishing the spectrum of toxicity associated with a particular agent, and the aim of research directed to such questions has been to develop adequate screening methods. Sensitivity is a secondary goal in these programs because extrapolation is only an implied, and not a direct, requirement or even role for such screens (e.g., Tilson et al., 1979).

A second type of validation embodies animal models of the kind discussed in this volume by Russell and by Overstreet. These models seek to mimic neurological diseases, such as parkinsonism and Alzheimer's disease, whose etiology currently is suspected by some to result from neurotoxic processes. Here, the validation theme takes the form of chemically induced lesions and behavioral endpoints that are assumed to be analogues of human function such as short-term memory.

The second theme, which I call amplification, addresses risk estimation directly and its ultimate goal of coupling exposure levels with risk incidence or severity. This goal would be essentially the next step, after hazard identification, in the risk assessment process. It may be undertaken either as an expansion of observations where humans served in the role of sentinels or as the successor to laboratory findings that have documented the existence of a hazard. No entirely new substance has yet passed through the defined phases of the conventional risk assessment process. Our literature is based almost exclusively on agents already defined by human exposure. Hazard identification has been pursued mainly as a validation process based on recognized toxicants. Dose-response (and dose-effect) phases have typically been conducted, in the laboratory, as programs to establish validity by demonstrating such relationships. Efforts to provide a basis for dose extrapolation to humans remain minimal.

The calculation of risk based on neurobehavioral criteria is complicated by the variety of prototypical situations in which adverse effects might appear. Exposures may be either acute or chronic; consequences may be either reversible or irreversible, progressive or stable. Some effects may remain latent, only to emerge with time, perhaps in advanced age, when the reserve capacity of the nervous system has been depleted. The anesthetic properties of volatile organic solvents, for example, represent an acute reversible situation, but consistent exposures may lead to progressive deterioration that eventually becomes irreversible. Delayed irreversible effects are associated, for example, with MPTP exposure in adults and methylmercury exposure in the fetus.

Clinical and Behavioral Criteria

For all these categories, our past evaluations of adverse effects were based largely on clinical endpoints, still the main basis for estimates of the hazards of systemic toxicants acting on organs other than the central nervous system. The adequacy of clinical criteria for risk assessment is questionable.

Clinical criteria are especially flawed when neurotoxicity is expressed by a gradual, progressive erosion of functional capacity. Consider the reasons for trying to develop and refine psychological test procedures sensitive to the early manifestations of Alzheimer's disease. By the time a patient comes to the attention of clinicians, he or she has already progressed to a stage that has captured the concerns of family members. At that point, an accurate diagnosis is not an especially formidable challenge. Even though the currently available test procedures cannot comfortably differentiate victims of the disease from controls, except in group designs, they remain vastly superior to the clinical examination in defining the areas and extent of functional deficit. Their precision is certain to improve now that the vast amount of research on the psychological deficits of Alzheimer's disease is being embodied in potential diagnostic procedures.

One of the most cogent examples of the difference between clinical standards and psychological test or experimental design standards is surely lead toxicity. I discussed earlier the novel way in which the risks of lead exposure should be formulated, an example of the way in which the amplification process works. Cory-Slechta (this volume) traces the progressive lowering, over the past four decades, of the blood levels accepted as hazardous to children and detectable in animal models. Such a progression is now evolving with methylmercury, which currently is undergoing an amplification process. Although it has many features in common with how our views of lead toxicity evolved, it has distinct features of its own that make it an appealing model. One important feature is our extensive knowledge of methylmercury neuropathology. Another is our ability to trace exposure history by the analysis of methylmercury in hair. The third is its specific effects on special systems, such as vision. The fourth is the narrow focus of its toxicity: unlike lead, which exerts significant effects on hematopoiesis and blood pressure, methylmercury exerts only minimal effects beyond the nervous system. The fifth feature is the often prolonged latency to overtly detectable effects during or following exposure.

Methylmercury as Prototype

Most current concerns about methylmercury arise from its potency in the developing human. Minamata suggested, and Iraq confirmed, that the fetus and neonate are far more sensitive than the adult. Clarkson and his colleagues, in a series of analyses based on the Iraq disaster (e.g., Clarkson et al., 1981), now suggest that the fetus may be as much as ten times more vulnerable than the adult to methylmercury. Such calculations are based on the appearance of paresthesias in adults with total body burdens of 25 mg and of retarded motor development, of a type leading to a diagnosis of cerebral palsy, in children whose mothers accumulated a body burden of 2.5 mg. These are clinical criteria based on examinations conducted in rural Iraq, not on the kind of careful neuropsychological evaluation possible in major medical centers.

It is provocative to consider what kind of results might have emerged from the application of what are considered to be more sensitive and specific tests currently found in neuropsychological testing centers, such the Bayley Scales of Infant Development used by Bellinger et al. (1987). Even such instruments are crude compared to the tools described in the current literature of child development, although they offer the virtue of standardization. Given our experience with lead, we might predict that reliance on even these imperfect instruments could amplify sensitivity by a factor of four or five. Research now in progress suggests that, in fact, the developmental neurotoxicity of methylmercury might have been underestimated, on the basis of clinical criteria, by almost such a factor.

It might be equally provocative to imagine the conclusions that would have been fostered by the kinds of schemes now envisaged for identifying neurotoxicity and then for extending it to risk estimates. Gross neurotoxicity in developing animals would surely have been identified at high doses by observations of developmental disorders. In rats, a massive study designed to evaluate reproducibility of behavioral observations between laboratories, the Collaborative Behavioral Teratology

Study (CTBS), chose 6 mg/kg, administered on gestation days 6–9 or 12–16, as the high dose on the basis of a preliminary study (Buelke-Sam et al., 1985). Most of the six participating laboratories would have selected that dose, a total of 24 mg/kg, as the LOAEL, on the basis of indices such as maternal and offspring weight gain together with a variety of behavioral indices.

Unfortunately, the protocols included neither sensitive morphological indices nor measurements of methylmercury tissue levels, so that a direct comparison with neurotoxic health risk estimates based on human data is not feasible, but crude parallels can be constructed from knowledge of levels prevailing in fish. The Food and Drug Administration action level is 1 ppm. Most swordfish exceed this level. Shark, an increasingly popular seafood, has an even higher content than swordfish. Freshwater pike from the Adirondacks, because of acid rain, typically exceed 1 ppm of methylmercury as well. Assume that a pregnant female consumes seven fish meals, within a one-month period, of a species at the FDA action level. If each meal consists of 240 grams of fish, she will accumulate a body burden of 1,680 grams.

For a body weight of 70 kg, this amounts to 24 μg/kg, or 1/1,000 of the rat-based LOAEL. Such a body burden is equivalent to what is now suggested to be the human LOAEL.

However, there is another provocative feature to methylmercury that might multiply our risk estimates even more. Spyker (1975) maintained mice, after prenatal treatment, for a lifetime. In mice that, until that time, had manifested no adverse effects, neurological disorders began to appear at about 15 months of age, and even in superficially healthy mice, behavioral testing revealed functional impairment. As the mice aged, they revealed more and more disorders. These observations are a powerful argument for longitudinal studies, but an even more powerful argument for including such possibilities in risk assessment. Spencer (this volume) offers a compelling argument that earlier cycad exposure may trigger the eruption of the amyotrophic lateral sclerosis/parkinsonism-dementia (ALS-PD) syndrome even decades later.

Individual Differences

More than other areas of toxicology, BT is sensitive to individual differences. Other disciplines typically model results solely as means, or even, as in carcinogenesis, within a stochastic model that relates total exposure in a population to number of tumors irrespective of the distribution of exposure. Some of our sensitivity perhaps stems from the historical junction of diagnosis with psychometrics; some of it may stem from our laboratory experiences with allegedly homogeneous groups of animals whose members all seem to exhibit unique experimental personalities, especially when we trace the development of a process such as learning. Despite our awareness (sometimes subliminal) of individual differences, most papers in BT, like most papers in toxicology, assume that subjects (animal and human) come from a uniform population and treat the data, as well as the design, accordingly. It is not the best approach to defining the characteristics of low-dose, chronic exposure. Physiologists are also now questioning the usefulness of group analyses without the data of individual subjects—a tradition exemplified by operant experimenters.

Even acute experiments may lead to wayward conclusions if individual differences are ignored. Ben F. Feingold was a pioneering pediatric allergist who formulated the hypothesis that some of the children labeled as hyperactive were actually responding to certain constituents of the diet (Feingold, 1975). Although he singled out synthetic colors and flavors, mostly because he doubted their nutritional value, his hypothesis had roots in an extensive allergy literature, but he never held that all children with that label suffered from excessive sensitivity to additives.

I have reviewed the experimental data generated by the Feingold hypothesis on several occasions (e.g., Weiss, 1982, 1986a). Two generalities arise from those data. First, Feingold was correct in principle: some children respond adversely to colors and perhaps to other additives. Second, the prevailing view that the Feingold hypothesis has been disproved is mostly attributable to the naive statistics practiced by experimenters and reviewers alike. Some flaws stem from the assumption of a uniform population and are illustrated in Figure 3. Assume a population comprised of 70 percent nonresponders and 30 percent responders (A). Then assume a treatment that shifts the responders by one standard deviation (B). The distribution (C) shows the results for the sample as a whole; the difference in means is hardly visible. The usual procedure for extrapolating animal data to human standards imposes a safety or uncertainty factor to compensate for wide individual differences in sensitivity. Although no one disputes that such differences exist, that recognition exercises little influence on experimental design and analysis. Even in the laboratory within a group of rats of the same age and strain, we see remarkable differences in the behavioral response to toxicants such as lead and have had to develop special statistical techniques to quantify these differences.

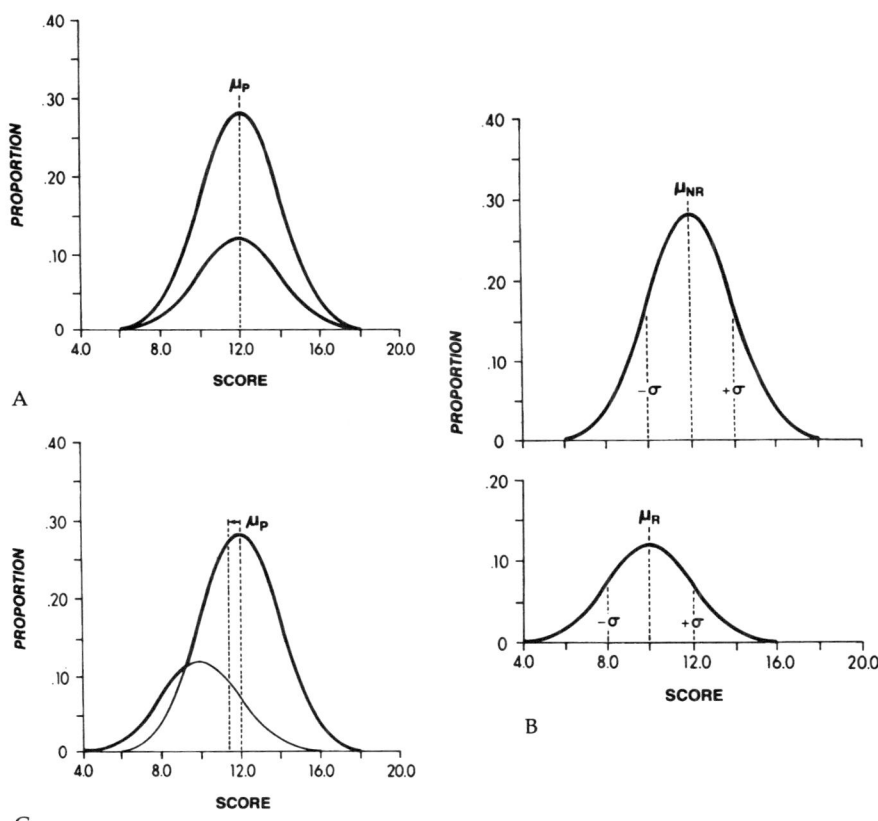

FIGURE 3 Hypothetical distributions showing interpretive difficulties arising from studies of populations comprised of both responders and nonresponders. (A) Distribution of scores, before toxic challenge, in a population consisting of 70 percent nonresponders (taller distribution) and 30 percent responders (shorter distribution). (B) Distribution of scores, shown separately for responders and nonresponders, to a toxic challenge that displaces the responders by one standard deviation. (C) Combined distribution, shown by heavy line, of nonresponders and responders. The difference in means, enclosed by the vertical dotted lines, indicates that even a significant displacement of the responders alters the mean of the distribution only slightly under these circumstances; only with large samples could such an effect be detected consistently.

THE REMOTE FUTURE

Behavioral toxicology first emerged as an alternative to traditional markers of toxicity such as tissue damage and as a potential reservoir of more sensitive methods for measuring toxicity. Impelled, perhaps, by regulatory questions, it veered from the sensitivity issue toward the development of techniques for the detection of neurotoxicity. Much of it remains clasped in the identification phase of risk assessment, a status that tends to isolate it from new advances in behavioral and neuroscience and that negates much of its early promise.

Risk assessment is viewed as the critical coupling of toxicological science and public policy. Behavioral toxicology surely has more to offer this process, and much more to extract from it, than a list of procedures. What other discipline is in the unique position of access to a technology for tracing a progression of toxicity from early, subtle effects to clear impairment? What other perspective on toxicology can integrate such a rich configuration of endpoints (Weiss, 1986b)? If BT abandons its early promise by confining itself to narrow questions of techniques for identification, it could rupture its close relationship with the major concerns of public health. The science will be the greater victim.

ACKNOWLEDGMENT

Preparation supported in part by grants ES01247, ES01248, and ESO44929 from the National Institute of Environmental Health Sciences.

REFERENCES

American Conference of Governmental Industrial Hygienists. 1974. Documentation of the Threshold Limit for Substances in Workroom Air, third edition. Cincinnati, Ohio: ACGIH.

Anger, W. K., and B. L. Johnson. 1985. Chemicals affecting behavior. Pp. 51–148 in Neurotoxicity of Industrial and Commercial Chemicals, J. O'Donoghue, ed. Boca Raton, Fla.: CRC Press.

Bellinger, D., A. Leviton, C. Waterhaus, H. Needleman, and M. Rabinowitz. 1987. Longitudional analysis of prenatal and postnatal lead exposure and early cognitive development. New England Journal of Medicine 316:1037–1043.

Buelke-Sam, J., C. A. Kimmel, and J. Adams, eds. 1985. Design considerations in screening for behavioral teratogens: Results of the collaborative behavior teratology study. Neurobehavioral Toxicology and Teratology 7:537–673.

Clarkson, T. W., C. Coc, D. O. Marsh, G. J. Myers, S. K. Al-Tikriti, L. Amin Zaki, and A. R. Dabbagh. 1981. Dose-response relationships for adult and prenatal exposures to methylmercury. Pp. 111–130 in Measurement of Risks, G. G. Berg and H. D. Maillie, eds. New York: Plenum.

Feingold, B. F. 1975. Why Your Child Is Hyperactive. New York: Random House.

Glass, R. I. 1975. A perspective on environmental health in the USSR. Archives of Environmental Health 30:391–395.
Grasso, P., M. Sharratt, D. M. Davies, and D. Irvine. 1984. Neurophysiological and psychological disorders and occupational exposure to organic solvents. Food and Cosmetic Toxicology 22:819–852.
Iversen, S. D., and L. L. Iversen. 1981. Behavioral Pharmacology, second edition. New York: Oxford University Press.
Klaassen, C. D., M. O. Amdur, and J. Doull, eds. 1986. Casarett and Doull's Toxicology, third edition. New York: Macmillan.
Maurissen, J. P. J. 1988. Quantitative sensory assessment in toxicology and occupational medicine: Applications, theory and critical appraisal. Toxicology Letters 43:321–343.
National Research Council. 1983. Risk Assessment in the Federal Government: Managing the Process. Washington, D.C.: National Academy Press.
Pavlenko, S. M. 1975. Methods for the study of the central nervous system in toxicological tests. Pp. 86–108 in Methods Used in the USSR for Establishing Biologically Safe Levels of Toxic Substances. Geneva: World Health Organization.
Spyker, J. M. 1975. Behavioral teratology and toxicology. Pp. 311–344 in Behavioral Toxicology, B. Weiss and V. G. Laties, eds. New York: Plenum.
Stebbins, W., ed. 1970. Animal Psychophysics. New York: Appleton-Century-Crofts.
Tepper, J. L., and B. Weiss. 1986. Determinants of behavioral response with ozone exposure. Journal of Applied Physiology 60:868–875.
Tepper, J. L., B. Weiss, and C. Cox. 1982. Microanalysis of ozone depression of motor activity. Toxicology and Applied Pharmacology 64:317–326.
Tilson, H. A., C. L. Mitchell, and P. A. Cabe. 1979. Screening for neurobehavioral toxicity: The need for and examples of validation of testing procedures. Neurobehavioral Toxicology 1(Suppl.):137–148.
Victory, W., A. J. Vander, J. M. Sherlock, P. Schoeps, and S. Julius. 1982. Lead, hypertension, and the renin-angiotensin system. Journal of Laboratory and Clinical Medicine 99:354–363.
Weiss, B. 1982. Food additives and environmental chemicals as sources of childhood behavior disorders. Journal of the American Academy of Child Psychiatry 21:144–152.
Weiss, B. 1984. Behavior as a measure of adverse response to environmental contaminants. Pp. 1–57 in Handbook of Psychopharmacology, Vol. 18, L. L. Iversen, S. D. Iversen, and S. H. Synder, eds. New York: Plenum.
Weiss, B. 1986a. Food additives as a source of behavioral disturbances in children. Neurotoxicity 7:197–208.
Weiss, B. 1986b. Emerging challenges to behavioral toxicology. Pp. 1–20 in Neurobehavioral Toxicology, Z. Annau, ed. Baltimore: Johns Hopkins University Press.
Weiss, B. 1988. Neurobehavioral toxicity as a basis for risk assessment. Trends in Pharmacological Science 9:52–62.
Weiss, B. 1989. Behavior as an endpoint for inhaled toxicants. Pp. 492–512 in Concepts in Inhalation Toxicology, R. O. McClennan and R. F. Henderson, eds. New York: Hemisphere.
Weiss, B., and V. G. Laties, eds. 1975. Behavioral Toxicology. New York: Plenum.
Weiss, B., J. Ferin, W. Merigan, S. Stern, and C. Cox. 1981. Modification of rat operant behavior by ozone exposure. Toxicology and Applied Pharmacology 58:244–251.
Wood, R. W., and C. Cox. 1986. A repeated-measures approach to the detection of the minimal acute effects of toluene. Toxicologist 6:221.

Appendix
Symposium Agenda

August 23-26, 1988

Australian National University
Canberra, Australia

TUESDAY, AUGUST 23, 5:00 P.M.
Welcome and Opening Remarks

Wayne Holtzman (President, IUPsyS)
University of Texas
Austin, Texas, U.S.A.

Roger Russell (Australian Workshop
Coordinator), The Flinders
University, Bedford Park, S.A., Australia

L.W. Nichol, Vice Chancellor,
Australian National University,
Canberra, Australia

Pamela Ebert Flattau (U.S. Workshop
Coordinator), National Research
Council, Washington, D.C., U.S.A.

Keynote Address: Peter Spencer, Center for Research on Occupational and Environmental Toxicology, Oregon Health Sciences University, U.S.A.

WEDNESDAY, AUGUST 24, 8:30 A.M.
Opening Remarks: Wayne Holtzman, University of Texas, U.S.A.

SESSION I
Assessment of Neurobehavioral Tests Now in Use

CHAIR: George Singer, Latrobe University, Australia

Helena Hanninen, Institute of Occupation Health, Helsinki, Finland	"Methods in Behavioral Toxicology: Current Test Batteries and Need for Further Development"
Ann Williamson, National Institute of Occupational Health and Safety, Sydney, Australia	"The Current Status of Test Development in Neurobehavioral Toxicology"
W. Kent Anger, National Institute for Occupational Safety and Health, Cincinnati, Ohio, U.S.A.	"Human Neurobehavioral Tests That Have Identified Effects of Both Short- and Long-Term Chemical Exposures"

DISCUSSANT: J. Graham Beaumont, University of Swansea, U.K.

11:00 A.M.
General Discussion

2:00 P.M.

SESSION II
Assessment of Animal Models: What Has Worked and What Is Needed

CHAIR: Roger Russell, The Flinders University, Australia

Hanna Michalek, Instituto Superiore di Sanita, Rome, Italy	"Neurochemical Tests for Assessment of Neurotoxicity of a Cholinesterase Inhibitor in Senescent Rats: Animal Model for Alzheimer's Treatment?"
David Overstreet, The Flinders University, Bedford Park, S.A., Australia	"Animal Models of Dementia: Relevance of Testing to Neurobehavioral Toxicology"
Deborah Cory-Slechta, University of Rochester, Rochester, New York, U.S.A.	"Bridging Experimental Animal and Human Behavioral Toxicology Studies"
Beverly Kulig, Medical Biology Laboratory TNO, Rijswick, The Netherlands	"Methods and Issues in Evaluating the Effects of Organic Solvents on Nervous System Function"

DISCUSSANT: Robert MacPhail, U.S. Environmental Protection Agency, U.S.A.

4:30 P.M.
General Discussion

APPENDIX   417

THURSDAY, AUGUST 25, 8:30 A.M.
Opening Remarks: Andrew Pope, National Research Council, U.S.A.

SESSION III
Chemical Time Bombs: Environmental Causes
of Neurodegenerative Diseases

   CHAIR: Robert MacPhail, U.S. Environmental Protection Agency, U.S.A.

Gerhard Winneke, Institute of Environmental Hygiene, Dusseldorf, West Germany	"Neuropsychological Toxicity of Environmental Chemicals: Clinical and Subclinical Aspects"
Deborah C. Rice, Health and Welfare Canada, Ontario, Canada	"The Health Effects of Environmental Lead Exposure: Closing Pandora's Box"
Norman Krasnegor, National Institute of Child Health and Human Development, Bethesda, Maryland, U.S.A.	"On the Identification and Measurement of Chemical Time Bombs: A Behavior Development Perspective"
Roger Russell, The Flinders University, Bedford Park, S.A., Australia	"Neurobehavioral 'Time Bombs': Their Nature and Their Mechanisms"

   DISCUSSANT: Peter Spencer, Center for Research on Occupational and
      Environmental Toxicology, Oregon Health Sciences University, U.S.A.

11:00 A.M.
General Discussion

2:00 P.M.

SESSION IV
Behavioral Aspects of Neurotoxicity: Regional Issues

   CHAIR: W. Kent Anger, National Institute for Occupational Safety
                           and Health, U.S.A.

Liang You-xin, Shanghai Medical University, The People's Republic of China	"The Development and Perspective of Neurobehavioral Toxicity Tests in China"
V.B. Saxena, P.P.N. College, Kanpur, India	"Regional Issues in the Late Development of Neurobehavioral Toxicity Tests"
R. Gilioli, University of Milan, Italy	"Regional Issues in the Development of Neurobehavioral Toxicity Tests"

   DISCUSSANT: Ann Williamson, National Institute of
      Occupational Safety and Health, Australia

4:30 P.M.
General Discussion

FRIDAY, AUGUST 26, 9:00 A.M.
Opening Remarks: Pamela Ebert Flattau,
National Research Council, U.S.A.

SESSION V
Recommendations for Further Research and Testing

Robert MacPhail, U.S. Environmental Protection Agency, Research Triangle Park, North Carolina, U.S.A.	"Environmental Modulation of Neurobehavioral Toxicity"
Anders Iregren, National Board of Occupational Safety and Health, Solna, Sweden	"Computerized Performance Testing in Neurotoxicology: Why, What, How and Where To? The SPES Example"
Bernard Weiss, University of Rochester, Rochester, New York, U.S.A.	"The Scope and Promise of Behavioral Toxicology"

11:00 A.M.
Concluding Remarks: Roger Russell,
The Flinders University, Australia

Index

A

Abalone divers, 64, 338
Acetylcholine (ACh)
    and behavior, 102–104, 108, 117
    and neurodegenerative disease, 209–210, 211, 212–215, 216
Acetylcholinesterase (AChE), 103, 106
    DFP and, 111, 127
    and neurodegenerative disease, 209–211, 213
Acetylcoenzyme A, 104
Acid rain, 409
Acrylamide, 78, 236, 270, 276
Active avoidance, 130
Additions Test, 368, 369, 373, 374, 375, 376, 377, 388
Affective behavior, 77, 81, 160, 175
AF64A, 124, 128, 133, 134
Age
    and cholinergic system, 115–118, 119
    lead and, 258–259
    in NCTB studies, 83, 302
    and neurochemical changes, 101–102, 110–111, 281
    and neurodegenerative disease, 11, 212, 227, 276, 282
    organophosphates and, 184
    in Swedish performance battery, 375, 381
Agricultural chemicals, 359
Air pollution, 293

Alcohol, 63
    fetal exposure, 199
    in Microtox battery, 60
    nervous system damage, 176, 276
    in Swedish performance battery, 360, 377–378, 380
    tolerance development, 164, 167
Aliphatic hydrocarbons, 236
Alkanes, 78
Aluminum
    and Alzheimer's disease, 16, 236, 270–271
    neurological damage, 15, 226, 228–229, 235–236
Alzheimer's disease, 13, 15, 32, 227
    aluminum and, 16, 236, 270–271
    in behavioral studies, 152–153, 154, 406, 407
    chemicals and, 281
    cholinergic system and, 124–125, 210–212, 213, 214–215
    lead and, 283
    neuronal degeneration, 219, 220
    treatment, 133, 209, 210, 269
American Conference of Governmental Industrial Hygienists, 79, 399
American Psychological Association, 365
Ames test, 70
Amitraz, 353, 355–356
Amphetamine, 130, 131, 150–151, 350
Amygdala, 276

Amyotrophic lateral sclerosis, 35, 283
ALS/P-D (western Pacific amyotrophic lateral sclerosis and parkinsonism-dementia complex), 12–16, 26, 27, 35–36
cycads and, 17, 18, 21, 22, 27–32, 34, 280–281, 409
ANADe (O-acetyl-N-animodeanol), 213
Anemia, 252, 253
Anesthetic gases, 78, 359, 360
Anger, W. Kent, 90, 314
Animal models, 184–188, 193, 194, 271–272
Alzheimer's disease, 213
barbiturate exposure, 195, 196
behavioral effects, 129–134, 152
brain implants, 220–221, 222
cholinergic system, 101, 102, 118, 120, 210, 212
dementia, 124–126, 127, 130, 134
extrapolation to humans, 7, 115, 129, 187, 199, 239, 406, 410
lead exposure, 139, 141, 146, 257, 258, 399, 407
mercury exposure, 408
nerve growth, 216
parkinsonism, 279–280, 406
repeated acquisition procedures, 148–155
solvent exposure, 160, 161–162, 166–174, 177, 178–179
startle response, 201
ANOVAs, 111, 114, 115, 117, 302
Anticholinesterases (antiChE)
and behavior, 103, 104, 106, 210
and neurodegenerative disease, 118, 209, 211, 269
tolerance development, 103, 128, 131–132, 210
Antihypertensive drugs, 274
Antimitotics, 278
Antipsychotic drugs, 275
Apparatus-driven research, 148
AP7, 33–34
Aromatic hydrocarbons, 236
Asian Conference on Behavioral Toxicology and Clinical Psychology, 331
Associative conditioning, 196–199
Ataxia, 80, 82
Attention impairment, 160

Australia, 338–339, 341
aborigines, 18–19, 30
cycadism in, 27
lead study, 248
Autocannibalism, 213–214, 215
Autografts, 221
Auyu people, 14, 27–28, 30
Avoidance behavior, 129–130, 131, 134, 178
Axonal degeneration, 273, 274, 275–276, 278, 280

B

Bailey, Elaine L., 185
Bangladesh, 12, 24
Banks, Joseph, 19–20
Barbiturates, 168, 194–196
Bayley Infant Behavioral Record, 248
Bayley Mental Development Index, 247–248
Bayley Psychomotor Development Index, 247–248
Bayley Scales of Infant Development, 239, 403, 408
Behavior, 1–3, 64, 119, 125
age and, 102
animal models, 118, 129–134
brain implants and, 220–222
chemical effects on, 76, 77, 79, 81, 91, 206–209
cholinergic system and, 102–104, 210, 214–215, 218
and definition of toxicity, 401–403
environmental effects on, 347–348, 350–353
lead effects on, 247–250
measurement of, 4–5, 6–7, 185
neurotransmitters and, 138, 154
performance tests, 149–150, 161, 302–304, 359–390
Behavioral pharmacology, 137, 141, 152, 397–399
Bell, E. A., 23
Bender Gestalt Test, 230, 238
Benton Visual Retention Test, 40, 47, 230, 302, 303, 320, 326, 328
Benzene exposure, China, 287, 288
Betz cells, 24, 26
BHMH (butylazo-2-hydroxy-5-methylhexane), 169

Bhopal, India, 324, 341
Biochemical markers, 307–308
  lead exposure, 254–255
Blacks, 260
Block Design Test, 41
Blood-brain barrier, 102, 118, 277
Blood lead concentration, 229, 230, 231, 244, 260, 291, 407
  and intellectual performance, 249, 250–253, 257, 294
  maternal, 254–255
  and spontaneous abortion, 247
Blood pressure, lead and, 254, 258–259
BMAA (beta-N-methylamino-L-alanine), 23–24, 25–27, 32–35
BOAA (beta-N-oxalylamino-L-alanine), 23, 24, 25, 33, 34, 274, 276
Bolla-Wilson and Bleecker battery, 44, 45
Bone lead, 254, 255, 263, 283
Bourdon-Wiersma Vigilance Test, 40, 326, 327, 328
Brain
  age and, 110, 118
  chemicals and, 102
  damage, 41, 93, 218–219
  nerve growth, 216
  neurocircuitry, 200–201
  solvent effects, 174–175, 176
  tissue implants, 219–222, 279–280
Brazelton Neonatal Assessment Scale, 234
British Ability Scales, 230, 249
British National Adult Reading Test, 41
British Psychological Association, 365
Bulgaria, 317
Butyl acetate, 236

C

Caffeine, 193
Calcium, 255
  deficiency, 15–16
  homeostasis, 253, 254, 259, 263
California, 8
Calories, 283
Cancer, 281, 395, 396–397, 403
Carbaryl, 78, 152
Carbon disulfide, 40, 78, 80, 268–269, 270, 338
  Chinese studies, 289–291, 304
  neurological effects, 168, 171–174, 175, 176, 236, 237, 282

Carbon monoxide, 60, 78
Carbon tetrachloride, 78
Case control studies, 237
Cassava, 268
Cassitto, Maria G., 315
Catecholamine, 178, 274, 308
Cattle, cycadism in, 27
Caudal nerve, 173, 174
Cavanagh, J. B., 173
Celebes, 19
Centers for Disease Control, 230, 261
Central nervous system, 3, 4, 41, 124–125, 147, 227–228, 312, 401
  age and, 102
  conditioned learning, 200–201
  development of, 192, 216–217, 218–219
  lathyrism, 24
  lead effects, 230, 231
  mercury effects, 232–233
  solvent effects, 160–161, 164, 167, 169, 175, 236–237
  test batteries, 80, 82, 91, 359, 360
Cerebellum, 200
Cerebral cortex, 110, 113, 114, 116, 117, 276
Cerebral palsy, 408
Chamorros, 13, 14, 15, 21, 22, 23, 30, 31–32
Chelation, 263
Chemicals
  animal testing, 184, 186
  in China, 287–288
  and dementia, 212
  in India, 323
  neurobehavioral effects, 74, 75–78, 80–83, 101–102, 206–209, 226–240, 273–278, 312–313
  neurotoxicity criteria, 268–272, 278–284
  performance test assessment, 359–360
  regulation of, 79, 222, 403
Chemomyelotomy, 198
Chemotherapy, 397
Children, 239, 410
  eyelid conditioning, 200
  lead exposure, 138, 146, 229–231, 243, 245–246, 247–257, 260, 262–263, 292–293, 294, 295, 308, 403–404, 407
  mercury exposure, 233, 408

PCB exposure, 234–235
solvent exposure, 167
China, 12
  Academy of Preventive Medicine, 297
  carbon disulfide, 289–291
  lead exposure, 291–293
  magnetic field exposure, 296–297
  toxicity testing, 287–289, 297, 298–300, 304–310, 314, 337, 341
  video terminal operation, 293–296
Chlordane, 186
Chlordimeform, 353–356
Chloroform, 78
Choice Reaction Time, 88
  in Swedish performance battery, 366, 368, 369, 373, 374, 375, 376, 377–378, 380, 385, 387
Choline, 212–215
Choline acetyltransferase (ChAT), 104, 106, 109, 110, 111, 112, 114, 115, 118–119, 211, 213, 216
Cholinergic system, 101, 124–125, 128, 184–185
  and behavior, 102–104, 200–201
  DFP and, 104–119
  false transmitters, 212–216
  in neurodegenerative disease, 209–212, 218, 219, 227, 274
Cholinesterase (ChE)
  in Alzheimer's disease, 211
  and behavior, 103–104
  DFP and, 104–110, 111, 112, 113, 115, 117–119
Chromaffin cells, 221
Chromium oxide, 328
Chronic toxic encephalopathy, 174–179
Ciguatera, 268
Clean Air Act (1970), 8, 401
Climatic conditions, 361
Clinical Interview Schedule, 89
Clostridium bacilli, 268
Cognitive componential models, 93–95, 96
Cognitive Failures Questionnaire, 89
Cognitive function, 3, 6–7
  animal models, 124, 125, 133–134, 239
  brain implants and, 220–221
  chemical effects, 77, 81
  cholinergic system and, 210, 215
  lead effects, 231, 283
  nervous system development and, 217
  PCB and, 235

performance tests, 60, 64, 74, 80, 87, 90, 91, 92–93, 97, 238, 362
solvent effects, 175, 177, 178, 187
taxonomy, 70–71
Cohen, Stanley, 217
Collaborative Behavioral Teratology Study, 408–409
Collaborative Perinatal Project, 195
Color Word Vigilance Test, 88
  in Swedish performance battery, 369, 373, 374, 375, 376, 377, 378, 379, 385, 387
Committee on Environmental Hazards, 260
Communication, 8–9, 335
Complex Reaction Time, 367, 368
Computerized testing, 5, 59–60, 65, 89, 90, 92–96, 238, 339, 361–365
  China, 297, 298–300, 305, 309
  Milan Automated Neurobehavioral System, 317–320
  Neurobehavioral Core Test Battery, 87
  Neurobehavioral Evaluation System, 41, 50, 51, 73–74, 80–82, 83, 89, 92, 318
  Swedish Performance Evaluation System, 363, 366–390
Conduction velocity of slow fiber (CVSF), 290
Confounding factors, 57, 306, 312–313
Control groups, 6, 57, 82–83, 177
Cortical cells, 216
Cory-Slechta, Deborah A., 186, 407
Costa Rica, 19
Cree Indians, mercury exposure, 233
Cresol, 78
Cross-sectional studies, 6, 57
  lead exposure, 231, 246, 248–249, 254
  solvent exposure, 175, 176, 177
Cultural differences, 90–91
Cycadism, 20, 27, 35
Cycads, 12, 15, 16–23, 24, 27–35, 277, 280–281, 409
Cycasin, 23, 27, 32, 33
Czechoslovakia, 317, 319

D

Dementia, 7, 91, 134, 270
  animal models, 124–126, 127, 130, 185–186

dialysis, 16, 235–236, 271
presenile, 12, 35, 237
progressive degenerative, 4, 6, 206, 209–212, 220
senile, 11, 36, 212, 277
Dendrites, 273
Denmark, 237, 319
Denver Developmental Screening Test, 233–234
Depression, 126
Developing countries, 59, 72, 312, 315, 317, 322, 329–331, 334–335, 338
Developmental behavioral toxicology, 192–193, 194, 199–200, 201, 269
Developmental psychology, 238–239
Diagnostic and Statistical Manual (DSM-III), 175–176, 340
Dialysis dementia, 16, 235–236, 271
Dichloroacetylene, 169
Dicrotophos, 106
Diethylstilbestrol, 191
Digit Classification Test, 368, 369, 388
Digit Learning Test, 87
Digit Serial Test, 320
Digit Span Test, 41, 47, 52, 58, 87, 302, 303, 320, 362
  in Swedish performance battery, 368, 369, 373, 374, 376, 378, 379, 385, 388
Digit Symbol Test, 41, 47, 58, 302, 303, 320
Diisopropyl fluorophosphate (DFP), 103, 104–119, 126–127
Dinitro-*o*-cresol, 78
Diphtheria toxin, 278–279
Disinhibition, 131–132
Disulfoton, 108
Dopamine, 138, 178, 275, 279
Dopaminergic neurons, 227
Dose-effect relationship, 3, 49, 142–143, 193, 194, 270, 406
Drugs, 126, 128, 176
  central nervous system, 141, 398
  recreational, 167, 280, 333
  therapeutic, 4, 8–9, 269, 275–276
  tolerance development, 350
Dutch East India Company, 19
Dyslexia, 93

E

Education, 339, 342–343, 375
Elderly, 11–12, 13, 91, 115, 282, 283

Electromagnetic fields, 296–297, 361, 366, 379
Elementary cognitive tasks, 60
Emotional changes, 52
Encephalopathy, 6–7, 164, 167, 174–179, 237, 230, 270
Environment
  in ALS/P-D, 12, 15
  chemicals in, 222, 226–227, 268, 403
  lead in, 244–246, 252–254, 255, 262–263
  and lethality, 350–351
  modulation of behavior, 347–350
  and operant behavior, 351–353
  and pesticides, 353–356
  pollution, 281, 287
  research design, 323–325, 357–358
  toxicants in, 125, 126, 134, 137
Environmental Protection Agency (EPA), 8, 70, 161, 169, 229, 261, 262, 403
Environmental toxicology, 8
Enzymes, 106
Epidemiological research, 40
Erythrocytes, 106
  free protoporphyrin, 250, 260
Erythropoiesis, 229–230
Esters, 236
Ethanol. *See* Alcohol
Ethiopia, 12, 24
Ethnic composition, 338–339
Ethylbenzene, 360
Ethylene dibromide, 78
Europe, 337, 338
  NCTB data, 83, 315, 316, 318–319
European Economic Community, 313, 386–387
"Expert systems," 94
Eyelid conditioning, 200–202

F

Fatigue, 61, 376
Feingold, Ben F., 410
Fenvalerate, 327
Fetus, 191, 201
  barbiturate exposure, 195–196
  brain tissue implants, 221
  DFP exposure, 104–108
  lead exposure, 231, 255, 263
  learning, 197–199
  mercury exposure, 232, 233, 407, 408
Finger Tapping Tests, 368, 369, 373, 374, 375, 376, 377, 389

Finland, 44, 83, 237, 319
  Institute of Occupational Health, 40, 44, 45, 58, 71–72, 80–82, 309, 329
Fischer 344 rats, 111–114, 115, 116, 184, 185
Fish
  mercury in, 232, 233, 271, 409
  PCB in, 234–235
Flicker fusion frequency, 305–306
Flinders rats, 114–115, 127–128
Florida, 20–21
Fluorocarbons, 78
Food
  mercury in, 232, 233
  PCB in, 234–235
Food and Drug Administration, 261, 409
Forced swim test, 126
Formaldehyde, 78
Formamidine, 353–356
Formetanate, 353–356
Fosberg, F. Raymond, 22
Frostig Scales, 230

G

Gajdusek, D. Carleton, 28
Gamma-aminobutyric acid (GABA) pathways, 274
Gamma-diketone pathway, 168
Gammexane, 327
Gasoline
  lead in, 243, 244–245, 256, 263, 403
  and memory, 326
Gehrig, Lou, 13
Gender
  ALS incidence and, 13
  and neurobehavioral performance, 302–304
  in Swedish performance battery, 373–375
Genetic factors, 4, 110–115, 184–185
Ginkgoaceae, 16
Glial cells, 277
Glutamate excitotoxins, 282
Glutamate pathways, 274
Glutamate receptors, 33
Greece, 317
Groote Eylandt, Australia, 18
Guam, 11, 12, 13–15, 16, 17, 18, 21–22, 30, 31, 32, 281

H

Hair mercury concentration, 232, 233, 408
Halogenated hydrocarbons, 236
Haloperidol, 350–351
Halstead Reitan Battery, 41
Hamsters, 194–195
Hand-eye coordination, 80
Hanninen, Helena, 89–90, 288, 315, 325
Harvard University, 59
Health effects, lead exposure, 252–254
Hearing, 170
  lead and, 250–252, 253, 257–258
Heavy metals, 167, 289
Hematopoietic system, 252, 262
Heme biosynthesis, 229–230, 252, 262
Hemicholinium-3, 214–215
Heroin, 350
Hexacarbons, 168, 236
Hexane, 236, 268–269, 270, 282
High-affinity choline transport (HAChT), 209–210
Hippocampus, 110, 111, 113, 114, 116, 117, 127, 128, 131, 133, 134, 200, 216
Hoffman, Howard, 201
Hogstedt Symptom Questionnaire, 89
Homeostasis, 208, 210
  calcium, 253, 254, 259, 263
Homovanillic acid, 308
Honduras, 19
Hormonal markers, 308
Hull, C. L., 119
Human studies
  aluminum toxicity, 235
  animal extrapolation, 7, 115, 129, 187, 199, 239, 406, 410
  barbiturate exposure, 195–196
  brain implants, 221, 279–280
  chemical effects, 101, 155, 226–227, 271–272
  cycad use, 18, 35
  eyelid conditioning, 200, 201
  lead exposure, 138–139, 146–147, 254
  neurobehavioral tests, 90–91
  organophosphate exposure, 102, 126, 185
  repeated acquisition, 153–154
  schedule-controlled behavior, 146
  solvent exposure, 80, 160–161, 162–163, 164, 166, 168, 169–170, 176–178, 237
  toluene exposure, 398

Hungary, 317, 319
Huntington's disease, 220
Hydrocarbons, 236
Hydrogen cyanide, 78
Hydrogen sulfide, 78
Hydrolysis
  ACh, 104, 209, 211
  ANADe, 213
Hyperactivity, 134
Hyperparathyroidism, 15
Hypertension, lead and, 398–399
Hypothalamus, 178

I

IBM computers, 73, 367
Ibotenic acid, 124
Idria mines, 399
Illiteracy, 332–333, 339
India
  barriers to test development, 331–333, 338, 341
  environmental research, 322–325
  lathyrism in, 12, 24
  neurobehavioral research, 325–331, 337
Indonesia, 7
Industrial chemicals, 79, 167, 317–318
  neurobehavioral effects, 228–240
  pollution, 293
  solvents, 7, 159, 175, 179, 272, 360
Inescapable shock, 126
Inferential systems, 94–96
Information-processing theory, 60, 61, 62
Intellectual functions, 49
Intelligence, 239
  lead exposure and, 138–139, 245, 249, 256–257, 260, 283, 293, 403–404
  tests, 230, 298
International Commission on Occupational Health, 318
International Labor Organization, 335
International Symposium on Neurobehavioral Methods in Occupational Health, 313
Iraq, mercury poisoning, 232, 233, 408
Irian Jaya, 11, 12, 14, 17, 18, 23, 27–28, 30, 31
Israel, 317
Italy, 317, 318, 319

J

Japan
  ALS/P-D in, 12, 14, 15, 18, 23, 27–28, 29, 30, 31
  PCB exposure, 234
Jaqai people, 14, 27–28
Jaundice epidemic, India, 323
Johns Hopkins School of Medicine, 44

K

Kainate, 33
Kanpur University, 325–329
Karolinska Institute, 178
Ketones, 78, 236
Kii peninsula, Japan, 12, 14, 15, 18, 23, 27–28, 29, 30, 31
Kisby, Glen, 32
Krasnegor, Norman A., 269
Kulig, Beverly M., 186
Kurland, Leonard, 15, 22

L

Lactation, 255
Language differences, 339, 367
Larvicides, 327, 337
Lathyrism, 12, 23–27, 34, 268
Lead exposure, 78
  and aging, 258–259, 283
  China, 287, 288, 291–293, 294, 295
  effects on children, 138, 146, 229–231, 243, 245–257, 260, 262–263, 292–295, 308, 403–404, 407
  environmental, 243–246, 262–263
  and hypertension, 398–399, 408
  India, 328–329
  markers of, 254–255
  nerve cell damage, 276
  neurobehavioral effects, 138–147, 152, 176, 186, 226, 256–257, 270, 403
  occupational, 56–58, 80, 230, 291, 305
  test batteries and, 59, 61–62, 237–238, 239
Lead salts, 268–269
Learning, 146, 149, 150–152, 153, 239, 410. *See also* Cognitive function
  animal models of dementia, 124, 125, 126, 130, 185–186
  lead exposure and, 139

maze, 132–133
operant conditioning, 131–132
perinatal, 196–199, 202
practice effect bias, 306, 376
Leather industry, 328
Lethality, environment and, 350–351
Levi-Montalcini, Rita, 217–218
Lithium chloride, 197, 198–199
Locomotor activity, 134
Long Evans rats, 185
Longitudinal studies, 6, 192, 198, 409
Los Angeles, Calif., 8
Lowest-observed-adverse-effect level, 396
Lysergic acid diethylamide (LSD), 274

M

McCarthy Scales, 230, 247
Magnesium deficiency, 15
Malathion, 78
MAM (methylazoxymethanol), 30–31, 32, 33, 35
Manic-depressive psychosis, 175
Mariana islands, 12, 14, 17, 22–23, 32
Maximum allowable concentration (MAC), 288, 332
Maze learning, 132–133
Medicaid, 195
Melville Island, 18–19
Memory, 150–151, 153, 178, 239. See also Cognitive function
   in Alzheimer's disease, 218, 220
   animal models of dementia, 124, 125, 126, 129–130, 131–133, 134, 185–186
   lead exposure and, 328
   magnetic fields and, 296
   solvent exposure and, 160, 326–327
   test batteries, 41, 52, 64, 65, 74, 80, 298–299, 366
Memory Reproduction Test, 88
Mercury, 78. See also Fetus; Fish; Hair mercury concentration; Rats
   in China, 287, 288
   Minamata disease, 232, 233, 271, 283–284, 408
   mine poisoning, 399
   neurotoxicity, 226, 228–229, 231–234, 270, 276, 277
   test batteries and, 58, 61, 64, 80
Metals, 167, 226, 228–229, 268–269, 289, 359

Methyl alcohol, 78
Methyl chloride, 80
Methylchloroform, 78, 360
Methylene chloride, 78, 360
Methyl ethyl ketone, 168, 236
Methyl isobutyl ketone, 360
Methyl isocyanide, 324
Methylmercury, 232–234, 271, 283–284, 317, 407–409
Methyl n-butyl ketone, 168, 282
Methyl parathion, 78
Methylphenidate, 187
Methylxanthine, 193
Mice, 33–34, 106, 110, 194, 350, 409. See also Rodents
Michalek, Hanna, 184
Michigan, 195
Microtox battery, 59–60
Milan Automated Neurobehavioral System, 317–320
Mill Hill Vocabulary Test, 94
Minamata Bay, Japan, mercury poisoning, 232, 233, 271, 283–284, 408
Monkeys, 35, 143–144, 151, 152, 256–257. See also Primates
Monoamine oxidase inhibitors, 274
Mortality
   ALS, 13
   DFP, 115, 117–118, 119
   heroin, 350
   lead, 245
   organophosphate, 102
   phenobarbital, 194
Motoneuron diseases, 227
Motor conduction velocity (MCV), 290
Motor neurons, 227
Motor performance, 49, 96
   anticholinesterases and, 132
   brain implants and, 220
   chemicals and, 77, 81
   in test batteries, 80
MPTP (methylphenyltetrahydropyridine), 269, 276, 277, 279, 280, 407
Mulder, Donald, 22
Muscarinic ACh receptors (mAChR), 103, 104, 106, 108, 109, 110, 111–114, 115–119, 127–129, 210, 213
Muscle cells, 278–279
Myelin, 277, 280
Myelinating cells, 278–279

INDEX

N

NADe (N-animodeanol), 212–215, 214
Nanjing Medical College, 292
National Center for Toxicological Research, 70
National Health and Nutrition Examination Survey (NHANES), 250, 253–254, 262, 398–399
National Institute of Child Health and Human Development, 197
National Institute for Occupational Safety and Health (NIOSH), 59, 72, 75–76, 78, 79, 83, 159, 261, 297, 313
National Institutes of Health, 22, 212
National Occupational Environmental Survey, 75
National Occupational Hazard Survey, 75
National Research Council, 2, 9, 222
Neonates, 201
    barbiturate exposure, 195
    neuronal function, 102
    PCB exposure, 234–235
Nerve gases, 209
Nerve growth factor, 216–222
Nervous system. *See also* Central nervous system; Neurotoxicity; Peripheral nervous system
    adult, 235–237, 273–277
    age and, 11, 101–102
    ALS degeneration, 13
    and behavior, 1, 2, 118, 347–349
    chemical effects on, 74, 75, 76, 77, 78, 79, 81, 86, 226–227, 268–269
    cycads and, 33
    development of, 216–217, 218, 229–235, 277–278
    lead effects, 258
    mercury effects, 408
    neurobehavioral tests and, 63, 64, 70–71
    pesticide effects, 353
    solvent effects, 174, 179, 187, 236
Neuroanatomy, 119
Neurobehavioral Core Test Battery (NCTB), 47–48, 51, 59, 72–73, 238, 343
    applications in China, 297–304, 309
    computerization, 87–88, 92, 318, 319–320
    international validation, 50, 66, 83, 314–317
    language differences, 339

Neurobehavioral Evaluation System (NES), 41, 50, 51, 73–74, 80–82, 83, 89, 92, 318
Neurobehavioral toxicology, 2, 9, 119–120, 137–138, 186, 206, 405–412. *See also* Neurotoxicity
    developmental, 192–193, 194, 199–200, 201, 269
    environment and, 126–127, 347–350
    human testing, 69–83, 90–91
    regional requirements, 7–8, 331, 334–335, 337–343
    repeated acquisition procedures, 148–155
    research, 6–7, 65–66, 114, 147–148
    risk assessment, 395–397, 398
    test batteries, 39–53, 87–89, 91–92
    test development, 56–66, 90, 278–284, 322, 340–341
    tort law and, 400, 405
Neurobiology, 64
Neurodegenerative diseases, 11–13, 222, 227, 280, 283
    chemical exposure history, 36, 281–282
    progressive degenerative dementias, 4, 6, 206, 209–212, 220
Neuroendocrine system, 278–279
Neurons, 348–349
    age and, 102
    autocannibalism, 213–214, 215
    chemical effects on, 273–277
    cycad damage, 34
    degeneration of, 13, 258, 276, 278, 281
    development of, 216–217, 218–219, 277
Neuropsychology, 64
Neurotoxicity. *See also* Neurobehavioral toxicology
    adaptive changes, 128, 129
    aluminum, 226, 228–229, 235–236
    animal models, 124–125, 168, 184, 193, 220
    barbiturates, 195
    behavioral effects, 76, 86, 96, 119, 147–148, 154–155, 312–313
    behavioral endpoints and, 278–284, 401–403
    behavioral tests and, 1, 48–49, 50–52, 63–64, 74, 91–92, 96–97, 148–149, 179, 322, 340–341

chemicals, 75–78, 79, 81, 103–104, 147–148, 226–227, 268–269, 313–314, 317–318
  in China, 287–288
  classification of, 227–228, 269–272, 273
  cycads, 32–35
  data base, 69–70
  environmental effects, 357–358
  identification of, 2–3, 193–195
  in India, 323
  lead, 138, 141–142, 230, 231, 244–246, 258. *See also* Lead exposure
  measurement of, 4–6, 196–199, 359–362
  mechanisms and sites of action, 1–2, 3–4, 62
  mercury, 232–233, 408
  organophosphates, 185
  pesticides, 353
  risk assessment, 3, 9, 101, 397, 402–403, 412
  solvents, 160–161, 164, 167, 169, 174, 175, 179, 187, 236–237
Neurotransmitters, 103, 138, 154, 273, 274–275, 278, 280, 307–308. *See also* Cholinergic system
  false transmitters, 212–216
New Guinea, 32
New Zealand
  lead study, 249
  mercury exposure, 233–234
Nicotinic receptors, 103, 213
Nictitating membrane response, 200–202
Nitriles, 78
Niven, David, 13
NMDA (*N*-methyl-D-aspartate), 33, 34, 35
Noise, 78, 361
No-observed-effect level, 396
Norway, 237
Numerical ability, 366
Nunn, Peter, 25

O

Occupational exposure, 399–400
  carbon disulfide, 175, 282, 289, 338
  chemicals, 75–76, 80, 226, 229, 268, 275–276, 281, 313–314
  chemicals, India, 322, 323, 324–325, 326–327, 331–333, 334–335
  China, 287–288
  lead, 56–58, 80, 230, 291, 305

  magnetic fields, 296–297
  organophosphates, 102, 287, 327
  regional characteristics, 338
  solvents, 159, 160, 167, 175, 176–177, 179, 236, 282
  testing, 72, 97, 238, 329–330, 340, 359–361
  tort law, 400, 405
  video terminals, 295–296
Occupational Safety and Health Act (1970), 76
Occupational Safety and Health Administration, 261
Ohta, Masayuki, 29
Operant conditioning, 131–132, 348
  environment and, 351–353
Optacon, 400
"Organic affective syndrome," 176
Organophosphate-induced delayed neurotoxicity, 185
Organophosphates, 184, 185, 207, 281, 287, 288, 327
  and cholinergic system, 101, 102, 103, 106, 108, 110, 117
  and memory, 126
Organotin, 305
Osertesky Motor Scales, 230
Overstreet, David H., 185
Oxotremorine, 128
Ozone, 401–402

P

Pacific islands, 11–12, 32, 35, 268
Paint
  lead in, 245, 262–263
  and memory, 326–327
  solvents in, 360
Palmer, Valerie, 28, 29
Paracelsus, Philippus Aureolus, 3
Paraoxon, 103, 207
Parathion, 78, 103, 106, 207
Paresthesias, 408
Parkinsonism, 277
  ALS/P-D, 12–16, 22, 28, 30, 31, 35, 36
  animal models, 279–280, 406
  brain implants, 221–222, 279–280
  dopamine in, 138, 227
  therapy, 275
Passive avoidance, 129–130, 131, 134, 178
Pavlov, I. P., 401
Pavlovian conditioning, 196–199

Perceptual ability, 74
Periamygdaloid cortex, 276
Perikaryon, 273, 274
Peripheral nervous system, 227–228, 270, 312
   lead effects, 230
   organophosphate effects, 185
   solvent effects, 168–174, 236
Permissible exposure level (PEL), 288
Pesticides, 102, 167, 207, 287, 288, 289, 337, 353–356
Petroleum solvents, 78
Pharmacodynamic processes, 108
Pharmacokinetic factors, 106
Pharmacology, 7, 127–129
Phenobarbital, 194
Phenylalanine, 4
Phenylketonuria, 4
Phospholipid metabolism, 213
Physiology, 119
Physostigmine, 133, 210, 211
Picrotoxin, 274
Pigeons, 152, 353
Piperidine dicarboxylic acid (PDA), 33, 34
Pittsburgh Occupational Exposures Test Battery, 44, 45
Plasma, 106
Plastics, 169
Poland, 317, 319
Polychlorinated biphenyls (PCBs), 226, 228–229, 234–235
Positron emission tomography, 31, 279, 280
Prediction, 199
Pregnancy, 194, 255
   barbiturate exposure, 195–196
   and caffeine, 193
   lead exposure, 246–247, 248, 254–255, 263
   mercury exposure, 233, 409
Premarket testing, 69, 169
Presenile dementia, 12, 35, 237
Primary degenerative dementias, 124
Primates. *See also* Monkeys
   ALS/P-D models, 25, 27, 35
   lead exposure, 146–147
   learning studies, 150–151, 152
   parkinsonism, 279
   solvent exposure, 167
Profile of Mood States, 47, 51, 80–82, 300–302, 316, 319

Progressive degenerative dementias, 4, 6, 206, 209–212, 220
Prospective Memory Test, 89
Prospective studies, 6, 57–58, 66
   lead exposure, 231, 246–248, 254
Proteins, 283
   synthesis, 118, 218
Protoporphyrin, erythrocyte, 250, 260
Psychological function, 60, 96
Psychology, 39–40, 93, 179, 397
   in India, 332
   tests, 53, 70, 237–240, 304
   theory, 64–65, 90, 226
Psychometric theory, 93, 94
Psychomotor performance, 64, 74, 160, 233, 299–300
Psychomotor stimulants, 356
Psycho-organic syndromes, 237
Psychosis, 168, 175
Psychosomatics, 207
Purdue Pegboard, 230
Pursuit Aiming Test, 47, 302–304, 320
Pyridoxine, 269–270

Q

Quinuclidynyl benzilate (QNB)
   receptors, 105, 106, 107
Quisqualate, 33

R

Rabbits, 193, 200–201
Radial arm maze, 132–133, 178
Rats, 23, 30–31, 193, 194, 350–351. *See also* Rodents
   cholinergic system, 101, 104, 213
   dementia models, 128, 129–130, 131–133
   DFP exposure, 104–119, 126–127
   fetal learning, 196–199
   lead exposure, 139–141, 144–145, 146–147, 186
   mercury exposure, 408–409
   nerve growth factor, 216
   organophosphate exposure, 184–185
   ozone exposure, 401–402
   pesticide exposure, 353–356
   repeated acquisition, 178
   solvent exposure, 162, 163, 165, 166, 170, 171–174
   toluene exposure, 398

Rayon industry, 175, 289
Reaction time, 238, 305, 400
　lead and, 250, 252
　in Swedish performance battery, 380–384
Reading ability, 93
Regulation, 69
　environmental chemicals, 222
　food chemicals, 215–216
　lead, 243, 260–262
Reinforcement schedules, 131, 142–146, 397, 401
Renal function, 235–236
Repeated acquisition procedures, 148–155, 178
Reproductive effects, lead exposure, 246–247
Research, 6–7
　chemical exposure, 282
　developing countries, 341–343
　environmental effects, 357–358
　methodology, 89–90, 192–193, 257–258
　test development, 66
　test selection, 57–58
Retrospective studies. *See* Cross-sectional studies
Risk assessment, 3, 9, 101, 395–397, 402–403, 412
Rodents, 27, 194–195. *See also* Mice; Rats
　cholinergic system, 103, 115, 216
　learning, 132
　repeated acquisition procedures, 152
　solvent exposure, 162, 167–168, 169, 170
Ross, Stephen, 33
Rota, 12, 14, 21
Ruffin, Joseph, 69
Russell, Roger W., 268, 269

S

Santa Ana Dexterity Test, 40, 41, 47, 58, 80, 302, 303, 326, 327, 328
Santa Ana Rotation Test, 87
Saxitoxin, 274
Schedule-controlled behavior, 142–146, 152
Schwann cells, 219, 278–279
Scientific ethics, 126
Scopolamine, 127, 128, 130, 131
Screening, 2, 86–87, 186, 194
　solvent effects, 169–174
　test battery design, 61, 62–63, 65, 69–70, 82, 92
Search and Memory Test, 368, 369, 373, 374, 375, 376, 388
Self-reports, 304–305, 363
　in Swedish performance battery, 368, 369, 389–390
Seminole Indians, 20–21
Senile dementia, 11, 36, 212, 277
Sensory function, 80, 82
　chemicals and, 77, 81
　lead and, 61–62, 257–258
　solvents and, 169–174
Serotonergic pathways, 274
Sexual maturation, 194–195
Shanghai Development Center of Science and Technology, 305
Shanghai Medical University, 292–293, 297
Short-term exposure limit, 399
Simple Reaction Time, 47, 50, 88, 303, 319, 362, 363
　in Swedish performance battery, 365, 368, 369, 372, 373, 374, 375, 376, 377–378, 379, 380, 385–386, 387
Smotherman, William, 197
Socioeconomic status, 339–340
Sodium, 274, 278
Solvent exposure, 226, 228–229, 236–237, 268–269, 270, 271, 272, 282, 289
　acute effects, 160–168
　animal models, 160, 161–163, 168–169, 178–179, 186–187
　chronic effects, 168–174, 399
　encephalopathy, 174–179
　health standards, 56, 159–160
　performance test assessment, 7, 59, 71, 88, 359, 360, 366, 372, 385–386
　and reaction time, 63
　recreational abuse, 167–168
　studies in India, 326–327
　subclinical effects, 227–228
South America, 314
Soviet Union, 319, 400–403
Spanish Toxic Oil Syndrome, 272
Spatial resolution, 141–142
Sperry, R. W., 217
Spinal cord, 185
Spinal transection, 198
Spontaneous abortion, lead and, 247
Sprague-Dawley rats, 111, 118
Stanford-Binet test, 230

Startle response, 201
Statistical interpretation, lead studies, 255–257
Sternberg memory scanning test, 60
Stress, 126–127
Stress and Arousal Checklist, 89
Striatum, 113, 114, 116, 117, 127, 220
Styrene, 49, 78, 165, 170, 178, 187, 237, 360
Substantia nigra, 279, 280
Suicide, 175
Supranuclear palsy, 277
Sweden, 237, 319
  National Board of Safety and Health, 88
  National Institute of Occupational Health, 365, 372–373
Swedish Performance Evaluation System (SPES), 363, 366–390
Symbol Digit Test, 368, 369, 373, 374, 375, 376, 377, 378, 379, 380, 385, 388
Systemic toxicants, 395–397

T

Tardive dyskinesia, 280
Target organ effects, 74, 75–79, 80–83, 160
Test batteries
  barriers to development, 64–66, 331–333
  China, 289–290, 291–293, 295–296, 304–310
  computer-based, 92–96, 361–365
  current test batteries, 39–53, 87–89, 91–92
  design, 58–63
  developing countries, 329–331
  human testing, 69–83, 90–91
  methodology, 89–90
  rationales for, 70–74, 86–87
  target organ effects, 80–83
  test development, 56–66, 90
Testosterone, 195
Tetanus, 268
Tetrachloroethane, 78
Tetrodotoxin, 274
Thalidomide, 5, 191, 269, 277
Thallium, 176, 270, 276
Theory-based tests, 60–62, 65, 90
Thiols, 78
Threshold Limit Values (TLV), 1, 3, 79, 161, 329–330, 332, 341, 399, 400

Tiwi people, 18–19
T maze, 132
Tolerance development
  antiChE, 103, 128, 131–132, 210
  organophosphate, 185
  solvents, 164–165, 187
Toluene, 78, 236
  hearing loss, 170
  neurochemical effects, 162, 178, 304, 360, 377–378, 398
  recreational abuse, 167, 168
  tolerance development, 164–165
Tone discrimination, 305
Tooth lead concentration, 229, 230, 250, 251, 254
Tort law, 400, 405
Toxicants. *See* Neurotoxicity
Toxic Oil Syndrome, 272
Toxic Substances Control Act (1976), 70, 169, 403
"Toxicopsychology," 288
Trail Making Test, 41
Trauma, 91
Triadimefon, 187, 356
Trichloroethane, 78
Trichloroethylene, 165–166, 168–169, 187, 236, 237, 360
Triethylcholine, 212
Trimethyltin, 151–152, 282
Trinitrotoluene exposure, China, 287, 288
Tungsten, 78
TURBO Pascal, 368
Turkey, 317

U

United Kingdom, 318
  Institute of Occupational Health, 89
United States
  ALS incidence, 13
  behavioral toxicology, 397
  carbon disulfide use, 338
  chemical exposure, 75–76
  Department of Housing and Urban Development, 261
  elderly population, 11
  environmental movement, 403
  lead regulation, 243, 260–262, 263
  NCTB validation, 83, 314, 318
  organic solvent exposure, 159
  zamia poisoning in, 20–21
University of Milan, 87–88, 319, 342–343

University of North Carolina, 59–60
University of Wisconsin, 257
Uttar Pradesh, India, 323

V

Vanillylmandelic acid, 308
Verbal memory, 49
Verbal Reasoning Test, 369, 373, 374, 375, 376, 377, 385, 389
Video display terminals, 293–296
Visual function
  lead exposure and, 61–62, 139–141, 258
  perception tests, 299
  styrene exposure and, 165
Visual memory, 49
Visual Recognition Memory Test, 235
Visual Searching test, 41
Vitamin B_6, 269–270
Vitamin D, 253, 259, 262
Vlamingh, Willem de, 19
Vocabulary Test, 368, 369, 389

W

Weakness, 80, 82
Wechsler Adult Intelligence Scales (WAIS), 40, 41, 44, 45, 58, 59, 230, 304, 361
Wechsler Memory Scale (WMS), 40, 52, 59, 304
Weiss, Bernard, 4
West Germany, 318, 319
White spirits, 162–163, 236, 360, 365–366
Whiting, Marjorie, 22
Williamson, Ann M., 90
Winneke, Gerhard, 271
WISC-R, 89, 292–293, 294, 304
Wistar rats, 111–114, 115, 184
World Health Organization, 47, 80–82, 83, 169, 325–326, 329, 335, 342, 360, 361
  Office of Environmental Health Service, 317
  Office of Occupational Health, 313
World War II, 32, 97, 209
Worksites. *See* Occupational exposure

X

X-ray fluorescence, 254
Xylene, 78, 170, 360

Y

Yugoslavia, 317
Yusho, 234

Z

Zamia plants, 20–21
Zinc oxide, 78